# DIGITAL GASTRONOMY

## From 3D Food Printing to Personalized Nutrition

# World Scientific Series in 3D Printing

**Series Editor:** Chee Kai Chua *(Nanyang Technological University, Singapore)*

---

World Scientific Series in 3D Printing

# DIGITAL GASTRONOMY

## From 3D Food Printing to Personalized Nutrition

Chee Kai Chua, Hong Wei Tan, U-Xuan Tan,
Chen Huei Leo, Michinao Hashimoto,
Justin Jia Yao Tan & Aakanksha Pant

*Singapore University of Technology and Design, Singapore*

Wai Yee Yeong

*Nanyang Technological University, Singapore*

Yi Zhang

*University of Electronic Science and Technology of China, China*

Gladys Hooi Chuan Wong

*Khoo Teck Puat Hospital, Singapore*

**World Scientific**

NEW JERSEY · LONDON · SINGAPORE · BEIJING · SHANGHAI · HONG KONG · TAIPEI · CHENNAI · TOKYO

*Published by*

World Scientific Publishing Co. Pte. Ltd.
5 Toh Tuck Link, Singapore 596224
*USA office:* 27 Warren Street, Suite 401-402, Hackensack, NJ 07601
*UK office:* 57 Shelton Street, Covent Garden, London WC2H 9HE

**Library of Congress Cataloging-in-Publication Data**
Names: Chua, Chee Kai, author.
Title: Digital gastronomy : from 3D food printing to personalized nutrition / Chee Kai Chua,
    Hong Wei Tan, U-Xuan Tan, Chen Huei Leo, Michinao Hashimoto, Justin Jia Yao Tan &
    Aakanksha Pant, Singapore University of Technology and Design, Singapore,
    Wai Yee Yeong, Nanyang Technological University, Singapore,
    Yi Zhang, University of Electronic Science and Technology of China, China,
    Gladys Hooi Chuan Wong, Khoo Teck Puat Hospital, Singapore.
Description: Singapore ; Hackensack, NJ ; London : World Scientific, [2023] | Series:
    World Scientific series in 3D printing ; vol. 4 | Includes bibliographical references and index.
Identifiers: LCCN 2022030349 | ISBN 9789811255908 (hardcover) |
    ISBN 9789811257407 (paperback) | ISBN 9789811256592 (ebook for institutions) |
    ISBN 9789811256608 (ebook for individuals)
Subjects: LCSH: Three-dimensional food printing. | Gastronomy. | Nutrition.
Classification: LCC TP372.85 .C47 2023 | DDC 641.01/3--dc23/eng/20220725
LC record available at https://lccn.loc.gov/2022030349

**British Library Cataloguing-in-Publication Data**
A catalogue record for this book is available from the British Library.

For any available supplementary material, please visit
https://www.worldscientific.com/worldscibooks/10.1142/12824#t=suppl

Desk Editors: Gregory Lee/Amanda Yun

Typeset by Stallion Press
Email: enquiries@stallionpress.com

# Dedications

To my wife, Wendy, children Cherie, son-in-law Darren & grandchildren Hannah, Esther and Gabriel, Clement, daughter-in-law Lynette & granddaughter Abigail, and Cavell, whose prayers, support and motivation have made it possible for us to finish this.

Chee Kai

To my love and joy, Lim Tee Seng, Lim Bao Rong, and Lim Zi Kai.

Wai Yee

To my wife, Clarrisa, and my family for their unconditional love and support.

Hong Wei

To my wife, Li, for her continued support and love.

Zhang Yi

To my wife, Wen-Jun, and daughters, Min-Yi and Min-Yue, for their support.

U-Xuan

To my parents and wife, Christina Tan, for always being my pillar of support and strength, which made it possible for me to complete this book.

Chen Huei

To my wife and children — Katsuki, An, and Zentaro — for their love, support and inspiration.

Michinao

To my colleagues at Khoo Teck Puat Hospital for sharing this vision that 3D-printed food is the safe solution to consistent textured consistency for our patients with swallowing difficulties.

Gladys

To my beloved parents, Frederick Tan and Doris Tan, all my family members and loved ones for their love, support and encouragement.

Justin

To my loving parents, K. C. Pant and Jayshree Pant, brother, Ashish Pant, and the love and light of my life, Nikhil and Navya.

Aakanksha

# Foreword

Food is very important to people. Not just for sustenance and enjoyment, but it is also where people seek comfort, and express love and care for others. For many Singaporeans, it is even a passion, our heritage, and our way of life.

We can therefore imagine the anguish people have to go through when they have swallowing difficulties or dysphagia. These individuals are unable to consume regular meals that require chewing or are harder in texture. If the food is modified into softer texture, it may look unappetising. If they do not consume food properly, they may lose weight, develop malnutrition, and experience poorer health. Worse, they lose dignity, and enjoyment of life that are closely associated with food.

3D food printing tries to overcome these difficulties. It can personalise pureed food into more visually pleasing food to create an enjoyable dining experience to encourage diners to eat more. Khoo Teck Puat Hospital has been working with the Singapore University of Technology and Design (SUTD) on two 3D food printing research projects to provide patients with safe and palatable meals. In this book, you can find examples and images of how pleasant and appetising the food can look.

3D food printing is an important contribution towards creating a healthcare system that is ready for our ageing population in Singapore. Beyond helping patients, 3D food printing can incorporate different nutrients within the same food, and therefore has the potential of tailoring meals to anyone's nutritional preferences and needs, taste and convenience. This has a deep impact in the field of nutrition research and practice.

My heartiest congratulations to Professor Chua and his team for their efforts in pushing the boundaries of 3D food printing technology. In this area, Singapore is a leader in the world, and I hope SUTD will continue to play a pioneering role.

Ong Ye Kung
Minister for Health, Singapore

# Preface

In the last several decades, the food industry has seen many changes — new technologies have been introduced into the way we cook, manufacture, and purchase food. Digital gastronomy, which combines new computational abilities such as three-dimensional (3D) printing with traditional food preparation, has allowed consumers to design and manufacture food with personalized shapes, colours, textures, and even nutrition. Consequently, our eating experiences can extend beyond taste to include food preparation, culture, economy, and the sciences behind the food ingredients.

3D printing of food has attracted growing interest in recent years; it offers unique advantages not achievable by traditional means of food preparation — for instance, personalization of food properties such as shapes and textures to meet the unique requirements of individual consumers, automation in food preparation, and promoting food sustainability through 3D-printed cell-based meats and alternative proteins.

In general, 3D-printed foods are food products prepared by an automated additive process, layer by layer. Entire meals can be constructed just by 3D food printing alone. Development and understanding of 3D printing of food require a combination of various disciplines, specifically material science of the food inks, mechanical-electrical engineering for the manufacturing and processing of the 3D food printers, and computational engineering for the design of 3D-printed food models.

This textbook has been written to offer students, researchers, and engineers a thorough understanding of the concepts, processes, applications, and developments of 3D food printing. This textbook begins with an introduction to 3D food printing — its background, commercial trends, instrumentation, materials, and significance. A chapter is dedicated to the different commercially available food printers in the market. Four subsequent chapters are dedicated to the various classes of food ingredients employed in 3D food printing, which are desserts and snacks (comprising dairy products, chocolate, sugars, and dough), fruits and vegetables, meats and alternative proteins, and pharmaceuticals and nutraceuticals. The book next discusses the safety and regulations around 3D-printed foods, identifying and assessing their various food hazards and related safety management systems. Next, the book presents business model developments for two food ingredients, namely, chocolate and alternative proteins, and how 3D food printing has been used to impact the markets for these foods. The last chapter is entirely dedicated to identifying and discussing emerging technologies stemming from current 3D food printing and offering useful insights for current challenges and the outlook of 3D food printing in the future.

At the end of each chapter, a set of problems offer undergraduate and postgraduate students practice on the main ideas discussed within the chapter. For tertiary-level lecturers and university professors, the topic of 3D food printing can be associated with other subjects in food and nutrition, pharmaceutical and nutraceutical sciences, and food engineering.

Chua C. K.
Professor

Leo C. H.
Assistant Professor

Yeong W. Y.
Associate Professor

Hashimoto M.
Associate Professor

Tan H. W.
Research Fellow

Wong H. C. G.
Senior Principal Dietitian

Zhang Y.
Assistant Professor

Tan J. J. Y.
Research Fellow

Tan U.-X.
Associate Professor

Pant A.
Research Associate

# About the Authors

**Chee Kai CHUA** is Head of Pillar for Engineering Product Development and Cheng Tsang Man Chair Professor at the Singapore University of Technology and Design (SUTD). Dr Chua has extensive teaching and consulting experience in 3D Printing and Additive Manufacturing (3DP & AM). For over 31 years, he has been an active contributor to the AM (or 3D Printing) field, where his redesign of AM processes for innovative devices such as tissue engineering scaffolds are highly regarded by the scientific community. He is now active in 3D printing of electronics, food, metals, and polymers. He won the prestigious International Freeform and Additive Manufacturing Excellence (FAME) Award in 2018. As of 2021, he has contributed more than 400 technical papers and patents, generating more than 19,000 citations, and co-authored five books, including *3D Printing and Additive Manufacturing: Principles and Applications* (5th edition), *3D Printing and Additive Manufacturing of Electronics: Principles and Applications* (1st edition), and *Bioprinting: Principles and Applications*. In addition, he is the chief editor of *Virtual and Physical Prototyping*, as well as the chief editor of *International Journal of Bioprinting*. Professor Chua can be contacted by email at cheekai_chua@sutd.edu.sg.

**Wai Yee YEONG** is Associate Professor at the School of Mechanical and Aerospace Engineering (MAE), Nanyang Technological University, Singapore. She also serves as Associate Chair (Students) at MAE. She has published more than 150 papers, generating more than 6,100 citations with a current H-index of 40, and co-authored 3 textbooks (published by World Scientific and Elsevier, respectively). Her works have been featured on *CNA*, *The Straits Times*, and other media channels. Her portfolio also includes serving as Programme Director (Aerospace and Defence) at the Singapore Centre for 3D Printing and at HP-NTU Digital Manufacturing Corporate Labs. She has filed 6 patents applications and 14 know-hows. Her main research interests are in 3D printing, bioprinting, and the translation of advanced technologies for industrial applications. Her current research topics include 3D printing of new materials, hybrid electronic-mechanical structures, and bioprinting for tissue engineering. She was named the winner of the TCT Women in 3D Printing Innovator Award 2019. Associate Professor Yeong can be contacted by email at wyyeong@ntu.edu.sg, and more information can be found on her website: www.yeongresearch.com.

**Hong Wei TAN** received his B.Eng. (First-Class Honours) and Ph.D. degrees in Mechanical Engineering from Nanyang Technological University, Singapore. He has filed one PCT patent in sintering metallic nanoparticle inks for 3D printed electronics applications. He also has co-authored *3D Printing and Additive Manufacturing of Electronics: Principles and Applications* (1st edition) (published by World Scientific). His research interests are additive manufacturing processes for 3D-printed electronics and materials characterizations. Dr Tan can be contacted by email at hongweitan1990@gmail.com.

**Yi ZHANG** is currently Professor at the School of Electronic Science and Engineering, University of Electronic Science and Technology of China (UESTC). Prior to joining UESTC, he was Assistant Professor at the School of Mechanical and Aerospace Engineering at Nanyang Technological University (NTU), Singapore. He received his Ph.D. in Biomedical Engineering from Johns Hopkins University School of Medicine, US, in 2013, and his B.Eng. in Bioengineering from NTU in 2007. He received his postdoctoral training at the Institute of Bioengineering and Nanotechnology, Agency of Science, Technology and Research (A*STAR), Singapore, in 2015, and subsequently worked there as a research scientist until 2016. His research aims to develop micro- and nanobiomedical systems for diagnostics and therapeutics using advanced manufacturing technologies such as additive manufacturing. Professor Zhang can be contacted by email at yi_zhang@uestc.edu.cn.

**U-Xuan TAN** is currently Associate Professor at the Singapore University of Technology and Design (SUTD), Singapore, as well as the Programme Director for SUTD's Technology Entrepreneurship Programme. He completed his Ph.D. at Nanyang Technological University, Singapore, and subsequently received his postdoctoral training at the University of Maryland, College Park, US, from 2009 to 2011. In 2012, he joined SUTD. His research areas are in robotics, mechatronics, and 3D printing; more specifically, designing sensing and control approaches that are easily deployable for various applications covering the healthcare, defence, and aerospace sectors. Associate Professor Tan can be contacted by email at uxuan_tan@sutd.edu.sg.

**Chen Huei LEO** is currently Assistant Professor at the Singapore University of Technology and Design (SUTD), Singapore. He completed his Ph.D. in Pharmacology and Physiology from RMIT University, Australia, in 2012, and attained his Bachelor of Science (Hons) in Pharmacology from the University of Melbourne, Australia. Prior to joining SUTD, he was a research fellow at the University of Melbourne from 2012 to 2016. His current research focuses on nutritional pharmacology, food science technology, and green chemistry. Assistant Professor Leo can be contacted by email at chenhuei_leo@sutd. edu.sg.

**Michinao HASHIMOTO** is currently Associate Professor at Singapore University of Technology and Design, Singapore, and the principal investigator of Soft Fluidics Group. The overarching research interests of the Group are microfluidics and 3D printing, both fundamental and applied, with the current focus on stretchable devices, organ-on-a-chip, 3D bioprinting, and 3D food printing. Prior to his independent academic career, he received his Ph.D. in Chemical Physics from Harvard University, followed by postdoctoral training at the Massachusetts Institute of Technology and Boston Children's Hospital. Associate Professor Hashimoto can be contacted by email at hashimoto@sutd.edu.sg.

**Gladys Hooi Chuan WONG** is an Accredited Dietitian of the Singapore Nutrition & Dietetics Association. She trained and worked as a dietitian in New Zealand before relocating to Singapore in 1995 to pioneer the nutrition diploma at Temasek Polytechnic. She then returned to clinical/foodservice dietetics in 2000 at Alexandra Hospital as Chief Dietitian of Nutrition & Dietetics before leaving for the newly built Khoo Teck Puat Hospital in 2010. She relinquished her 17-year headship at the end of 2017 to concentrate on dietetic placement education and geriatric and psycho-geriatric dietetics, as well as innovative projects pertaining to e-learning, health promotion, food safety standards, sustainability, and 3D food printing. She strongly believes that 3D-printed food will be the future of safe and dignified feeding for individuals with swallowing and chewing difficulties. Ms Wong can be contacted by email at wong.gladys. hc@ktph.com.sg.

**Justin Jia Yao TAN** is currently a postdoctoral research fellow at Hashimoto Lab, Singapore University of Technology and Design, Singapore. He received his Ph.D. in the field of pharmaceutical sciences from the National University of Singapore, Singapore, in 2019, and his B.Sc. (Honours) in Pharmacy from the National University of Singapore, Singapore, in 2014. His research interest is in 3D food printing and its related applications for various functional purposes. Dr Tan can be contacted by email at justin_tan@sutd.edu.sg.

**Aakanksha PANT** is a research associate at the Singapore University of Technology and Design. She received her M.S. in Molecular Biology and Immunology from the University of Maryland Baltimore, US, in 2007, and her B.S. in Zoology from Delhi University, India, in 2003. From 2008 to 2013, she worked at the National University of Singapore, researching bacterial virulence and the development of nanodevices for transdermal delivery. Subsequently, from 2013 to 2017, she was with the Agency of Science, Technology and Research (A*STAR), Singapore, working on targeted photodynamic therapy in melanoma research. Her work at the Singapore Centre for 3D Printing at Nanyang Technological University, Singapore, focused on 3D-printed food for dysphagia patients. Ms Pant can be contacted by email at aakanksha_pant@sutd.edu.sg.

# Acknowledgements

First, we would like to thank God for granting us His strength throughout the writing of this book. Secondly, we are especially grateful to our respective spouses, Wendy and Tee Seng Lim, and our respective children, Cherie Chua, Clement Chua, Cavell Chua, son-in-law Darren (Cherie's husband) and daughter-in-law Lynette (Clement's wife), Bao Rong Lim, and Zi Kai Lim for their patience, support, and encouragement throughout the year it took to complete this book.

We wish to thank the valuable support from the administration of the Singapore University of Technology and Design (SUTD), Nanyang Technological University (NTU), and Khoo Teck Puat Hospital, especially to their respective departments, the Engineering Product Development (EPD) pillar and the School of Mechanical and Aerospace Engineering (MAE).

We would like to extend our special appreciation to Mr Ong Ye Kung, Minister for Health of Singapore, for his foreword. We would also like to express our special gratitude to Meydan Levy, Dov Ganchrow, Bogdan Sokol, Shay Maman, Esti Baranets, Emilya Galdon, and Alona Balona for providing us with the Neo Fruit pictures to use in our book.

The acknowledgements would not be complete without the contributions of the following companies for supplying and helping us with the information about the products they develop, manufacture, or represent:

1. Aprecia Pharmaceuticals, LLC
2. byFlow
3. Craft Health Pte Ltd
4. FabRx Ltd

5. Natural Machines
6. Nourished (www.get-nourished.com)
7. Shiyintech
8. Wiiboox

Lastly, we would like to thank you, our readers. Your suggestions, corrections, and contributions will be appreciated and reflected in the later edition of this book.

Chua C. K.
Professor

Yeong W. Y.
Associate Professor

Tan H. W.
Research Fellow

Zhang Y.
Assistant Professor

Tan U.-X.
Associate Professor

Leo C. H.
Assistant Professor

Hashimoto M.
Associate Professor

Wong H. C. G.
Senior Principal Dietitian

Tan J. J. Y.
Research Fellow

Pant A.
Research Associate

# Contents

# List of Abbreviations

| | |
|---|---|
| 2-ACBs | 2-alkylcyclobutanones |
| 3D | Three-Dimensional |
| 3DFP | Three-Dimensional Food Printing |
| 4D | Four-Dimensional |
| ABS | Acrylonitrile Butadiene Styrene |
| ADSCs | Adipose-Derived Stem Cells |
| AI | Artificial Intelligence |
| ALM | Additive-Layer Manufacturing |
| AM | Additive Manufacturing |
| API | Active Pharmaceutical Ingredient |
| APs | Alternative Proteins |
| ASTM | American Society for Testing and Materials |
| bADSCs | Bovine Adipose-Derived Stem Cells |
| BMC | Business Model Canvas |
| bSCs | Bovine Satellite Cells |
| BSF | Black Soldier Flies |
| BSFL | Black Soldier Flies Larvae |
| Ca | Capillary Number |
| CAGR | Compound Annual Growth Rate |
| CAD | Computer-Aided Design |
| CE | Conformitè Europëenne |
| CFR | Code of Federal Regulation |
| Ci3DP | Chocolate-Based Ink Three-Dimensional Printing |
| CJP | Colour Jet Printing |
| CNFs | Collagen Nanofiber Solution |

| | |
|---|---|
| CT | Centralized Tomography |
| CVD | Cardiovascular Diseases |
| dECM | Decellularized Extracellular Matrix |
| DIW | Direct Ink Writing |
| DIY | Do It Yourself |
| DLP | Digital Light Processing |
| DMEM | Dulbecco's Modified Eagle Media |
| DNA | Deoxyribonucleic Acid |
| DPE | Direct Powder Extrusion |
| DSP | Drawing Soy Protein |
| ECM | Extracellular Matrix |
| ECM | Extra Cellulose Matrix |
| EFSA | European Food Safety Authority |
| EMA | European Medicines Agency |
| ES | Embryonic Stem |
| ESC | Embryonic Stem Cell |
| EU | European Union |
| FABP4 | Fatty-Acid-Binding Protein 4 |
| FAO | Food and Agriculture Organization |
| FBS | Fetal Bovine Serum |
| FDA | Food and Drug Administration |
| FDM | Fused Deposition Modelling |
| FFF | Fused Filament Fabrication |
| $G*$ | Complex Modulus |
| $G'$ | Storage Modulus |
| $G''$ | Loss Modulus |
| GC | Ground Chicken |
| GelMA | Gelatin Methacryloyl |
| GFP | Green Fluorescent Protein |
| GG | Guar Gum |
| G-Gel | Granular Particles of Gelatin |
| G-GG | Granular Particles of Gellan Gum |
| GHG | Greenhouse Gas Emissions |
| GI | Gastrointestinal |
| GMO | Genetically Modified Organisms |
| GMP | Good Manufacturing Practice |
| GRAS | Generally Recognized as Safe |
| H&E | Hematoxylin and Eosin |
| HACCP | Hazard Analysis Critical Control Points |

| | |
|---|---|
| HACM | Heat Acid Coagulated Milk |
| HCAs | Heterocyclic Amines |
| HDMP | Heat Desiccated Milk Powder |
| HEK-293 | Human Embryonic Kidney-293 |
| hMPC | Human Muscle Progenitor Cell |
| HPC | Hydroxypropyl Cellulose |
| HPMC | Hydroxypropyl Methylcellulose |
| HSPC | High-Speed Printed Cheese |
| HUVECs | Human Umbilical Vein Endothelial Cells |
| ICCO | International Cocoa Organization |
| IDDSI | International Dysphagia Diet Standardization Initiative |
| iPSCs | Induced Pluripotent Stem Cells |
| ISO | International Organization for Standardization |
| ITOP | Integrated Tissue-Organ Printing |
| $K$ | Flow Consistency Index |
| KC | Kappa Carrageenan |
| LB | Locust Bean |
| LBG | Locust Bean Gum |
| LCA | Life Cycle Assessment |
| LSPC | Low-Speed Printed Cheese |
| LVR | Linear Viscoelastic Region |
| MHC | Major Histocompatibility Complex |
| MIT | Massachusetts Institute of Technology |
| mRNA | Messenger Ribonucleic Acid |
| MSC | Mesenchymal Stem Cell |
| $n$ | Flow Behaviour Index |
| NAMIC | National Additive Manufacturing Innovation Cluster |
| NCDs | Non-Communicable Diseases |
| NFG | Nanoemulsion-Filled Gel |
| NMR | Nuclear Magnetic Resonance |
| NSRDEC | Natick Soldier Research, Development and Engineering Center |
| O/W | Oil-in-Water |
| O/W/O | Oil-in-Water-in-Oil |
| OD | Oropharyngeal Dysphagia |
| ODT | Orally Disintegrating Tablet |
| PAH | Polycyclic Aromatic Hydrocarbons |
| PAT | Process Analytical Technology |
| PBMAs | Plant-Based Meat Alternatives |

| | |
|---|---|
| PCL | Poly(E-Caprolactone) |
| PDCAAS | Protein Digestibility-Corrected Amino Acid Score |
| PDMS | Polydimethylsiloxane |
| PE | Polyethylene |
| PEG | Poly(Ethylene Glycol) |
| PEGDA | Polyethylene Glycol Diacrylate |
| PEGMA | Poly(Ethylene Glycol) Dimethacrylate |
| PEH | Pseudoephedrine Hydrochloride |
| PET | Polyethylene Terephthalate |
| PP | Polypropylene |
| PPAR $\gamma$2 | Peroxisome Proliferator-Activated Receptor-$\gamma$2 |
| PPI | Pea Protein Isolate |
| PS | Phytosterol |
| PVA | Poly(Vinyl Alcohol) |
| PVP | Polyvinylpyrrolidone |
| RSM | Response Surface Modelling |
| RWF | Refined Wheat Flour |
| SEM | Scanning Electron Microscopy |
| SFA | Singapore Food Agency |
| SHASAM | Selective Hot Air Sintering and Melting |
| SLA | Stereolithography Apparatus |
| SLS | Selective Laser Sintering |
| SMP | Skimmed Milk Powder |
| SSE | Semi-solid Extrusion |
| SSMP | Semi-Skimmed Milk Powder |
| tan $\delta$ | Loss Tangent |
| TEC | Ethanolic Triethyl Citrate |
| Tg | Glass Transition Temperature |
| TIP | Tendon-Gel-Integrated Bioprinting |
| TNO | Netherlands Organisation for Applied Scientific Research |
| TPA | Texture Profile Analysis |
| TPO | Diphenyl (2,4,6-Trimethylbenzoyl) Phosphine Oxide |
| TSP | Textured Soybean Protein |
| UCL | University College London |
| UCLA | University of California, Los Angeles |
| UNICEF | United Nations Children's Fund |
| UV-C | Ultraviolet C |
| VPC | Value Proposition Canvas |
| W/O | Water-in-Oil |

| | |
|---|---|
| W/O/W | Water-in-Oil-in-Water |
| w/w | Weight by Weight |
| WC% | Water Content Percentage |
| WHO | World Health Organization |
| XG | Xanthan Gum |

# Chapter 1

# An Introduction to the Principles of 3D Food Printing

## 1.1 3D Food Printing: Background

Additive manufacturing, also known as three-dimensional (3D) printing, is the layer-by-layer production of structures using a variety of technologies, such as extrusion-based printing of *food inks*. In recent years, 3D food printing (3DFP) has begun receiving particular attention both commercially and academically, as it can harness 3D printing technology to create food items personalized in both content and design. Examples of foods that have been printed include not only decorative structures of chocolate or sugar, but also nutritious foods such as vegetables and dairy products. The targeted segment includes both healthy people (e.g., astronauts, soldiers) and patients.

The health condition of healthy subjects can be critical, especially if their job requires them to maintain a high level of health. Special efforts are put in to ensure a high level of health of the space crew due to the demands of the mission. However, nutrition actually gets challenging when they are away from Earth and out in space. In addition, a space mission can take years, and the food is packed beforehand. It is a challenge to provide precise amounts of needed nutrients as well as minimize the amount of trash. Hence, efforts have been made to explore a 3D-printed food system for long-duration space missions [1].

There has also been an increasing interest in utilizing 3DFP for healthy subjects like soldiers. Similar to space crew, soldiers need to maintain a high level of health, and nutrition becomes challenging when they are out in the field for a long duration. Food technologists at the US Army Natick Soldier Research, Development and Engineering Center (NSRDEC) have been looking into how 3DFP can be leveraged to enable personalized nutrition to meet the requirements of the individual soldier while in the field. The individual nutritional requirements of the soldiers can be sent to a 3D food printer, which will print out meals with proper amounts of vitamins and minerals [2]. With such motivation, studies have also been conducted to explore the attitudes towards 3D-printed food in a real-life military setting. The initial findings demonstrate that repeated consumption of 3D-printed foods increases consumers' acceptance of 3D-printed food [3].

There has also been an increasing interest in a personalized diet for prevention of diseases [4]. Diet is an important component for a healthy life, and 3DFP has the potential to provide customized amounts of protein, sugar, vitamins, and minerals into the printed food for consumption. This will allow one to have a customized diet to meet individual nutrition requirements.

With the rapidly ageing population, there is an increasing demand around the world (especially among developed countries) to address age-ing issues that are arising, and one of the issues that a number of the elderly are facing is the loss of ability to enjoy food, which is caused by an illness known as dysphagia (dysphagia patients can include the mentioned elderlies as well as stroke patients who lose control of their muscles). Food is critical to provide the elderly with energy and nutrients. Nutrient management for the elderly is a growing market with increasing importance as keeping them healthy is much lower in cost for society. Unfortunately, as elderlies age, they start to lose muscles. This results in dysphagia, which is a challenge to eating due to a difficulty in swallow-ing. In this area, 3DFP has surfaced repeatedly as a potential solution [5–8]. Various dishes such as pork [5], vegetables [6], and black fungus [7] have been investigated and shown to be printable. Given the increas-ing interest as illustrated in this research, there is a huge potential for 3DFP to become a possible solution for dysphagic patients.

Other than nutritional consideration, 3DFP has also gained traction in other considerations. One of them is food sustainability, and food wastage is a global concern. Waste as well as the by-products of the food industry contribute significantly to food waste, and 3DFP has shown its potential to reduce food waste [9]. 3DFP can achieve this by utilizing undesired food products like seeds and distorted fruits and vegetables. Jagadiswaran *et al.* have investigated the utilization of food industry waste streams, and 3D printing was used to prepare grape pomace powder-incorporated broken wheat (flour) cookies [10]. Using printing technologies, foods can be printed on demand, which will reduce the amount of food waste generated [11].

In parallel, there has been an increasing interest in exploring alternative ingredients like insects and algae as they are rich sources of protein, dietary fibre, and other benefits. Studies have shown that using 3D printing technologies to render ingredients such as insects is more acceptable and appealing to consumers [12].

Cafés and restaurants have also gained interest in exploring 3DFP for food decorations. With the availability of commercial 3D food printers, interesting customized edible 3D models can be printed to serve as decorations in a dish. Similarly, personalized print can be done on the top layer of drinks like coffee. More details and examples will be mentioned in Chapter 2.

With these motivations, 3DFP has been gaining popularity since the first academic paper published in 2007. Since then, there has been an increasing demand for 3DFP and the market for 3DFP has been growing worldwide.

## 1.2 Mechanism and Instrument of 3D Food Printing Techniques

### 1.2.1 *Material Extrusion*

There are several 3DFP techniques, and extrusion is currently the most common technique among commercial printers. The general principle of the extrusion approach is that the food ink is pushed through a nozzle and extruded out during the printing. The actuating mechanism can be realized by using a piston to squeeze the food ink out or by using compressed air, as illustrated in Figure 1.1.

There is a range of nozzle diameter sizes, and a balance between printing speed and resolution is required. The smaller diameter size is good for better resolution, but the printing time will be longer. After printing, some dishes will require additional cooking processing such as baking. Examples of food ink will be introduced in the later part of this chapter.

Force/pressure applied
to push the food ink

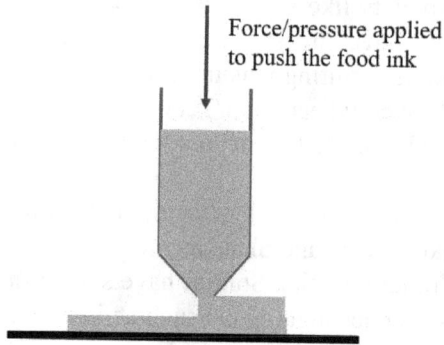

Figure 1.1.   Extrusion process.

## 1.2.2 *Selective Laser Sintering*

Selective laser sintering (SLS) of powdered food materials is another 3DFP technique. In this process, the food material can be in powdered form and is selectively heated using a laser as the heat source, as illustrated in Figure 1.2. The heat in the targeted region will bind the powder together. After the current layer is prepared, a new layer of powder is applied. The levelling roller or brush is used to ensure the thickness of the next layer. This process is then repeated layer by layer until the whole print is done.

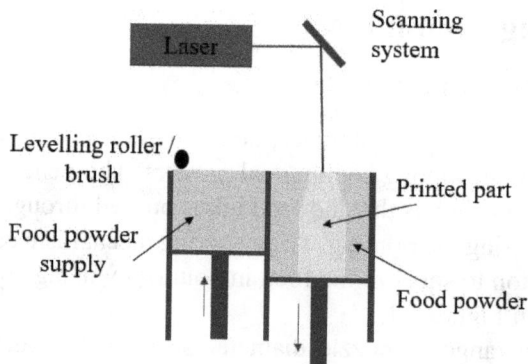

Scanning
system

Laser

Levelling roller /
brush

Printed part

Food powder
supply

Food powder

Figure 1.2.   Selective laser sintering process.

The SLS approach can be used to print complex shapes with good resolutions. There are limited food printers using this technique because research in the food materials still needs further investigation and development.

### 1.2.3 *Binder Jetting*

The food materials used for binder jetting are typically powder based. Instead of using heat sources like a laser to bind the selected region, a liquid binder is used to bind the selected region of the food powder together. The general idea is illustrated in Figure 1.3.

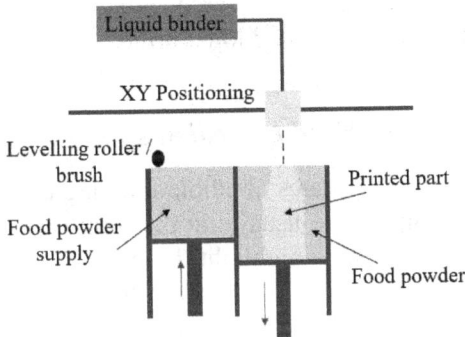

Figure 1.3.   Binder jetting process.

A *XY* positioning system is placed inside the printer to deposit the binder at the desired location to bind the food material. Similar to selective laser sintering, a new layer of powder is spread over it when the current layer is completed. After that, a levelling roller is used to ensure the consistency of each applied layer. This process is repeated until the whole print is completed. Again, similar to selective laser sintering, research in the food materials (i.e., powdered ingredients) and compatible binders still needs further investigation and development.

## 1.3 Materials of 3D Food Printing Techniques

3DFP is a recent application of additive manufacturing, with the first academic articles published in the field around 2007. Applicable materials depend on the category of 3D printing. ASTM International has defined seven additive manufacturing categories, four of which have been effectively employed in 3DFP, namely, binder jetting, material extrusion, material jetting, and powder bed fusion using selective laser sintering [13]. Material extrusion has been by far the most widely utilized in 3DFP research because a broad range of food ingredients, from chocolate to vegetables, can be processed. Extrudable food materials can be formed using simple processes such as melting and blending.

3D-printed foods in academic research were mostly based on materials that can be easily extruded, including frosting, cheese, chocolate, edible hydrogels, and cookie dough; a few outliers such as pureed celery combined with hydrocolloids have been demonstrated [14]. Furthermore, there has been a significant increase in the number of original research over the last two years, substantially broadening the range of food ingredients that have been demonstrated to be printed.

### 1.3.1  *Classification of Food Ingredients for 3D Food Printing*

Food inks are the printable states of edible food ingredients consisting of one or more raw components, usually in the form of a paste or gel. The term *food inks* is used here to highlight the popularity of extrusion food printing from a printer nozzle, while other forms of raw materials are required according to the method of printing.

We can define five kinds of food inks based on the ingredients: (1) confectionary inks such as chocolate or dough, (2) dairy inks obtained from milk or its derivatives, (3) edible hydrogel inks, (4) plant-based inks, and (5) meat-based inks [14]. Based on the general objective of 3D food printability, food inks can be divided into two major categories: natively extrudable (confectionery, dairy, hydrogels) and non-natively extrudable (plants, meat) [14]. The distinction between natively printable and nonprintable food materials has been briefly mentioned in previous publications [15]. Some food inks by this classification may not be mutually exclusive. For instance, potato starch suspensions would be classified as hydrogels, whereas mashed potato would be classified as a plant-based ink. While this method of classification may bring some arbitrariness for the individual food inks, the goal of this classification is to confer an intuitive approach that is easily applied to most of the food inks. Other classification methods, such as traditional (e.g., chocolate, cheese, fruit, and vegetables) and non-traditional food materials (e.g., insects, seaweeds, plant, and animal by-products) [16], and nutrition-based classification (e.g., carbohydrate, fat, and protein) [17], have been reported in the literature.

## 1.3.1.1 *Food inks for extrusion-based printing*

(1) Natively extrudable food inks

Natively extrudable food inks are inks with desirable viscoelastic characteristics that allow direct extrusion without the use of matrix ingredients like gum or binder. However, these additives may still be employed to enhance print quality. This category of inks is easily extruded via a nozzle and has been the subject of most of the academic research in 3DFP. These inks are classified into three types: confectionery (i.e., chocolate, dough), dairy, and hydrogels [14].

• Confectionery: Chocolate, dough

Chocolate was one of the first materials employed in 3DFP, as demonstrated by Schaal in 2007 [14]. Following this work, a group at the University of Exeter designed the ChocALM prototype printer in 2010 [18]. This printer was eventually refined and commercially launched as the Choc Creator V1 under the Choc Edge label in 2012 [14]. A few more chocolate printers were launched in the following years: Choc Creator V2 and V2.0 Plus, CocoJet from 3D Systems and Hershey, and Chocobot from Global 3D Labs [14].

Despite the emergence of these commercial printers, there were few studies on chocolate printing until the publication of three papers in 2017 addressing, respectively, the determination of printing parameters and the influence of supports and additives [19], the creation and optimization of a low-cost printer utilizing open-source parts [20], and the use of 3D-printed chocolate treats as a self-monitoring system of an individual physical activity [21].

The improvements to the printing process and platform in chocolate 3D printing have remained the focus of recent research [14]. Karyappa *et al.* demonstrated the direct printing of chocolate-based inks (i.e., chocolate paste and syrup formulated to have appropriate rheological properties) using a direct ink writing (DIW) 3D printer to produce complex 3D shapes without the use of temperature control [22] (Figure 1.4). This research also focused on the printing of chocolate-based models containing different ink types, including those with features of a semi-solid enclosure and liquid filling using multiple nozzles. In another approach, Takagishi *et al.* created an electrostatic inkjet printer, which employs a nozzle with a thin fibre of ABS resin at the tip to reduce droplet size that allows the printing of chocolate with a precision of below 50 μm [23].

Figure 1.4.   (a)–(e) Optical images of 3D chocolate models printed with direct ink writing 3D printing (Scale bar = 1 cm) [11]. Reproduced from [11], which is licensed under CC BY 4.0.

Electrostatic inkjet printing overcame the challenges inherent in the printing of highly viscous foods such as molten chocolate.

While commercial dark chocolate and milk chocolate are widely used as food inks, the inclusion of additives has been shown as an effective route to fortify nutrients or to enhance 3D printability. Milk chocolate containing vitamin C or cranberry powder and methylcellulose [24], and dark chocolate with magnesium stearate or plant sterol powder [19, 25], have been reported. The addition of magnesium stearate and plant sterol powder has been shown to improve the hardness of the printed chocolate [25]. Moreover, magnesium stearate has been shown to delay the crystallization of chocolate, thus increasing the extrudability of the chocolate ink during printing. In addition, magnesium stearate has shown anti-sticking and lubricating properties [19]. Despite these advantages, Hao *et al.* noted that the addition of too high a quantity of dry powders could lower the fat ratio, which leads to overly fluid chocolate that compromises the printability [24]. This limit was determined to be about 1% by weight, after which the addition of methylcellulose was required to avoid the collapse of the printed chocolate. Further increase in the methylcellulose concentration, however, again resulted in the collapse [24]. Mantihal *et al.* managed to print chocolate containing up to 5% additive content without the use of methylcellulose or other thickeners [26]. However, it is uncertain whether this difference is attributed to the type of chocolate or additive used, and more studies are required to understand the effects of interactions between food ingredients on the chocolate printing performance.

Cookie dough is another confectionery product under active investigations in 3DFP. Cookie dough usually comprises common ingredients such as water, butter, and sugar [27, 28]. In some cases, additives such as probiotics [29] and ground mealworm powder [30, 31] were added. Furthermore, post-processing of dough (such as baking) necessitates the development of a formulation that is both extrudable and immune to deformation after baking. Such a formulation is usually achieved by changing the ratios of its components experimentally [27–29]. Some research groups have reported employing rapid cooling between printing and baking [32], and using texture modifiers (e.g., calcium caseinate [29], xanthan gum [33], and pea protein [34]) to enhance the printability and processability of the 3D-printed dough. Ideally, these modifications enable the dough to be 3D-printed and baked while retaining a similar flavour and texture to traditional cookies without prominent changes to the quality of the foods.

- Dairy

A variety of dairy-based inks have been explored for 3D printing. Such inks contain primarily semi-skimmed milk powder (SSMP), skimmed milk powder (SMP) [35], milk protein [36, 37], and processed/semihard model cheese [38, 39]. Research in dairy-based inks has mostly focused on identifying the influence of composition on material characteristics and printability. For example, adding whey protein isolate [36] and increasing the total protein content [37] were shown to enhance the printability of the milk protein paste. Other dairy products and derivatives have also been studied, including heat-desiccated milk powder and heat acid coagulated milk. Joshi *et al.* found that heat-desiccated milk powder and semi-skimmed milk powder mixed in the ratio of 55:5 respectively (60% by weight) yielded the highest print fidelity at the extrusion temperature of 40 °C [40]. Bareen *et al.* showed that 4% w/w whey protein isolate, 2% w/w maltitol, and 94%

w/w heat acid coagulated milk gave the best printability and structural stability of the printed constructs [41]. A variety of dairy products and derivatives have also been utilized as components in various food inks, including milk in dough [33], dried non-fat milk in fruit snacks [42], and sodium caseinate in gels [43]. Apart from experimenting with the type of constituent within the milk ink formulation to influence its printability, changes to the milk powder concentration have also been demonstrated as a facile way to improve the milk ink printability. Lee *et al.* demonstrated that milk ink can be printed successfully into complex shapes at room temperature by only changing the concentration of the milk powder without the use of rheological modifiers [44] (Figure 1.5). The exclusion of multiple additives reduces the complexity involved in optimizing the pre-identified concentration of each component to achieve the overall printability of the milk ink. Additionally, researchers have also demonstrated multi-material printing with milk and other edible inks, including chocolate, coconut, maple syrup, and blueberry inks using several syringes and nozzles [44]. This demonstration highlighted the capability to fabricate 3D milk models with a rigid enclosure containing different fillings or different layered structures to offer a variety of tastes and textures.

(a)                                                                      (b)

Figure 1.5.    Two different designs of 3D-printed milk models using direct ink writing 3D printing: (a) wheel and (b) cloverleaf.

3D printing for cheese is another interesting class of demonstration. The levels of calcium and pH were found to influence the gel-sol transition temperature and the critical shear rate (going above would result in unstable flow and melt fracture). The gel-sol transition temperature and critical shear rate would then define the printing temperature for hot-melt extrusion printing of cheese [39]. Additionally, the melting and shearing involved in the printing have been shown to influence the texture of processed cheese. Tohic *et al.* reported a lower hardness of 3D-printed cheese than untreated commercially processed cheese samples. The shearing during extrusion 3D printing has an additional impact on the textural properties of processed cheese [38]. Differences in textural properties between 3D-printed cheese and unprinted (i.e., processed in a regular way of production) cheese have also been reported by Ross *et al.* [45]. Additionally, cheese samples printed at higher temperature (65 °C) were more resilient, less adhesive, gummier, harder, and springier than those printed at a lower temperature (40 °C) [45].

These findings emphasize the importance of understanding the impact of dairy product characteristics and their composition on the 3D printing process, and vice versa; the knowledge gained in these studies would allow us to exploit the abilities of dairy-based food inks to form structures. Such considerations apply beyond 3DFP. For example, pharmaceutical industries employ pure lactose as a bulking agent in tablets [45].

• Hydrogels

In early work, Cohen *et al.* demonstrated 3D printing of edible hydrogels with a wide variety of textures (i.e., weak to firm) and mouth feels (i.e., smooth to granular). They used combinations of hydrocolloids (i.e., xanthan gum and gelatin) infused with food-grade flavour concentrates such as raspberry, strawberry, banana, and chocolate [46]. While they have proposed using a minimal number of food materials with only two hydrocolloids to simulate different food textures and mouth feels, the range of materials applicable in hydrogel-based 3DFP has only risen over the years [14]. Given the multiple research works in hydrogel printing (bioprinting, precision medicine in addition to food printing), such trends are indeed expected.

Current developments in 3D printing of edible hydrogels aim to accomplish one or more of the following objectives: (i) finding new hydrogel inks, (ii) finding the effect of constituents and their concentrations, (iii) identifying the printing parameters for selected hydrogel inks, (iv) investigating novel printing technologies, and (v) developing theoretical simulations or models for the ink and/or the process of printing [14].

(i) Finding new hydrogel inks

Given the diversity in the available hydrogels, research in the development of 3D printable hydrogel inks has prioritized breadth over depth. A large number of hydrogel inks and their composites have been demonstrated for 3D printing [14]. Current research has favoured formulating complex inks with multiple constituents, including pectin [47], plant-based proteins [48], sodium caseinate [43], hydrocolloids [49], and egg white [50].

(ii) Finding the effect of constituents and their concentrations

The composition of hydrogels is one of the most crucial variables that determine the printability of hydrogel inks. Most literature has recorded its optimization to evaluate the influence of various compositions on hydrogel rheology and printability, while other variables including micro-structure and gelation temperature are also related. For instance, Chen *et al.* examined suspensions of several starches at varying concentrations to determine their associated effects on the rheological characteristics of the ink. The inks were subsequently printed using hot extrusion to evaluate their printability and the quality of the printed models, which were correlated with their rheological data. Rice starch suspensions with concentrations of 15–25% w/w printed at 80 °C exhibited excellent shape stability [51]. Liu *et al.* conducted a similar study on the printability and rheology of a multi-component carrageenan-xanthan gum-potato starch gel by adjusting the concentrations of the hydrocolloids; inks having 1% w/w κ-carrageenan (KC), 0.25–0.5% w/w xanthan gum (XG), and 2% starch produced the best results for printing [52]. These experiments suggest that the printability of hydrogel inks is highly dependent on the material composition.

(iii) Optimizing printing parameters for selected hydrogel inks

Some research has aimed to identify printing parameters, including the movement speed, height of the nozzle, the diameter of the nozzle, and the rate of extrusion [53]. In addition, some parameters specific to the printing platforms and materials may be critical in hydrogel printing; for instance, gelation temperature, mode of gelation, and the arrangement of multiple types of hydrogels are the key to achieving printability. For example, Warner *et al.* examined the consistency and dimensional accuracy of 3D printing of gelatin-carrageenan gel based on its printing temperature, which was set at or above its gelation temperature [54]. Liu *et al.* studied dual-extrusion printing of strawberry juice gel and mashed potato; they also identified the extruder offset and retraction value to ensure replicability of the printed model with its intended design. They also demonstrated the prevention of ink dripping when the extruders were switched or paused during printing [55]. Overall, this research highlights the identification of printing parameters for hydrogel inks to ensure printability and shape retention.

### (iv) Identifying novel printing technologies

Research efforts are made towards improving and creating novel hardware for improved functions. The desired functions of hydrogel printing include reducing the print duration, facilitating the ink preparation, and so on. Vancauwenberghe *et al.* created a co-axial extrusion print head, which facilitated the dispensing of a pectin-based ink within the inner flow and calcium chloride cross-linking solution within the outer flow, removing the need for the post-treatment step, which required the incubation of the 3D-printed pectin structure to be cross-linked within a calcium chloride bath. This capability offered improved control over the texture of the printed objects [56]. Similarly, Diañez *et al.* demonstrated the modification of an existing 3D printer to perform *in situ* gelations of 3D-printed KC inks within several minutes; the developed process generated comparable rheological properties to those prepared in the conventional process of gelation [57]. Another instance of hardware modification was demonstrated by Schutyser *et al.* when they configured the dispensing nozzle to have a side inlet to facilitate the flow of an olive oil phase into sodium caseinate gel ink to modify the texture of the printed objects [43]. Although regulating the pressure for simultaneous deposition was challenging, this method was able to generate 3D-printed structures with particular spatial distributions of the oil phase, which is not possible by pre-mixing the oil into the gel before printing [43].

### (v) Developing theoretical simulations or models for the ink and/or the process of printing

Recent research has employed quantitative models to determine experimental outcomes. For example, response surface modelling (RSM) enables the identification of the desired printing speed to achieve a specific gel strength, using the experimental data involving relevant variables (e.g., time and temperature) [57]. Other examples include numerical models, for instance, modelling the effects of printing parameters such as shear rate, velocity, and pressure fields in the flow channel from barrel to nozzle [58].

The last two objectives (i.e., printing technologies and theoretical models) may be of particular relevance in future research. Although the discovery of new hydrogel inks may lay the foundation for developing conventional and novel food types, advances in printing technologies and the development of theoretical models may be advantageous for 3DFP in general. These advances will also contribute to scaled-up food production, testing, and standardization [14].

## (2) Non-natively extrudable food inks

Not all types of foods can form extrudable pastes in their original form. For such foods, additional processes are necessary to create a printable food ink to ensure the extrusion from the 3D printing nozzles. In this section, we discuss examples of food inks that fall into such a category.

- Plants: Fruits and vegetables

This section discusses food inks produced from solid plant components in the form of pastes or slurries. The pastes derived from plants can be printed in varying shapes (Figures 1.6 and 1.7). Concentrated fruits or fruit juices derived from plants have been added to hydrogels to form extrudable inks, but they are not discussed in this section. In the 3D printing of plant inks, five stages have been recognized: (1) selecting materials,

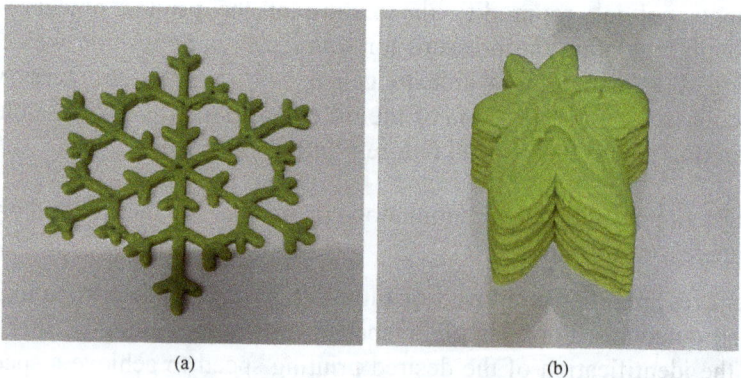

Figure 1.6. Two different designs of 3D-printed garden pea models using direct ink writing 3D printing: (a) snowflake and (b) flower.

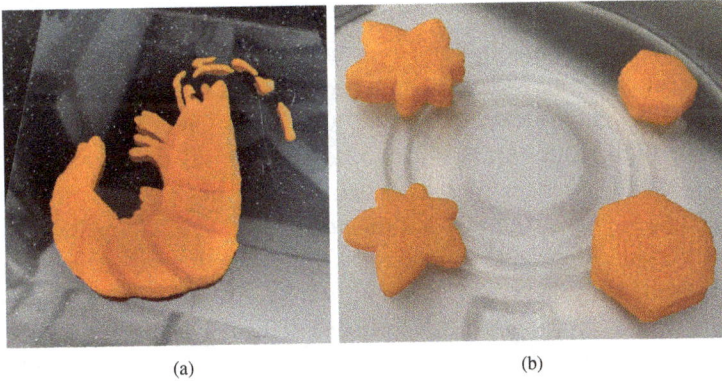

(a) (b)

Figure 1.7. Different designs of 3D-printed carrot ink (containing 0.3% w/w XG) models using direct ink writing 3D printing: (a) prawn and (b) two different designs including flower and hexagonal shapes.

(2) determining the component composition, (3) producing food inks, (4) deciding the conditions for printing, and (5) increasing the shelf life of the finished structures [59]. First, differences in characteristics across batches should be reduced or controlled. Small differences in the characteristics of inks might result in a pronounced departure in printing outcomes [60]. Second, the ink viscosity can be modified using thickeners by separating the solid and liquid phases [61]. Lastly, given that most fruits and vegetables are perishable, the storage and stability of 3D-printed foods need to be addressed unless the printed foods are consumed immediately. Inks that require drying after printing may benefit from a higher dry matter composition to decrease water loss. Deformation of shapes due to drying can also take place [35].

Mashed potato has received a lot of attention in 3DFP. It may be obtained directly from raw potatoes and gelatinized potato flakes or powders [62, 63]. Even though mashed potato is extrudable on its own, the addition of additives such as agar and alginate [62], or hydrocolloids such as KC and XG [63] at low concentrations (1% w/w), has been shown to enhance the printing outcomes in terms of its surface quality, shape stability, and texture. Researchers have compared the microstructure and texture between printed and casted mashed potatoes [63] and performed dual-nozzle extrusion using mashed potato and strawberry juice gel [55]. Kim *et al.* demonstrated other uses of additives for plant inks; they tested the printability of inks produced from three different freeze-dried vegetable powders (i.e., broccoli, spinach, and carrot) with four different types of hydrocolloids (i.e., XG, guar gum [GG], locust bean [LB] powder, and hydroxypropyl methylcellulose) [64]. Only XG gave 3D-printed structures with high resolution without a collapse in all the three vegetable powders tested, even at high concentrations. These observations were attributed to the high water hydration capacity of XG, which allowed it to prevent variances induced by powder swelling and resulted in inks with good extrudability and shape stability [64]. Although all the above research studies employed fresh or processed plant materials as a major part of the ink, other approaches, such as incorporating live plant cells into edible bio-inks, can also be used to generate artificial plant tissue with excellent cell viability within a 3D-printed structure [48].

• Meat

In recent years, meat production has increased steadily with the rising global demand for meat consumption, and its continued growth is expected in the near future [65]. This increase in conventional meat production poses issues relating to food security, public health, as well as loss of biodiversity and animal welfare stemming from livestock production [66]. 3DFP offers an excellent opportunity to salvage meat by-products by modifying them into customized meat products exhibiting similar or even better nutritional content, texture, and taste than high-value meat cuts. By combining a variety of food materials such as meat paste and fat slurry, recombined meats can be created. 3D printing of meats contributes to improving the appearance, nutritional value, and flavour of meat products while addressing environmental sustainability [67].

Fibrous ingredients of such meats and by-products are considered to be natively non-extrudable [68]. They require the addition of flow enhancers such as hydrocolloids and fats to modify their rheological properties to yield an extrudable paste-like ink for printing. Binders such as the enzyme transglutaminase can be added to ensure that adjacent printed layers stay adhered to one another upon printing. For instance, it has been shown that the printability of meat slurry containing chicken, pork, and fish can be improved by adding gelatin [68]. In the same way, before printing, transglutaminase can be added to turkey and scallop pastes as a binder to hold the printed meat structures together [69].

A number of research groups have attempted to 3D-print a variety of meats, including chicken nuggets [70], *surimi* [71–73], beef [74, 75], and pork [5]. The effects of additives on the printability of the meat inks were investigated. For instance, increasing the sodium chloride concentration has been shown to increase the printability of *surimi* gels [71]. For beef, the addition of hydrocolloids such as XG and GG increased the viscosity of the pork pastes and ensured solid-like behaviour with a self-supporting capacity over time after printing [5]. *Surimi* inks with 12% w/w potato starch produced the least deformed and most stable samples by co-axial extrusion-based 3D printing [72]. Because post-processing may result in weight loss and size shrinkage, the influence of the post-processing method on the cooking loss and texture of post-processed meats has been evaluated. Relevant process parameters (e.g., infill density [74], transglutaminase concentration [73], hydrocolloid addition [75, 76]) were evaluated.

An important application of 3DFP is the preparation of customized 3D-printed meals for individuals with chewing and swallowing difficulties (i.e., dysphagia) [5]. To reduce choking risks, texture-modified diets are suggested for dysphagic patients. These diets are made of thickened fluids and minced and pureed foods; such foods can be unappetizing and distressing to these individuals consuming them [5]. In addition to the aesthetics of the foods, some studies suggested that diets comprising 3D-reformed pureed and minced foods elevate the energy levels, nutrient intake, and weight of patients compared to the conventional texture-modified diets served at the institutions [77–79]. With 3DFP, custom-made meals could be produced to resemble the original shape of the food while retaining the textural characteristics suitable for consumption by dysphagic individuals.

In this aspect, the texture of 3D-printed post-processed meat was investigated by various research groups to ensure that they meet the requirements of the International Dysphagia Diet Standardization Initiative (IDDSI) framework. The IDDSI framework contains eight levels (0–7), with levels 0–4 consisting of drinks and levels 3–7 consisting of foods. Foods can be further classified into the following categories: liquidized (level 3), pureed (level 4), minced and moist (level 5), soft and bite-sized (level 6), and easy to chew/regular (level 7) [75]. Among the cooked 3D-printed meats, a level 6 pork product has been successfully printed for dysphagic individuals [5], while levels 4 [75], 6, and 7 [74] have been achieved for beef.

## 1.3.1.2 *Non-extrusion-based printing*

In addition to extrusion, three other printing methods have been demonstrated for food printing: (1) inkjet printing, (2) selective laser sintering, and (3) binder jetting [80]. In selective laser sintering and binder jetting, suitable food precursors (or food inks) are frequently constrained to powdered ingredients (together with a liquid binder in binder jetting). In inkjet printing, low-viscosity ingredients are used. Multiple printing methods have been demonstrated with commercial printers. For instance, CandyFab uses a hot air beam to sinter powdered sugar. ChefJet Pro uses water as a binding agent to create 3D models. FoodJet uses a solidifier in food inks to merge the printed layers and prevent structural collapse [14]. Apart from commercial printers, hobbyists have conducted do it yourself (DIY) trials with various food inks and reported their findings on digital platforms. For example, sake (Japanese rice wine) and other alcohol-based preparations were demonstrated as binders in the binder jetting, while sugar or sugar mixtures were still primarily used as powder formulations [14]. In academic research, no advanced study beyond early testing on these printing platforms has been documented. For example, researchers at the TNO, the Netherlands, demonstrated laser sintering on drinking chocolate powder Nesquik, fabricating chain links and corporate logos due to its relatively high sugar and fat composition. However, detailed characterization of the ingredients and the processes is not available [14]. A few research publications on this topic include electrostatic chocolate inkjet printing [23] and binder jetting 3D printing [81]. Since a majority of food printing research is conducted with extrusion-based printing, we will discuss the physical properties of food inks, which influence the outcome of their printing.

### 1.3.2 *Physical Properties of Food Inks*

In this section, we will discuss the physical properties of food inks used in extrusion-based printing. In extrusion-based 3D printing, the food inks are extruded smoothly and retain their structure post-printing. Both phenomena depend on the rheological properties of the inks. Other considerations concerning the physical properties of the food inks would be the texture of its eventual 3D-printed product and its resistance to deformation due to post-processing steps such as cooking and baking. This section describes some of the key parameters in defining the physical properties of food inks and their printed products.

## 1.3.2.1 *Rheological properties of food inks*

Rheology is the study of the flow of matter [82]. It applies to a range of materials, including foods, pastes, and polymers, all of which are eligible for 3D printing. Understanding the rheological characteristics of food materials is essential for evaluating and enhancing their printing outcomes. For example, shear-thinning properties are desired for ink extrusion from the narrow nozzle, while yield stress properties are essential in creating self-supporting structures [17].

- Viscosity

The viscosity of a fluid is the measure of its resistance to deformation (strain) at a particular rate [83]. As a result, viscosity is the opposition to flow when a shear force is applied. This resistance is analogous to the internal frictional force that exists between two adjacent planes of liquid that are in relative motion during deformation. As a result, shear stress refers to the force required to overcome the friction between adjacent planes of a liquid. Shear stress is represented as $\tau$ and has the unit Pa (Pascal).

Shear stress causes the flow of matter. The shear parameters of stress and strain rate are derived from torque and flow rate measurements [84]. Shear strain describes material deformation in such a way that parallel internal surfaces within the material slide over each other. Shear rate is defined as the rate of change in velocity as one layer slides over its adjacent layer divided by the distance between them [85]. Shear rate is represented by $\dot{\gamma}$ and has the unit s$^{-1}$.

The apparent (or shear) viscosity of a fluid is defined as the viscosity calculated from the shear stress and shear rate. It is represented by $\mu$ and has the unit Pa.s (Pascal-second). Equation 1.1 defines the relationship between shear viscosity, shear stress, and shear rate:

$$\tau = \mu\,\dot{\gamma} \qquad\qquad (1.1)$$

Previous studies have suggested that DIW extrusion printers can print inks with $\mu > 100$ Pa.s [86, 87].

When a linear relationship exists between the shear stress and shear rate of a fluid, the fluid is called a Newtonian fluid [88]. As a result, the viscosity of a Newtonian fluid is unaffected by shear stress or shear rate and remains constant. Water and mineral oil are examples of Newtonian fluids. For non-Newtonian fluids, shear stress and shear rate do not exhibit a linear relationship [89]. Honey is an example of a non-Newtonian fluid [90]. Non-Newtonian fluids are divided into two categories: time-independent fluids and time-dependent fluids. The time refers to the duration of time that the fluids are sheared. For time-dependent fluids, the viscosity is independent of the duration of shear; for time-independent fluids, the viscosity is dependent on the shear duration. Time-independent fluids can be classified into shear-thinning and shear-thickening fluids. The shear stress of time-independent non-Newtonian fluids changes non-linearly with shear rate, as shown in Equation 1.2 [91]:

$$\tau = K\dot{\gamma}^n \tag{1.2}$$

where $\tau$ is the shear stress (Pa), $\dot{\gamma}$ is the shear rate (s$^{-1}$), $K$ is the consistency coefficient (Pa.s$^n$), and $n$ is the flow behaviour index (unitless).

The flow behaviour index ($n$) demonstrates how the shear stress changes with the shear rate. When $n > 1$, the ink displays shear-thickening (dilatant) behaviour; its viscosity increases with an increase in shear stress. For instance, Oobleck (i.e., a mixture of corn starch and water) is a shear-thickening fluid. On the other hand, when $0 < n < 1$, the ink displays shear-thinning (pseudoplastic) behaviour; its viscosity decreases with an increase in shear stress. Ketchup is an example of a shear-thinning fluid [92].

Some shear-thinning fluids possess a yield shear stress, which needs to be overcome before it can flow. A common model known as the Herschel–Bulkley model considers the yield shear stress, which can be mathematically expressed as

$$\tau = \tau_y + K\dot{\gamma}^n \tag{1.3}$$

where $\tau$ is the shear stress (Pa), $\dot{\gamma}$ is the shear rate (s$^{-1}$), $K$ is the consistency coefficient (Pa.s$^n$), $n$ is the flow behaviour index (unitless), and $\tau_y$ is the yield shear stress (Pa).

The Herschel–Bulkley model is a mathematical model containing two parameters: yield stress and the power-law index [93]. This model describes the shear-thinning inks that behave like a solid at rest and begins to flow only when the applied stress exceeds the yield stress for that ink. This model is often used to assess the rheological behaviour of suspensions, emulsions, or pastes in which the material exhibits the power-law behaviour after yielding [26]. In 3D printing, the yield stress is commonly used to assess the extrudability of food inks [33]. Yield stress is related to the mechanical strength of the food inks. A greater extrusion force correlates to a higher mechanical strength of the food ink [28].

Time-dependent fluids can be classified into two categories: rheopectic and thixotropic fluids. Rheopectic fluids display an increase in viscosity when they are sheared over time. Cream is an example of a rheopectic fluid. On the contrary, thixotropic fluids display a reduction in viscosity when they are sheared over time. Honey is an example of a thixotropic fluid. The longer the length of time during which a force is applied to the fluid, the larger the change in the viscosity of these fluids.

The viscosities of different fluid types can be measured using stationary tests. The inks are exposed to a certain stress and the resultant strain is determined (or vice versa). A rheometer or viscometer can be used to conduct these simple rotational tests. The inks are exposed to a constant shear rate and the viscosity of the ink is measured from the amount of torque (i.e., force) required to turn the spindle. The shape of the curve obtained from the results determines the type of fluid. For time-dependent fluids, their viscosities can be observed over a time duration at a single shear rate to determine whether the fluid is thixotropic (i.e., viscosity decreases with time) or rheopectic (i.e., viscosity increases with time).

Food materials used in 3DFP should be pseudoplastic fluids demonstrating acceptable shear-thinning behaviour. In addition, they must be extruded from the printer nozzle at a suitable shear stress, possessing quick structural recovery and solidification after extrusion [17].

- Viscoelasticity: Oscillatory rheology

Apart from the viscosity, the viscoelastic behaviour of the fluid is also an important requirement for printability. Viscoelasticity refers to the property of materials that exhibit both viscous and elastic properties when undergoing deformation. In the event that the stress is applied to a completely viscous medium, they resist the shear flow and strain in a linear manner. Elastic materials, on the other hand, experience strain when stretched but return to their normal shape once the stress is removed.

Oscillatory tests employing rheometers can evaluate the elastic and viscous response of the ink under two operating conditions: controlled stress and controlled rate.

The stress and strain happen one after another in stages in fully elastic samples. In fully viscous materials, however, a phase difference between the stress and strain is seen, with the strain lagging by a 90° (or $\pi/2$ radian) phase difference. Viscoelastic samples exhibit intermediate behaviours between fully elastic and fully viscous materials, with the phase lag in strain. The phase angle ($\delta$) reflects the phase lag between shear stress and strain. This parameter can be computed using the loss tangent equation as follows [94]:

$$\text{Loss tangent} = \tan\delta = \frac{G''}{G'} \tag{1.4}$$

where $\tan\delta$ is the ratio of $G''$ to $G'$. Loss tangent, $\tan\delta$, can be used to assess the viscoelastic characteristic of the printing ink. $\tan\delta < 1$ suggests that the material has a mostly elastic behaviour, whereas $\tan\delta > 1$ suggests that the material has a more viscous behaviour [70]. The phase angle may be used to determine the complex modulus ($G*$), which measures the overall resistance of the deformation of the material. The complex modulus is also an appropriate marker for the stiffness or flexibility of the material [38]. The complex modulus is composed of the following [95]:

$$G' = G*\cos\delta \tag{1.5}$$

$$G'' = G*\sin\delta \tag{1.6}$$

$G'$ is the storage modulus and measures the elastic response of the material to stress. The applied energy is retained by the material, and the material returns to its original condition once the stress is removed. $G''$, however, indicates the loss modulus and defines the viscous response of the material to stress. The applied energy is released as the material flows.

Amplitude sweep is a form of oscillatory test that characterizes these parameters. The amplitude of deformation is gradually raised over several cycles to identify the point at which the internal structure of the material starts to break down. Within the linear viscoelastic region (LVR) of the material, $G'$, $G''$, and $\delta$ are unaffected by the applied stress or strain [96]. The yield point is the point at which $G'$ deviates from linearity. This yield point established by oscillatory testing measurements is referred to as static yield stress (i.e., the lowest shear stress or strain required to cause the material to flow), whereas dynamic yield stress is measured by flow curves (i.e., the shear stress required to keep the flow at low shear rates) [96]. The frequency sweep is another form of oscillatory test, in which the amplitude of deformation stays constant as the frequency of oscillation increases. The frequency sweep is performed to determine the time-dependent behaviour of materials using the values obtained from the LVR of the material (as predetermined by amplitude sweep). Higher frequencies mimic quick motion over short timescales, whereas lower frequencies mimic slow motion over long timescales or at rest. These measurements give information on (1) the inner structure of the materials, (2) the time-dependent shear behaviour, and (3) the long-term stability of dispersions. High-frequency sweep, for instance, exhibits short-term shear behaviour such as extruding or mixing, whereas low-frequency sweep exhibits long-term shear behaviour such as settling within dispersions over a long time.

- Rheological properties of food inks for extrusion-based 3D printing.

Rheological properties such as yield stress, storage modulus ($G'$), loss modulus ($G''$), flow behaviour index ($K$), and flow characteristic index ($n$) are frequently employed to predict the effect of extrusion from the nozzle. The yield stress and storage modulus are essential parameters to predict the ability of the inks to self-support. The flowability of the food inks can be achieved by (1) adjusting the concentration of the components within the ink, (2) changing the printing temperature and/or (3) including additives to the inks. For instance, Lee *et al.* showed that the milk ink can be successfully printed into complex structures at room temperature by only changing the concentration of the milk powder without the use of rheological modifiers [44]. Similarly, Karyappa *et al.* demonstrated the direct printing of chocolate-based inks to fabricate complex 3D shapes without the use of temperature control by varying the concentrations and composition of the inks [22]. Mantihal *et al.* maintained a nozzle temperature of 32 °C to support the extrusion of the melted chocolate and simultaneously retained the stable β crystals within the chocolate that are essential for chocolate quality [26].

Each material has a distinct thermal property, and the extrusion temperature depends on the printed materials [13]. Materials such as chocolate undergo temperature-induced phase transitions; molten chocolate can be extruded as a low-viscosity liquid and subsequently cooled rapidly post-deposition to allow the material to solidify. Like chocolate, hydrocolloids that create low-viscocity solutions at high temperatures and gel after cooling can be used in 3D printing [97], although careful temperature control is necessary for both situations. Printing at room temperature is also possible with formulations that exhibit shear rate-dependent flow behaviour (i.e., shear-thinning inks). Previous studies have demonstrated that shear-thinning materials can be effectively used in extrusion-based 3D printing, such as hydrocolloid gels [98, 99], mashed potatoes [100], and dough [29]. Overall, rheological characterization is essential in predicting the appropriateness of the food formulations for extrusion-based 3DFP [99, 101, 102].

## 1.3.2.2 *Textural properties of 3D-printed foods*

Food texture refers to all the rheological and structural (geometric and surface) attributes of the product that are perceptible by means of mechanical, tactile, and where appropriate, visual and auditory receptors [103]. It is determined by the physical and physicochemical qualities of the product, as well as the distinctive, complicated features of the human senses [104]. Bourne stated that the structure of foods determines their texture [105]. Since 3D printing determines the structure of 3D-printed foods, it can be used to control the texture of the 3D-printed food product. In this section, we describe the assessments which have been employed to determine food texture, and some examples where textural measurements have been conducted for 3D-printed food products.

Food texture can be assessed using descriptive sensory (subjective) or instrumental (objective) assessments [106]. Because of the time and cost involved in sensory analysis, empirical mechanical tests can be correlated to sensory analysis of food texture. A variety of tests have been used to measure food texture in both research and industry throughout the years.

Subjective texture assessment, often known as *sensory perception* or *sensory evaluation*, covers all techniques for measuring, analyzing, and interpreting the human responses to the characteristics of foods and materials experienced by the five senses: taste, smell, touch, sight, and hearing [107, 108]. Bourne categorized these approaches into *oral* and *non-oral* [105]. The measurement of the texture serves as an instrument to understand human reactions to external stimuli [109, 110]. Traditionally, sensory characteristics are evaluated by a panel of ~10 trained individuals [110–112]. However, for the assessment of hedonic liking, preference, or purchase intent, a large number of untrained panelists (>70 people) is required [107]. Many research works, however, have been based on a smaller number of untrained panelists, especially when sensory perception data are combined with instrumental analysis [106]. Typically, panelists describe food texture using certain popular sensory words. The texture of potato chips, for example, is frequently characterized in terms of crispness, hardness, and crunchiness [113]. For the assessment of 3D-printed foods, honeycomb-shaped chocolate with varying infill percentages of 25, 50 and 100% has been 3D-printed and compared by panelists based on their preferences towards the appearance and hardness of the chocolate samples [114].

Objective assessment utilizing instrumental techniques has been established to address the limits of sensory perception on food textures [115]. The texture felt in the mouth is primarily determined by the mechanical behaviour of the food, which in turn, determines the dynamics of the breakdown of the food materials during eating [116]. The mechanical and rheological characteristics of the food remain the focus of most of these objective assessments.

The most typical methods to characterize food textural qualities, including 3D-printed foods, are puncture and compression tests. These tests are often simple to conduct using low-cost equipment and operation [117]. Both solid and semi-solid foods can be tested. Probes for compression and puncture are typically cylindrical with heads of varying sizes. The heads for compression tests were reported to be in the range of 10–150 mm in diameter [118–121]; the heads of puncture tests are typically smaller than those of compression tests, which range from 1 to 11 mm in diameter [122–124].

The head speeds used for puncture and compression tests differ as well. For puncture tests, the testing speed is estimated to be a few millimetres per second (~4 mm/s) [122]. For compression tests, the head speed can be as high as 30 mm/s [125, 126]. Additionally, the extents of material deformation required for compression and puncture tests differ. For compression tests, deformation of the original sample height by 50% [127] and 75% [128] have been reported at the maximum compression force. On the other hand, the puncture depth of the material tested in puncture tests is typically small, within the range of a few millimetres [106]. The puncture test is similar to the compression test, except that the contact area of the probe is considerably smaller than the sample size. For instance, a needle-shaped probe is used for the puncture test. The probe compresses the sample to a certain strain to measure the material characteristics such as the breaking strength, maximum force, and penetration depth of the food sample [129].

Single compression tests are usually carried out in the form of an axial compression between two flat plates [130]. The samples tested must be smaller in size than the contact area of the used probe. Samples are compressed to the point of failure, or often to a predetermined amount of deformation. However, double-compression tests, also called texture profile analysis (TPA), are often conducted. TPA mimics mastication with double-compression cycles. It is also known as the two-bite test, mimicking the motion of the mouth to bite in order to study the response of the food upon chewing.

A large variety of food textural properties can be determined with TPA, including hardness, adhesiveness, cohesiveness, resiliency, springiness, gumminess, and chewiness in a single set of experiments [127, 131, 132]. These properties are some of the most common textural properties evaluated among 3D-printed foods. The parameters obtained from TPA are classified as primary (hardness, adhesiveness, springiness, and cohesiveness) or secondary (gumminess, chewiness, and resilience) [133]. The resultant force/time graph obtained from TPA can be used to directly determine the primary parameters; the secondary parameters can be obtained from the primary parameters [129]. The positive peak of the first compression cycle was used to calculate hardness, which represents the firmness of the food samples [74]. Hardness was described as the force applied by the molar teeth to compress the food [134] or the force needed to compress a sample [135]. In 3DFP, the hardness of the material was correlated to the extrusion force needed to push it through the print head, where the print head possessed a changing cross-sectional area from the syringe feeder to a narrow aperture of the extrusion nozzle [70].

The adhesiveness of food is the ability of food to stick to the teeth during chewing [134]. Adhesiveness measures the rate at which the food detaches from the probe or the roof of the mouth and teeth [136]. Adhesiveness is also an important attribute that aids in the binding between layers of the 3D-printed foods [70]. The adhesiveness of a food sample can be determined from the negative area of the force-time curve following its first compression in TPA.

The springiness of a food sample is related to its elasticity; it can be defined as how the food sample is reformed after pressure is applied. Springiness is associated with the height recovered by the food sample within the duration between the end of the first bite and the beginning of the second bite [137]. The higher the springiness of the food, the more energy is needed for chewing in the mouth.

The cohesiveness (consistency) of food has been described as a measure of how the food holds itself together [136]; this property relates to the ability to withstand deformation [72]. The cohesiveness of a food sample implies the strength of the internal bonding that makes up the food as well as the extent to which a food may be deformed before it breaks [138]. Therefore, the cohesiveness of food reflects its capacity to hold itself together [137]. Cohesiveness can be determined by taking the ratio of the positive force area under the second compression to the positive force area under the first compression of the force-time graph obtained from TPA [137].

Gumminess is the property of semi-solid foods that exhibits low hardness and high cohesiveness. It is a more essential textural characteristic for semisolids than solids [137]. Gumminess is obtained by the multiplication of the hardness and cohesiveness values from TPA [137]. Chewiness is a measure of the amount of energy needed to chew the food, which is often reported for solid foods; it is derived from the multiplication of the gumminess and springiness values, which is the same as the product of hardness, cohesiveness, and springiness values [137]. Resilience is a measure of how quickly and forcefully a sample recovers from deformation [139]. In other words, it refers to the elastic recovery of the food sample [137]. Resilience is determined by dividing the area of work (i.e., force times displacement) during the upstroke of the first compression by the area of work during the downstroke of the first compression from the force-time graph of the TPA [140].

Multiple 3D-printed food samples have been assessed for their texture based on the parameters described above. For instance, Le Tohic *et al.* have reported that the hardness of the 3D-printed cheese was significantly lower than the untreated commercially processed cheese. The difference was attributed to the shearing during printing [38]. Previous studies have also demonstrated that the changes in the texture of 3D-printed products may occur by varying the printing temperature [45], pH [45], infill pattern [25], infill percentages [25, 141], and food ink formulation [27, 36, 41, 142]. 3DFP, therefore, allows the control of the textural qualities of the food. However, the texture is a multimodal sensory characteristic. The use of one type of mechanical testing is unlikely to capture all the subtleties of food texture that a human perceives during eating [110]. The measurement by physical instruments permits quantifying only some textural properties, which must be translated to sensory experience [143]. Overall, instrumental approaches should be combined with sensory assessments to improve the effectiveness and accuracy of food texture determination [106].

### 1.3.2.3 *Post-processing and physical properties of 3D-printed foods*

In 3D printing, the change in dimensions of the food sample may occur due to the post-processing of the food, including cooking or baking [144]. Examples of such food inks include meat and dough inks, which need to be cooked before their consumption. Post-processing at high temperatures may result in the deformation of the printed patterns [144].

Several parameters have been used to quantify the physical and dimensional changes of 3D-printed food. One way to quantify the changes in 3D-printed foods is cooking loss; cooking loss is the loss of the weight described in Equation 1.7 [74]:

$$Cooking\,loss\,(\%) = \frac{(Raw\,weight - cooked\,weight)}{Raw\,weight} \times 100 \qquad (1.7)$$

Another method to quantify the physical changes is the shrinkage of the food printed as a cube/cuboid; the shrinkage is characterized by the percentage of the change in the lateral dimensions [74]:

$$Shrinkage\,(\%) = \frac{(Sum\,of\,differences\,in\,L,W,and\,H\,between\,raw\,and\,cooked\,samples)}{(Sum\,of\,L,W,and\,H\,of\,raw\,sample)} \times 100$$

$$(1.8)$$

where $L$, $W$, and $H$ refer to the length, width, and height of the sample, respectively.

For example, the effect of baking (at 150 °C for 25 min) was determined for 3D-printed cookies by measuring the percentage changes in their dimensions (i.e., diameter and height) before and after baking in Equation 1.9 [145]:

$$Variation\,(\%) = \frac{[(Difference\,between\,baked\,and\,raw\,cookie\,dimensions) \times 100]}{Raw\,cookie\,dimension} \qquad (1.9)$$

In another work, the deformation caused by baking 3D-printed cookie dough was calculated using the dimensions (i.e., width, height, and thickness) of the samples, as described in the following equation [28]:

$$\Delta X\,(\%) = \frac{(X_{bb} - X_{ab})}{X_{bb}} \times 100 \qquad (1.10)$$

where $X_{bb}$ and $X_{ab}$ represent the dimensions of the printed sample before and after baking, respectively. $\Delta X > 0$ (or $X_{bb} > X_{ab}$) indicates the shrinkage of the samples, while $\Delta X < 0$ (or $X_{bb} < X_{ab}$) indicates a collapse or spreading of 3D-printed cookie samples after baking.

To ensure the dimensional stability of the food structures, changes in the food ink composition and additives, as well as the post-processing method itself, are often investigated [146]. Adding XG to wheat flour-based cookie dough has been shown to enhance its dimensional stability during baking at 170 °C by improving the mechanical strength of the 3D-printed cookie dough during baking [33]. To improve the dimensional stability of 3D-printed food; rapid freezing has been explored. For example, 3D-printed samples made from low-gluten flour-based cookie dough at −65 °C before baking at 190 °C resulted in the improvement of shape stability [32]. Another study reported that heating by convection resulted in higher shrinkage and cooking loss than by heating by steam convection and microwave; of those, steam convection resulted in the least dimensional deviation post-reheating [76]. By using steam convection, 3D-printed beef samples can retain their original printed shape post-reheating, thereby improving their presentation towards consumers.

## 1.4 Applications and Significance of 3D Food Printing

3DFP may have a variety of applications and significance, including (1) personalization of foods, (2) automation in food processing, (3) design and fabrication of novel and complex food shapes, and (4) improvements in food sustainability [11, 46, 146–148].

### 1.4.1 *Personalization of Food*

3DFP allows the customization of food products to meet the unique requirements of individual consumers or specific groups of consumers.

Appearance, texture, and nutrition are the key properties of the food amendable by 3DFP.

A study conducted by Severini *et al.* reported that the appearance of 3D-printed smoothie paste consisting of a blend of fruits and vegetables (i.e., carrots, pears, kiwi, broccoli raab leaves, and avocado) was preferred over non-printed samples. These results suggested that 3DFP can be used to enhance the visual appearance of food [61]. Such visually appealing food may enhance the palatability and acceptability of oral intake of food for children. For instance, 3D-printed candy-like pediatric tablets in the shape of a heart, ring, lion, bottle, and bear were created by hot-melt extrusion and fused deposition modelling (FDM) 3D printing. The research also suggested that these 3D-printed dosages help mask the bitterness of the drug [149]. In another study, 3D-printed chewable isoleucine tablets of varying colour and flavour were well accepted among four pediatric patients aged between 3 and 16 years, who also displayed their preferences for certain colours and flavours [150]. Chewable chocolate-based oral dosage forms (containing paracetamol or ibuprofen) were also 3D-printed into various shapes such as a bone, rocket, and star to improve the acceptability of orally administering drugs within the pediatric population [142]. Apart from creating visually appealing dosages, 3DFP offers the flexibility to change the weight and other parameters to modify the dose based on the pathophysiology of each patient [142].

Food texture is an important sensory feature that can impact an individual experience during oral intake. Studies have been conducted to evaluate consumer preferences for 3D-printed foods. For instance, Manithal *et al.* reported that 3D-printed chocolate at 100% infill was preferred by a sensory panel than those printed at 25% and 50% infill due to its smoother surfaces [114]. In this way, the shape and texture of 3D-printed foods can be personalized to achieve desired outcomes of individual consumers.

One application of 3DFP in food personalization is the preparation for 3D-printed meals for people suffering from chewing and swallowing difficulties (i.e., dysphagia). Each meal exposes them to the risk of choking [5]. Dysphagia is a medical condition affecting ~14% of the population over 50 years old, and around 40–50% of residents in aged care facilities [5]. Dysphagia can be a result of several acquired or congenital medical conditions. For example, physical abnormalities of the oesophagus, or neurological disorders resulting from Parkinson's disease, multiple sclerosis, cancer and its treatment with radiation therapy [75]. To reduce the risk of choking, texture-modified foods are proposed for dysphagic patients. These meals are usually made of thickened fluids, minced, and pureed foods, which can be unappetizing and distressing to these individuals consuming them [5]. Patients on these texture-modified diets were shown to exhibit 17–37% lower energy levels than those on regular diets [151], causing malnourishment and a weakened immune system [152]. On the contrary, it has been reported that diets consisting of 3D-reformed pureed and minced foods have increased the energy levels, nutrient intake, and weight in patients, as compared to the conventional texture-modified diets served at the institutions [77–79]. With 3DFP, custom-made meals resembling the original shape of the food can be produced while keeping the textural properties favourable for consumption by dysphagic individuals.

In 3DFP, the food ingredients are made into a paste before printing. The food ink can be tailored with desired healthy ingredients or added with selected nutrients to improve the nutritional content of the 3D-printed food product. This approach helps to address the dietary and nutritional needs of the consumers. In this aspect, Derossi *et al.* experimented with a fruit-based formulation consisting of fresh bananas, mushrooms, white beans, non-fat milk and lemon juice to 3D-print a snack for children aged 3–10 years. This formulation is selected to overcome the insufficient intake of calcium, iron, and vitamin D for children [42]. In another work, the cookie dough fortified with microalga *Arthrospira platensis* (and its antioxidant extracts) was 3D-printed and baked to produce a functional food containing a rich source of antioxidants having anti-cancer, anti-diabetes, and anti-inflammatory properties [145]. Finally, in the work done by Liu *et al.*, the probiotic strain *Bifidobacterium animalis subsp. lactis BB-12* was added to mashed potato for the 3D printing of ready-to-eat meals to improve the gastrointestinal health and immune function of the consumers [153].

### 1.4.2 *Automation in Food Production*

3DFP also enables automated food production, creating foods with unique shapes, textures, and improved nutritional content. This is made possible through the development of 3D food printers. Some 3D food printers are meant for domestic use, while others are for professional use. Most commercially available printers are extrusion-based [154]. Examples of 3D food printers include Focus (byFlow, Netherlands) and Choc

Creator V2.0 Plus (Choc Edge Ltd., UK), both of which are employed in the bakery and confectionery industry, respectively [154]. Another 3D food printer, known as Foodini (Natural Machines, Spain), is a desktop printer (with an open capsule system for loading fresh food ingredients) designed for both professional and domestic uses [154]. Lipson *et al.* have established the Fab@Home project to develop an open-source mass-collaboration personal 3D printer [154]. The open-source nature of this hardware allows the whole community working on this project all around the world to access the printer [154]. Overall, 3DFP offers a quick and reproducible method to make customizable food products with automation.

### 1.4.3 *Novel and Complex Food Shapes and Structures*

The ability of 3DFP to make objects with geometric complexity enables the creation of sophisticated 3D food designs that may be utilized for aesthetic presentations as well as complex textures [17]. For instance, 3DFP allows constructing intricate interior and exterior patterns of the 3D-printed foods that would be difficult to achieve using traditional methods such as moulding [74]. Multiple studies have shown that infill designs and patterns have a critical influence on the overall texture of 3D-printed foods. Examples include potato starch, skimmed milk, and chocolate [19, 25, 141]. Furthermore, the precision and accuracy of 3D printing aid in the reduction of structural heterogeneity among complex 3D-printed samples, which would negatively impact their reproducibility. As a result, 3DFP can be employed in the texture and taste development of the food. Importantly, it serves as a unique tool to study the relationship between the texture and taste of the food.

Because of the constraints of current 3D printers, most research uses only one type of food ingredient or a blend containing several food ingredients (such as fruits and vegetable blend), which is then deposited using a single nozzle-extruder type 3D printer. However, in recent years, multi-material food printing is used to achieve complex food shapes and distribution of materials. For example, a dual-nozzle extrusion system was used to print strawberry juice gel and mashed potato in varying 3D configurations (such as alternating layers of the two inks) with different infill percentages [55]. In another study, milk and other edible inks, including chocolate, coconut, maple syrup, and blueberry inks, were extruded from multiple syringes and nozzles to fabricate 3D milk models enclosing different fillings [44]. In a similar way, multiple nozzles can be used to print chocolate-based models consisting of different chocolate-based inks (e.g., viscosities) with features of a semi-solid enclosure and liquid filling [22].

### 1.4.4 *Improvements in Food Sustainability*

Malnutrition and undernutrition persist to be major challenges all over the world, despite a reduction in the percentage of people experiencing these issues since the beginning of the 21st century [155]. Nonetheless, UNICEF reported that almost half of all fatalities in children under the age of five were linked to undernutrition. This situation caused the loss of nearly three million young lives each year, with Africa and Asia being the worst impacted regions [155]. Furthermore, over 700 million people are still malnourished; the United Nations Food and Agricultural Organization (FAO) projects that the global population will continue to expand, rising from roughly 7.4 billion in 2017 to 9 billion in 2048 [155], exacerbating the problem of malnutrition globally.

Food wastage has been identified as one of the major causes of global food shortage. Large amounts of food products become waste because they were either lost during the production or distribution without consumption [155]. It is estimated that $1.3 \times 10^{12}$ kg of food waste is produced every year. This amount accounts for a third of all the food generated for human consumption [156]. Another cause of food shortage is the growing global population. According to the United Nations, the world population is expected to exceed 10 billion by 2050. Under this estimate, we need to deliver 60–70% more food than we presently produce to fulfil the demand [157]. Moreover, the growth in global meat production is widely deemed to be unsustainable. Conventional meat production is associated with several issues of food security, public health, loss of biodiversity, and animal welfare, all of which stem from livestock production [66]. Additionally, the production of greenhouse gas emissions by global livestock leads to climate change, which has been identified as a major concern around the world.

3DFP may potentially aid in alleviating the global food shortage by upcycling and recycling food wastes and food by-products. Approximately a third of the food produced ends up as waste. For instance, in the production of meats, high-value steaks only represent 7.2% of the weight of a cattle carcass [158]; the remainder of the meat cuts are either sold as low valued by-products or deemed as waste [159]. Additionally, by-products from livestock, such as their skin, intestines, fat, and feet, account for 66% and 52% of the pig's live weight and cattle's live weight, respectively, even though they are rich sources of carbohydrates, amino acids, fatty acids, and vitamins [160]. This example highlights that more than half of the weight of live animals is deemed unfit for eating. Therefore, alternative methods enable us to properly utilize these meat by-products [161]. As a solution to this problem, 3DFP may offer a route to salvage meat by-products by transforming them into customized meat products exhibiting similar or even better nutritional content, texture, and taste than high-value meat cuts. Recombined meats may be constructed by 3D printing utilizing diverse food ingredients such as meat paste and fat slurry to mimic the look, nutritional content, and flavour of a steak.

Additionally, based on a recent study, fresh vegetables and salads make up 25% of the total household edible food waste [156]. In an attempt to salvage edible food wastes, Upprinting Food, a start-up in the Netherlands, recycles unsellable but edible foods such as bread and ugly/overripe vegetables and fruits by 3D printing them into snacks [154]. Another study utilized grape pomace and broken wheat, which are considered by-products from wineries and the wheat milling industry, respectively, as ingredients for 3D-printed cookies with customized shapes — an approach to improve consumer perception towards foods derived from waste and by-products [10]. Similarly, non-edible food wastes were also recycled for application in 3D printing. For instance, mussel shell wastes (which is commonly discarded but a rich source of calcium carbonate) were recycled as 3D printable inks, suitable for extrusion 3D printing [162]. Coconut shells were recycled as polymer composite filaments, used in 3D printing to replace hazardous non-degradable plastics [163]. Components derived from fish scales and eggshells were used in the fabrication of composite filaments for the 3D printing of bone implants and other biomedical appliances [164]. These examples highlight the potential benefits of applying 3D printing to transform wastes into products.

3DFP also helps to promote the consumption of alternative food sources and ingredients. Alternative ingredients are novel proteins and fibres derived from sources such as algae, fungus, insects, and seaweed [165]. These alternative components are gaining popularity as possible additions to a balanced diet, supplementing conventional food sources such as cattle and crops, and can be made into a paste or powder appropriate for 3D printing inside a meat paste to provide meals [159]. One of the sustainable sources of high-value animal proteins is edible insects. Edible insects can help meet the growing demand for meat products [166]. Insects, with their diverse nutritional profiles, might also serve as substitutes for shellfish, nuts, and pulses [167]. While approximately two billion people globally consume insects in their diet, negative perception towards insect consumption remains in parts of the world where insect consumption is not traditional (e.g., Europe and North America) [31]. However, the acceptability of insect consumption can be improved if the insects are incorporated invisibly into familiar foods [168]. In this aspect of designing suitable insect-containing products with no sensory expectations, 3D printing has been acknowledged as an essential tool by the edible insect industry [169]. In one of the earlier works involving 3D printing of insects, Insect Au Gratin project, insect-protein-based flour created from crickets, mealworms, and silkworm pupae was mixed with familiar ingredients such as chocolate, cream cheese, icing butter, and spices to produce an extrudable paste that was later printed into insect-inspired designs [159]. In a similar work, dough-containing blends of wheat flour and yellow mealworm (*Tenebrio molitor*) powder in varying compositions were used as printing inks, which were then 3D-printed into cylindrical shapes and baked in a convection oven to produce protein enriched snacks [31].

Another practical alternative source of protein is cultured meat, which is produced from healthy animal muscle cells; this method does not diminish its nutritional content, nor does it require the slaughtering of animals [170]. Over the years, cultured meat is also known by other names such as *clean meat* and *cell-based meat* [171]. Muscle cells may be grown on plastics with reasonable ease as a normal *in vitro* model of the cell culture. This approach has been utilized for decades in biological laboratories and is widely accepted as a model for investigating the molecular principles of muscle growth and degradation [172]. In recent years, this method has been recognized as a viable alternative to animal meat production. The single layer of cultivated muscle cells, on the other hand, varies greatly from muscle tissue, which is a highly complex organ. Adipose and connective tissues, as well as the vascular bed, are found in skeletal muscle [172]. The individual components give rise to the flavour, fragrance, and nutritional content of the meat in a distinct way. However, because this approach cannot be used to manufacture highly structured meats such as steaks, the scope of this technology is yet limited [173]. While printing of muscle cells is possible, an extracellular matrix supporting cells and lipids inside the scaffold that stimulates cell development is also essential in cell growth and development. Because of this complexity, 3D printing cultured meat is more advanced than culturing muscle cells alone [174]. In a recent paper, Kang *et al.* created a tendon-gel-integrated bioprinting method for the generation of cell fibres and differentiation into skeletal muscle, adipose, and blood capillary fibres. These building blocks are assembled to build engineered steak-like meats by mimicking the histological features of a real beef steak [175].

By decreasing the need for cattle rearing and eliminating the necessity for traditional meat processing, 3D printing of cultured meat has the potential to enhance human health and the environment [176, 177]. While evidence for sustainability advantages is yet limited [170], 3D printing of cultured meat adds unquestionable diversity to meat products by allowing them to be manufactured with nutritional value for specific groups of customers and constructed into novel textures and forms [177, 178]. 3D-printed cultured meat has the potential to improve food intake in specific categories of people, notably those who struggle to obtain enough nutrition or those suffering from medical disorders such as dysphagia [179]. As a result, 3D printing can create textured and appealing meats for the elderly or individuals who have swallowing and chewing difficulties [180]. Because there has been a tendency of children and adolescents not getting sufficient vitamins and minerals, printed cultured meat may also be developed with better nutritional content, such as containing certain types of vitamins [177, 180]. People with allergies would also benefit from 3D printing of cultured meat since they would no longer have to worry about dietary restrictions if meat could be modified to be free of certain allergens [174]. On a global scale, 3D-printed cultured meat has the potential to tackle challenges such as famine by optimizing the available resources and combining them into a personalized meat structure [180]. Lastly, it may also address ethical and religious concerns of the food supply by providing alternative routes to create foods for consumers who require specific attention.

# References

[1]   Irvin, D. (2013). 3D printed food system for long duration space missions, *Advanced Food Systems Technology*.

[2]   US army investigating 3D printers for food production. Retrieved from https://www.engineering.com/story/us-army-investigating-3d-printers-for-food-production

[3] Caulier, S., Doets, E. and Noort, M. (2020). An exploratory consumer study of 3D printed food perception in a real-life military setting, *Food Quality and Preference*, 86, p. 104001.

[4] Sarwar, M. H., Sarwar, M. F., Khalid, M. T. and Sarwar, M. (2015). Effects of eating the balance food and diet to protect human health and prevent diseases, *American Journal of Circuits, Systems and Signal Processing*, 1, pp. 99–104.

[5] Dick, A., Bhandari, B., Dong, X. and Prakash, S. (2020). Feasibility study of hydrocolloid incorporated 3D printed pork as dysphagia food, *Food Hydrocolloids*, 107, p. 105940.

[6] Pant, A., Lee, A. Y., Karyappa, R., Lee, C. P., An, J., Hashimoto, M., Tan, U.-X., Wong, G., Chua, C. K. and Zhang, Y. (2021). 3D food printing of fresh vegetables using food hydrocolloids for dysphagic patients, *Food Hydrocolloids*, 114, p. 106546.

[7] Xing, X., Chitrakar, B., Hati, S., Xie, S., Li, H., Li, C., Liu, Z. and Mo, H. (2022). Development of black fungus-based 3D printed foods as dysphagia diet: effect of gums incorporation, *Food Hydrocolloids*, 123, p. 107173.

[8] Lee, A. Y., Pant, A., Pojchanun, K., Lee, C. P., An, J., Hashimoto, M., Tan, U.-X., Leo, C. H., Wong, G. and Chua, C. K. (2021). Three-dimensional printing of food foams stabilized by hydrocolloids for hydration in dysphagia, *International Journal of Bioprinting*, 7.

[9] Nachal, N., Moses, J., Karthik, P. and Anandharamakrishnan, C. (2019). Applications of 3D printing in food processing, *Food Engineering Reviews*, 11, pp. 123–141.

[10] Jagadiswaran, B., Alagarasan, V., Palanivelu, P., Theagarajan, R., Moses, J. and Anandharamakrishnan, C. (2021). Valorization of food industry waste and by-products using 3D printing: a study on the development of value-added functional cookies, *Future Foods*, p. 100036.

[11] Dankar, I., Haddarah, A., Omar, F. E., Sepulcre, F. and Pujolà, M. (2018). 3D printing technology: the new era for food customization and elaboration, *Trends in Food Science & Technology*, 75, pp. 231–242.

[12] Lupton, D. and Turner, B. (2018). Food of the future? Consumer responses to the idea of 3D-printed meat and insect-based foods, *Food and Foodways*, 26, pp. 269–289.

[13] Liu, Z., Zhang, M., Bhandari, B. and Wang, Y. (2017). 3D printing: printing precision and application in food sector, *Trends in Food Science & Technology*, 69, pp. 83–94.

[14] Voon, S. L., An, J., Wong, G., Zhang, Y. and Chua, C. K. (2019). 3D food printing: a categorised review of inks and their development, *Virtual and Physical Prototyping*, 14, pp. 203–218.

[15] Sun, J., Peng, Z., Yan, L., Fuh, J. Y. H. and Hong, G. S. (2015). 3D food printing an innovative way of mass customization in food fabrication, *International Journal of Bioprinting*, 1, pp. 27–38.

[16] Feng, C., Zhang, M. and Bhandari, B. (2019). Materials properties of printable edible inks and printing parameters optimization during 3D printing: a review, *Critical Reviews in Food Science and Nutrition*, 59, pp. 3074–3081.

[17] Jiang, H., Zheng, L., Zou, Y., Tong, Z., Han, S. and Wang, S. (2019). 3D food printing: main components selection by considering rheological properties, *Critical Reviews in Food Science and Nutrition*, 59, pp. 2335–2347.

[18] Hao, L., Mellor, S., Seaman, O., Henderson, J., Sewell, N. and Sloan, M. (2010). Material characterisation and process development for chocolate additive layer manufacturing, *Virtual and Physical Prototyping*, 5, pp. 57–64.

[19] Mantihal, S., Prakash, S., Godoi, F. C. and Bhandari, B. (2017). Optimization of chocolate 3D printing by correlating thermal and flow properties with 3D structure modeling, *Innovative Food Science & Emerging Technologies*, 44, pp. 21–29.

[20] Lanaro, M., Forrestal, D. P., Scheurer, S., Slinger, D. J., Liao, S., Powell, S. K. and Woodruff, M. A. (2017). 3D printing complex chocolate objects: platform design, optimization and evaluation, *Journal of Food Engineering*, 215, pp. 13–22.

[21] Khot, R. A., Aggarwal, D., Pennings, R., Hjorth, L. and Mueller, F. F. (2017). Edipulse: investigating a playful approach to self-monitoring through 3D printed chocolate treats. In *Proceedings of the 2017 CHI Conference on Human Factors in Computing Systems*, pp. 6593–6607.

[22] Karyappa, R. and Hashimoto, M. (2019). Chocolate-based Ink Three-dimensional Printing (Ci3DP), *Scientific Reports*, 9, p. 14178.

[23] Takagishi, K., Suzuki, Y. and Umezu, S. (2018). The high precision drawing method of chocolate utilizing electrostatic ink-jet printer, *Journal of Food Engineering*, 216, pp. 138–143.

[24] Hao, L., Li, Y., Gong, P. and Xiong, W. (2019). *Fundamentals of 3D Food Printing and Applications*, "Material, process and business development for 3D chocolate printing" (Elsevier), pp. 207–255.

[25] Mantihal, S., Prakash, S. and Bhandari, B. (2019). Textural modification of 3D printed dark chocolate by varying internal infill structure, *Food Research International*, 121, pp. 648–657.

[26] Mantihal, S., Prakash, S., Godoi, F. C. and Bhandari, B. (2019). Effect of additives on thermal, rheological and tribological properties of 3D printed dark chocolate, *Food Research International*, 119, pp. 161–169.

[27] Yang, F., Zhang, M., Prakash, S. and Liu, Y. (2018). Physical properties of 3D printed baking dough as affected by different compositions, *Innovative Food Science & Emerging Technologies*, 49, pp. 202–210.

[28] Pulatsu, E., Su, J.-W., Lin, J. and Lin, M. (2020). Factors affecting 3D printing and post-processing capacity of cookie dough, *Innovative Food Science & Emerging Technologies*, 61, p. 102316.

[29] Zhang, L., Lou, Y. and Schutyser, M. A. (2018). 3D printing of cereal-based food structures containing probiotics, *Food Structure*, 18, pp. 14–22.

[30] Azzollini, D., Derossi, A., Fogliano, V., Lakemond, C. and Severini, C. (2018). Effects of formulation and process conditions on microstructure, texture and digestibility of extruded insect-riched snacks, *Innovative Food Science & Emerging Technologies*, 45, pp. 344–353.

[31] Severini, C., Azzollini, D., Albenzio, M. and Derossi, A. (2018). On print-ability, quality and nutritional properties of 3D printed cereal based snacks enriched with edible insects, *Food Research International*, 106, pp. 666–676.

[32] Yang, F., Zhang, M., Fang, Z. and Liu, Y. (2019). Impact of processing parameters and post — treatment on the shape accuracy of 3D — printed baking dough, *International Journal of Food Science & Technology*, 54, pp. 68–74.

[33] Kim, H. W., Lee, I. J., Park, S. M., Lee, J. H., Nguyen, M.-H. and Park, H. J. (2019). Effect of hydrocolloid addition on dimensional stability in post-processing of 3D printable cookie dough, *LWT*, 101, pp. 69–75.

[34] Chuanxing, F., Qi, W., Hui, L., Quancheng, Z. and Wang, M. (2018). Effects of pea protein on the properties of potato starch-based 3D printing materials, *International Journal of Food Engineering*, 14.

[35] Lille, M., Nurmela, A., Nordlund, E., Metsä-Kortelainen, S. and Sozer, N. (2018). Applicability of protein and fiber-rich food materials in extrusion-based 3D printing, *Journal of Food Engineering*, 220, pp. 20–27.

[36] Liu, Y., Liu, D., Wei, G., Ma, Y., Bhandari, B. and Zhou, P. (2018). 3D printed milk protein food simulant: improving the printing performance of milk protein concentration by incorporating whey protein isolate, *Innovative Food Science & Emerging Technologies*, 49, pp. 116–126.

[37] Liu, Y., Yu, Y., Liu, C., Regenstein, J. M., Liu, X. and Zhou, P. (2019). Rheological and mechanical behavior of milk protein composite gel for extrusion-based 3D food printing, *LWT*, 102, pp. 338–346.

[38] Le Tohic, C., O'Sullivan, J. J., Drapala, K. P., Chartrin, V., Chan, T., Morrison, A. P., Kerry, J. P. and Kelly, A. L. (2018). Effect of 3D printing on the structure and textural properties of processed cheese, *Journal of Food Engineering*, 220, pp. 56–64.

[39] Kern, C., Weiss, J. and Hinrichs, J. (2018). Additive layer manufacturing of semi-hard model cheese: effect of calcium levels on thermo-rheological properties and shear behavior, *Journal of Food Engineering*, 235, pp. 89–97.

[40] Joshi, S., Sahu, J. K., Bareen, M. A., Prakash, S., Bhandari, B., Sharma, N. and Naik, S. N. (2021). Assessment of 3D printability of composite dairy matrix by correlating with its rheological properties, *Food Research International*, 141, p. 110111.

[41] Bareen, M. A., Joshi, S., Sahu, J. K., Prakash, S. and Bhandari, B. (2021). Assessment of 3D printability of heat acid coagulated milk semi-solids "soft cheese" by correlating rheological, microstructural, and textural properties, *Journal of Food Engineering*, 300, p. 110506.

[42] Derossi, A., Caporizzi, R., Azzollini, D. and Severini, C. (2018). Application of 3D printing for customized food. A case on the development of a fruit-based snack for children, *Journal of Food Engineering*, 220, pp. 65–75.

[43] Schutyser, M., Houlder, S., de Wit, M., Buijsse, C. and Alting, A. (2018). Fused deposition modelling of sodium caseinate dispersions, *Journal of Food Engineering*, 220, pp. 49–55.

[44] Lee, C. P., Karyappa, R. and Hashimoto, M. (2020). 3D printing of milk-based product, *RSC Advances*, 10, pp. 29821–29828.

[45] Ross, M. M., Crowley, S. V., Crotty, S., Oliveira, J., Morrison, A. P. and Kelly, A. L. (2021). Parameters affecting the printability of 3D-printed processed cheese, *Innovative Food Science & Emerging Technologies*, p. 102730.

[46] Cohen, D. L., Lipton, J. I., Cutler, M., Coulter, D., Vesco, A. and Lipson, H. (2009). Hydrocolloid printing: a novel platform for customized food production. In *Solid Freeform Fabrication Symposium*, Austin, TX, pp. 807–818.

[47] Vancauwenberghe, V., Katalagarianakis, L., Wang, Z., Meerts, M., Hertog, M., Verboven, P., Moldenaers, P., Hendrickx, M. E., Lammertyn, J. and Nicolaï, B. (2017). Pectin based food-ink formulations for 3-D printing of customizable porous food simulants, *Innovative Food Science & Emerging Technologies*, 42, pp. 138–150.

[48] Vancauwenberghe, V., Mbong, V. B. M., Vanstreels, E., Verboven, P., Lammertyn, J. and Nicolai, B. (2019). 3D printing of plant tissue for innovative food manufacturing: encapsulation of alive plant cells into pectin based bio-ink, *Journal of Food Engineering*, 263, pp. 454–464.

[49] Azam, R. S., Zhang, M., Bhandari, B. and Yang, C. (2018). Effect of different gums on features of 3D printed object based on vitamin-D enriched orange concentrate, *Food Biophysics*, 13, pp. 250–262.

[50] Liu, L., Meng, Y., Dai, X., Chen, K. and Zhu, Y. (2019). 3D printing complex egg white protein objects: properties and optimization, *Food and Bioprocess Technology*, 12, pp. 267–279.

[51] Chen, H., Xie, F., Chen, L. and Zheng, B. (2019). Effect of rheological properties of potato, rice and corn starches on their hot-extrusion 3D printing behaviors, *Journal of Food Engineering*, 244, pp. 150–158.

[52] Liu, Z., Bhandari, B., Prakash, S., Mantihal, S. and Zhang, M. (2019). Linking rheology and printability of a multicomponent gel system of carrageenan-xanthan-starch in extrusion based additive manufacturing, *Food Hydrocolloids*, 87, pp. 413–424.

[53] Yang, F., Zhang, M., Bhandari, B. and Liu, Y. (2018). Investigation on lemon juice gel as food material for 3D printing and optimization of printing parameters, *LWT*, 87, pp. 67–76.

[54] Warner, E., Norton, I. and Mills, T. (2019). Comparing the viscoelastic properties of gelatin and different concentrations of kappa-carrageenan mixtures for additive manufacturing applications, *Journal of Food Engineering*, 246, pp. 58–66.

[55] Liu, Z., Zhang, M. and Yang, C.-h. (2018). Dual extrusion 3D printing of mashed potatoes/strawberry juice gel, *LWT*, 96, pp. 589–596.

[56] Vancauwenberghe, V., Verboven, P., Lammertyn, J. and Nicolaï, B. (2018). Development of a coaxial extrusion deposition for 3D printing of customizable pectin-based food simulant, *Journal of Food Engineering*, 225, pp. 42–52.

[57] Diañez, I., Gallegos, C., Brito-de La Fuente, E., Martínez, I., Valencia, C., Sánchez, M., Diaz, M. and Franco, J. (2019). 3D printing *in situ* gelification of κ-carrageenan solutions: effect of printing variables on the rheological response, *Food Hydrocolloids*, 87, pp. 321–330.

[58] Yang, F., Guo, C., Zhang, M., Bhandari, B. and Liu, Y. (2019). Improving 3D printing process of lemon juice gel based on fluid flow numerical simulation, *LWT*, 102, pp. 89–99.

[59] Ricci, I., Derossi, A. and Severini, C. (2019). *Fundamentals of 3D Food Printing and Applications*, "3D printed food from fruits and vegetables" (Elsevier), pp. 117–149.

[60] An, Y. J., Guo, C. F., Zhang, M. and Zhong, Z. P. (2019). Investigation on characteristics of 3D printing using Nostoc sphaeroides biomass, *Journal of the Science of Food and Agriculture*, 99, pp. 639–646.

[61] Severini, C., Derossi, A., Ricci, I., Caporizzi, R. and Fiore, A. (2018). Printing a blend of fruit and vegetables. New advances on critical variables and shelf life of 3D edible objects, *Journal of Food Engineering*, 220, pp. 89–100.

[62] Dankar, I., Pujolà, M., El Omar, F., Sepulcre, F. and Haddarah, A. (2018). Impact of mechanical and microstructural properties of potato puree-food additive complexes on extrusion-based 3D printing, *Food and Bioprocess Technology*, 11, pp. 2021–2031.

[63] Liu, Z., Bhandari, B., Prakash, S. and Zhang, M. (2018). Creation of internal structure of mashed potato construct by 3D printing and its textural properties, *Food Research International*, 111, pp. 534–543.

[64] Kim, H. W., Lee, J. H., Park, S. M., Lee, M. H., Lee, I. W., Doh, H. S. and Park, H. J. (2018). Effect of hydrocolloids on rheological properties and

printability of vegetable inks for 3D food printing, *Journal of Food Science*, 83, pp. 2923–2932.

[65] Ritchie, H. and Roser, M. (2017). Meat and dairy production, *Our World in Data*.

[66] De Bakker, E. and Dagevos, H. (2012). Reducing meat consumption in today's consumer society: questioning the citizen-consumer gap, *Journal of Agricultural and Environmental Ethics*, 25, pp. 877–894.

[67] Lupton, D. and Turner, B. (2017). *Digital Food Activism*, "'Both fascinating and disturbing': consumer responses to 3D food printing and implications for food activism" (Routledge), pp. 151–167.

[68] Liu, C., Ho, C. and Wang, J. (2018). The development of 3D food printer for printing fibrous meat materials. In *IOP Conference Series: Materials Science and Engineering*, IOP Publishing, p. 012019.

[69] Lipton, J., Arnold, D., Nigl, F., Lopez, N., Cohen, D., Norén, N. and Lipson, H. (2010). Multi-material food printing with complex internal structure suitable for conventional post-processing. In *Solid Freeform Fabrication Symposium*, pp. 809–815.

[70] Wilson, A., Anukiruthika, T., Moses, J. and Anandharamakrishnan, C. (2020). Customized shapes for chicken meat–based products: feasibility study on 3D-printed nuggets, *Food and Bioprocess Technology*, 13, pp. 1968–1983.

[71] Wang, L., Zhang, M., Bhandari, B. and Yang, C. (2018). Investigation on fish surimi gel as promising food material for 3D printing, *Journal of Food Engineering*, 220, pp. 101–108.

[72] Kim, S. M., Kim, H. W. and Park, H. J. (2021). Preparation and characterization of surimi-based imitation crab meat using coaxial extrusion three-dimensional food printing, *Innovative Food Science & Emerging Technologies*, 71, p. 102711.

[73] Dong, X., Pan, Y., Zhao, W., Huang, Y., Qu, W., Pan, J., Qi, H. and Prakash, S. (2020). Impact of microbial transglutaminase on 3D printing quality of Scomberomorus niphonius surimi, *LWT*, 124, p. 109123.

[74] Dick, A., Bhandari, B. and Prakash, S. (2019). Post-processing feasibility of composite-layer 3D printed beef, *Meat Science*, 153, pp. 9–18.

[75] Dick, A., Bhandari, B. and Prakash, S. (2021). Printability and textural assessment of modified-texture cooked beef pastes for dysphagia patients, *Future Foods*, 3, p. 100006.

[76] Dick, A., Bhandari, B. and Prakash, S. (2021). Effect of reheating method on the post-processing characterisation of 3D printed meat products for dysphagia patients, *LWT*, p. 111915.

[77] Cassens, D., Johnson, E. and Keelan, S. (1996). Enhancing taste, texture, appearance, and presentation of pureed food improved resident quality of life and weight status, *Nutrition Reviews*, 54, p. S51.

[78] Farrer, O., Olsen, C., Mousley, K. and Teo, E. (2016). Does presentation of smooth pureed meals improve patients consumption in an acute care setting: a pilot study, *Nutrition & Dietetics*, 73, pp. 405–409.

[79] Germain, I., Dufresne, T. and Gray-Donald, K. (2006). A novel dysphagia diet improves the nutrient intake of institutionalized elders, *Journal of the American Dietetic Association*, 106, pp. 1614–1623.

[80] Liu, Z. and Zhang, M. (2019). *Fundamentals of 3D Food Printing and Applications*, "3D food printing technologies and factors affecting printing precision" (Elsevier), pp. 19–40.

[81] Holland, S., Tuck, C. and Foster, T. (2018). Selective recrystallization of cellulose composite powders and microstructure creation through 3D binder jetting, *Carbohydrate Polymers*, 200, pp. 229–238.

[82] Buschow, K. J. (2001). *Encyclopedia of Materials: Science and Technology* (Elsevier).

[83] Symon, K. R. (1971). *Mechanics*, Third Edition (Addison-Wesley Publishing Company).

[84] Struble, L. J. and Ji, X. (2001). *Handbook of Analytical Techniques in Concrete Science and Technology* (William Andrew).

[85] Folayan, J. A., Anawe, P. A. L., Abioye, P. O. and Elehinafe, F. B. (2017). Selecting the most appropriate model for rheological characterization of synthetic based drilling mud, *International Journal of Applied Engineering Research*, 12, pp. 7614–7649.

[86] Truby, R. L. and Lewis, J. A. (2016). Printing soft matter in three dimensions, *Nature*, 540, pp. 371–378.

[87] Aguado, B. A., Mulyasasmita, W., Su, J., Lampe, K. J. and Heilshorn, S. C. (2012). Improving viability of stem cells during syringe needle flow through the design of hydrogel cell carriers, *Tissue Engineering Part A*, 18, pp. 806–815.

[88] Panton, R. (2013). Lubrication approximation, *Incompressible Flow*, 4th edition (John Wiley & Sons), pp. 650–668.

[89] Irgens, F. (2014). *Rheology and Non-Newtonian Fluids* (Springer).

[90] Witczak, M., Juszczak, L. and Gałkowska, D. (2011). Non-Newtonian behaviour of heather honey, *Journal of Food Engineering*, 104, pp. 532–537.

[91] Guo, B. and Liu, G. (2011). *Applied Drilling Circulation Systems: Hydraulics, Calculations and Models* (Gulf Professional Publishing).

[92] Berta, M., Wiklund, J., Kotzé, R. and Stading, M. (2016). Correlation between in-line measurements of tomato ketchup shear viscosity and extensional viscosity, *Journal of Food Engineering*, 173, pp. 8–14.

[93] Venkatesan, J., Sankar, D., Hemalatha, K. and Yatim, Y. (2013). Mathematical analysis of Casson fluid model for blood rheology in stenosed narrow arteries, *Journal of Applied Mathematics*, 2013.

[94] Tan, J. J., Tee, J. K., Chou, K. O., Yong, S. Y. A., Pan, J., Ho, H. K., Ho, P. C. and Kang, L. (2018). Impact of substrate stiffness on dermal papilla aggregates in microgels, *Biomaterials Science*, 6, pp. 1347–1357.

[95] McKenna, B. and Lyng, J. (2003). Introduction to food rheology and its measurement, *Texture in Food*, 1, pp. 130–160.

[96] Steffe, J. F. (1996). *Rheological Methods in Food Process Engineering* (Freeman Press).

[97] Tan, C., Toh, W. Y., Wong, G. and Li, L. (2018). Extrusion-based 3D food printing: Materials and machines, *International Journal of Bioprinting*, 4.

[98] Vancauwenberghe, V., Delele, M. A., Vanbiervliet, J., Aregawi, W., Verboven, P., Lammertyn, J. and Nicolaï, B. (2018). Model-based design and validation of food texture of 3D printed pectin-based food simulants, *Journal of Food Engineering*, 231, pp. 72–82.

[99] Kim, H. W., Bae, H. and Park, H. J. (2017). Classification of the printability of selected food for 3D printing: development of an assessment method using hydrocolloids as reference material, *Journal of Food Engineering*, 215, pp. 23–32.

[100] Liu, Z., Zhang, M., Bhandari, B. and Yang, C. (2018). Impact of rheological properties of mashed potatoes on 3D printing, *Journal of Food Engineering*, 220, pp. 76–82.

[101] Zhu, S., Stieger, M. A., van der Goot, A. J. and Schutyser, M. A. (2019). Extrusion-based 3D printing of food pastes: correlating rheological properties with printing behaviour, *Innovative Food Science & Emerging Technologies*, 58, p. 102214.

[102] Nijdam, J. J., Agarwal, D. and Schon, B. S. (2021). Assessment of a novel window of dimensional stability for screening food inks for 3D printing, *Journal of Food Engineering*, 292, p. 110349.

[103] Lawless, H. T. and Heymann, H. (2010). *Sensory Evaluation of Food: Principles and Practices* (Springer).

[104] Di Monaco, R., Cavella, S. and Masi, P. (2008). Predicting sensory cohesiveness, hardness and springiness of solid foods from instrumental measurements, *Journal of Texture Studies*, 39, pp. 129–149.

[105] Bourne, M. (2002). *Food Texture and Viscosity: Concept and Measurement* (Elsevier).

[106] Chen, L. and Opara, U. L. (2013). Texture measurement approaches in fresh and processed foods — A review, *Food Research International*, 51, pp. 823–835.

[107] Civille, G. V. and Oftedal, K. N. (2012). Sensory evaluation techniques — Make "good for you" taste "good", *Physiology & Behavior*, 107, pp. 598–605.

[108] de Liz Pocztaruk, R., Abbink, J. H., de Wijk, R. A., da Fontoura Frasca, L. C., Gavião, M. B. D. and van der Bilt, A. (2011). The influence of auditory

and visual information on the perception of crispy food, *Food Quality and Preference*, 22, pp. 404–411.

[109] Kealy, T. (2006). Application of liquid and solid rheological technologies to the textural characterisation of semi-solid foods, *Food Research International*, 39, pp. 265–276.

[110] Foegeding, E., Daubert, C., Drake, M., Essick, G., Trulsson, M., Vinyard, C. and Van de Velde, F. (2011). A comprehensive approach to understanding textural properties of semi- and soft-solid foods, *Journal of Texture Studies*, 42, pp. 103–129.

[111] Drake, M. (2007). Invited review: sensory analysis of dairy foods, *Journal of Dairy Science*, 90, pp. 4925–4937.

[112] Brookfield, P. L., Nicoll, S., Gunson, F. A., Harker, F. R. and Wohlers, M. (2011). Sensory evaluation by small postharvest teams and the relationship with instrumental measurements of apple texture, *Postharvest Biology and Technology*, 59, pp. 179–186.

[113] Salvador, A., Varela, P., Sanz, T. and Fiszman, S. (2009). Understanding potato chips crispy texture by simultaneous fracture and acoustic measurements, and sensory analysis, *LWT*, 42, pp. 763–767.

[114] Mantihal, S., Prakash, S. and Bhandari, B. (2019). Texture-modified 3D printed dark chocolate: sensory evaluation and consumer perception study, *Journal of Texture Studies*, 50, pp. 386–399.

[115] Costa, F., Cappellin, L., Longhi, S., Guerra, W., Magnago, P., Porro, D., Soukoulis, C., Salvi, S., Velasco, R. and Biasioli, F. (2011). Assessment of apple (Malus × domestica Borkh.) fruit texture by a combined acoustic-mechanical profiling strategy, *Postharvest Biology and Technology*, 61, pp. 21–28.

[116] Foegeding, E. A., Çakır, E. and Koç, H. (2010). Using dairy ingredients to alter texture of foods: implications based on oral processing considerations, *International Dairy Journal*, 20, pp. 562–570.

[117] Claus, J. R. (1995). Methods for the objective measurement of meat product texture. In *Reciprocal Meat Conference Proceedings*, pp. 96–101.

[118] Sasikala, V., Ravi, R. and Narasimha, H. (2011). Textural changes of green gram (Phaseolus aureus) and horse gram (Dolichos biflorus) as affected by soaking and cooking, *Journal of Texture Studies*, 42, pp. 10–19.

[119] De Roeck, A., Mols, J., Duvetter, T., Van Loey, A. and Hendrickx, M. (2010). Carrot texture degradation kinetics and pectin changes during thermal versus high-pressure/high-temperature processing: a comparative study, *Food Chemistry*, 120, pp. 1104–1112.

[120] Farris, S., Gobbi, S., Torreggiani, D. and Piergiovanni, L. (2008). Assessment of two different rapid compression tests for the evaluation of texture differences in osmo-air-dried apple rings, *Journal of Food Engineering*, 88, pp. 484–491.

[121] Takahashi, T., Hayakawa, F., Kumagai, M., Akiyama, Y. and Kohyama, K. (2009). Relations among mechanical properties, human bite parameters, and ease of chewing of solid foods with various textures, *Journal of Food Engineering*, 95, pp. 400–409.

[122] Ioannides, Y., Seers, J., Defernez, M., Raithatha, C., Howarth, M. S., Smith, A. and Kemsley, E. K. (2009). Electromyography of the masticatory muscles can detect variation in the mechanical and sensory properties of apples, *Food Quality and Preference*, 20, pp. 203–215.

[123] Nguyen, L. T., Tay, A., Balasubramaniam, V., Legan, J., Turek, E. J. and Gupta, R. (2010). Evaluating the impact of thermal and pressure treatment in preserving textural quality of selected foods, *LWT*, 43, pp. 525–534.

[124] Tsukakoshi, Y., Naito, S. and Ishida, N. (2007). Probabilistic characteristics of stress changes during cereal snack puncture, *Journal of Texture Studies*, 38, pp. 220–235.

[125] Moreira, R., Chenlo, F., Chaguri, L. and Fernandes, C. (2008). Water absorption, texture, and color kinetics of air-dried chestnuts during rehydration, *Journal of Food Engineering*, 86, pp. 584–594.

[126] Varela, P., Salvador, A. and Fiszman, S. (2008). On the assessment of fracture in brittle foods: the case of roasted almonds, *Food Research International*, 41, pp. 544–551.

[127] Jaworska, G. and Bernaś, E. (2010). Effects of pre-treatment, freezing and frozen storage on the texture of *Boletus edulis* (Bull: Fr.) mushrooms, *International Journal of Refrigeration*, 33, pp. 877–885.

[128] Farahnaky, A., Azizi, R. and Gavahian, M. (2012). Accelerated texture softening of some root vegetables by Ohmic heating, *Journal of Food Engineering*, 113, pp. 275–280.

[129] Schreuders, F. K., Schlangen, M., Kyriakopoulou, K., Boom, R. M. and van der Goot, A. J. (2021). Texture methods for evaluating meat and meat analogue structures: a review, *Food Control*, 127, p. 108103.

[130] Barbut, S. (2015). *The Science of Poultry and Meat Processing* (University of Guelph).

[131] De Huidobro, F. R., Miguel, E., Blázquez, B. and Onega, E. (2005). A comparison between two methods (Warner–Bratzler and texture profile analysis) for testing either raw meat or cooked meat, *Meat Science*, 69, pp. 527–536.

[132] Guiné, R. P. and Barroca, M. J. (2012). Effect of drying treatments on texture and color of vegetables (pumpkin and green pepper), *Food and Bioproducts Processing*, 90, pp. 58–63.

[133] Novaković, S. and Tomašević, I. (2017). A comparison between Warner–Bratzler shear force measurement and texture profile analysis of meat and meat products: a review. In *IOP Conference Series: Earth and Environmental Science* (IOP Publishing), p. 012063.

[134] Paula, A. M. and Conti-Silva, A. C. (2014). Texture profile and correlation between sensory and instrumental analyses on extruded snacks, *Journal of Food Engineering*, 121, pp. 9–14.

[135] Aguirre, M., Owens, C., Miller, R. and Alvarado, C. (2018). Descriptive sensory and instrumental texture profile analysis of woody breast in marinated chicken, *Poultry Science*, 97, pp. 1456–1461.

[136] Szczesniak, A. S. (1963). Classification of textural characteristics, *Journal of Food Science*, 28, pp. 385–389.

[137] Chandra, M. and Shamasundar, B. (2015). Texture profile analysis and functional properties of gelatin from the skin of three species of fresh water fish, *International Journal of Food Properties*, 18, pp. 572–584.

[138] Radočaj, O. F., Dimić, E. B. and Vujasinović, V. B. (2011). Optimization of the texture of fat-based spread containing hull-less pumpkin (Cucurbita pepo L.) seed press-cake, *Acta Periodica Technologica*, pp. 131–143.

[139] Fermin, B. C., Hahm, T., Radinsky, J. A., Kratochvil, R. J., Hall, J. E. and Lo, Y. M. (2005). Effect of proline and glutamine on the functional properties of wheat dough in winter wheat varieties, *Journal of Food Science*, 70, pp. E273–E278.

[140] Chiabrando, V., Giacalone, G. and Rolle, L. (2009). Mechanical behaviour and quality traits of highbush blueberry during postharvest storage, *Journal of the Science of Food and Agriculture*, 89, pp. 989–992.

[141] Yang, F., Cui, Y., Guo, Y., Yang, W., Liu, X. and Liu, X. (2021). Internal structure and textural properties of a milk protein composite gel construct produced by three-dimensional printing, *Journal of Food Science*, 86, pp. 1917–1927.

[142] Karavasili, C., Gkaragkounis, A., Moschakis, T., Ritzoulis, C. and Fatouros, D. G. (2020). Pediatric-friendly chocolate-based dosage forms for the oral administration of both hydrophilic and lipophilic drugs fabricated with extrusion-based 3D printing, *European Journal of Pharmaceutical Sciences*, 147, p. 105291.

[143] Szczesniak, A. S. (2002). Texture is a sensory property, *Food Quality and Preference*, 13, pp. 215–225.

[144] Lille, M., Kortekangas, A., Heiniö, R.-L. and Sozer, N. (2020). Structural and textural characteristics of 3D-printed protein-and dietary fibre-rich snacks made of milk powder and wholegrain rye flour, *Foods*, 9, p. 1527.

[145] Vieira, M. V., Oliveira, S. M., Amado, I. R., Fasolin, L. H., Vicente, A. A., Pastrana, L. M. and Fuciños, P. (2020). 3D printed functional cookies fortified with Arthrospira platensis: evaluation of its antioxidant potential and physical-chemical characterization, *Food Hydrocolloids*, 107, p. 105893.

[146] Lipton, J. I., Cutler, M., Nigl, F., Cohen, D. and Lipson, H. (2015). Additive manufacturing for the food industry, *Trends in Food Science & Technology*, 43, pp. 114–123.

[147] Yang, F., Zhang, M. and Bhandari, B. (2017). Recent development in 3D food printing, *Critical Reviews in Food Science and Nutrition*, 57, pp. 3145–3153.

[148] Holland, S., Foster, T., MacNaughtan, W. and Tuck, C. (2018). Design and characterisation of food grade powders and inks for microstructure control using 3D printing, *Journal of Food Engineering*, 220, pp. 12–19.

[149] Scoutaris, N., Ross, S. A. and Douroumis, D. (2018). 3D printed "Starmix" drug loaded dosage forms for paediatric applications, *Pharmaceutical Research*, 35, pp. 1–11.

[150] Goyanes, A., Madla, C. M., Umerji, A., Piñeiro, G. D., Montero, J. M. G., Diaz, M. J. L., Barcia, M. G., Taherali, F., Sánchez-Pintos, P. and Couce, M.-L. (2019). Automated therapy preparation of isoleucine formulations using 3D printing for the treatment of MSUD: first single-centre, prospective, crossover study in patients, *International Journal of Pharmaceutics*, 567, p. 118497.

[151] Miles, A., Liang, V., Sekula, J., Broadmore, S., Owen, P. and Braakhuis, A. J. (2020). Texture-modified diets in aged care facilities: nutrition, swallow safety and mealtime experience, *Australasian Journal on Ageing*, 39, pp. 31–39.

[152] Broz, C. C. and Hammond, R. K. (2014). Dysphagia: education needs assessment for future health-care foodservice employees, *Nutrition & Food Science*, 44, pp. 407–413.

[153] Liu, Z., Bhandari, B. and Zhang, M. (2020). Incorporation of probiotics (*Bifidobacterium animalis* subsp. Lactis) into 3D printed mashed potatoes: effects of variables on the viability, *Food Research International*, 128, p. 108795.

[154] Baiano, A. (2020). 3D printed foods: a comprehensive review on technologies, nutritional value, safety, consumer attitude, regulatory framework, and economic and sustainability issues, *Food Reviews International*, pp. 1–31.

[155] McClements, D. J. (2019). *Future Foods: How Modern Science Is Transforming the Way We Eat* (Springer).

[156] Cooper, K. A., Quested, T. E., Lanctuit, H., Zimmermann, D., Espinoza-Orias, N. and Roulin, A. (2018). Nutrition in the bin: a nutritional and environmental assessment of food wasted in the UK, *Frontiers in Nutrition*, 5, p. 19.

[157] Hoy, A. Q. (2018). Agricultural advances draw opposition that blunts innovation, *Science*, 360, pp. 1413–1414.

[158] Conroy, S., Drennan, M., Kenny, D. and McGee, M. (2010). The relationship of various muscular and skeletal scores and ultrasound measurements in the live animal, and carcass classification scores with carcass composition and value of bulls, *Livestock Science*, 127, pp. 11–21.

[159] Dick, A., Bhandari, B. and Prakash, S. (2019). 3D printing of meat, *Meat Science*, 153, pp. 35–44.

[160] Ramachandraiah, K. (2021). Potential development of sustainable 3D-printed meat analogues: a review, *Sustainability*, 13, p. 938.

[161] Jayathilakan, K., Sultana, K., Radhakrishna, K. and Bawa, A. (2012). Utilization of byproducts and waste materials from meat, poultry and fish processing industries: a review, *Journal of Food Science and Technology*, 49, pp. 278–293.

[162] Sauerwein, M. and Doubrovski, E. (2018). Local and recyclable materials for additive manufacturing: 3D printing with mussel shells, *Materials Today Communications*, 15, pp. 214–217.

[163] Umerah, C. O., Kodali, D., Head, S., Jeelani, S. and Rangari, V. K. (2020). Synthesis of carbon from waste coconutshell and their application as filler in bioplast polymer filaments for 3D printing, *Composites Part B: Engineering*, 202, p. 108428.

[164] Wu, C.-S., Wang, S.-S., Wu, D.-Y. and Shih, W.-L. (2021). Novel composite 3D-printed filament made from fish scale-derived hydroxyapatite, eggshell and polylactic acid via a fused fabrication approach, *Additive Manufacturing*, 46, p. 102169.

[165] Sun, J., Zhou, W., Huang, D., Fuh, J. Y. and Hong, G. S. (2015). An overview of 3D printing technologies for food fabrication, *Food and Bioprocess Technology*, 8, pp. 1605–1615.

[166] Van Huis, A. (2013). Potential of insects as food and feed in assuring food security, *Annual Review of Entomology*, 58, pp. 563–583.

[167] Raubenheimer, D., Rothman, J. M., Pontzer, H. and Simpson, S. J. (2014). Macronutrient contributions of insects to the diets of hunter–gatherers: a geometric analysis, *Journal of Human Evolution*, 71, pp. 70–76.

[168] Tan, H. S. G., van den Berg, E. and Stieger, M. (2016). The influence of product preparation, familiarity and individual traits on the consumer acceptance of insects as food, *Food Quality and Preference*, 52, pp. 222–231.

[169] Payne, C. L., Dobermann, D., Forkes, A., House, J., Josephs, J., McBride, A., Müller, A., Quilliam, R. and Soares, S. (2016). Insects as food and feed: European perspectives on recent research and future priorities, *Journal of Insects as Food and Feed*, 2, pp. 269–276.

[170] van der Weele, C., Feindt, P., van der Goot, A. J., van Mierlo, B. and van Boekel, M. (2019). Meat alternatives: an integrative comparison, *Trends in Food Science & Technology*, 88, pp. 505–512.

[171] Ong, S., Choudhury, D. and Naing, M. W. (2020). Cell-based meat: current ambiguities with nomenclature, *Trends in Food Science & Technology*, 102, pp. 223–231.

[172] Orzechowski, A. (2015). Artificial meat? Feasible approach based on the experience from cell culture studies, *Journal of Integrative Agriculture*, 14, pp. 217–221.

[173] Gaydhane, M. K., Mahanta, U., Sharma, C. S., Khandelwal, M. and Ramakrishna, S. (2018). Cultured meat: state of the art and future, *Biomanufacturing Reviews*, 3, pp. 1–10.

[174] K. Handral, H., Hua Tay, S., Wan Chan, W. and Choudhury, D. (2020). 3D printing of cultured meat products, *Critical Reviews in Food Science and Nutrition*, pp. 1–10.

[175] Kang, D.-H., Louis, F., Liu, H., Shimoda, H., Nishiyama, Y., Nozawa, H., Kakitani, M., Takagi, D., Kasa, D. and Nagamori, E. (2021). Engineered whole cut meat-like tissue by the assembly of cell fibers using tendon-gel integrated bioprinting, *Nature Communications*, 12, pp. 1–12.

[176] Rorheim, A., Mannino, A., Baumann, T. and Caviola, L. (2016). Cultured meat: an ethical alternative to industrial animal farming, *Policy paper by Sentience Politics*, 1, pp. 1–14.

[177] Bhat, Z. F., Kumar, S. and Fayaz, H. (2015). *In vitro* meat production: challenges and benefits over conventional meat production, *Journal of Integrative Agriculture*, 14, pp. 241–248.

[178] Ben-Arye, T. and Levenberg, S. (2019). Tissue engineering for clean meat production, *Frontiers in Sustainable Food Systems*, 3, p. 46.

[179] Sher, D. and Tutó, X. (2015). Review of 3D food printing, *Temes de disseny*, 31, pp. 104–117.

[180] Portanguen, S., Tournayre, P., Sicard, J., Astruc, T. and Mirade, P.-S. (2019). Toward the design of functional foods and biobased products by 3D printing: a review, *Trends in Food Science & Technology*, 86, pp. 188–198.

# Problems

1. Describe the mechanism of extrusion-based 3D food printing.
2. What are the current challenges in deploying selective laser sintering printers for commercial usage?
3. How are food inks classified for extrusion-based 3D food printing?
4. What are the types of foods applicable to 3D printing? Classify them into different categories.
5. What are the physical properties of food inks important for 3D food printing?
6. What are the parameters measured for food texture, and what are the tests which can be used to measure them?
7. Explain the significance of 3D food printing.

# Chapter 2

# Food Printers and Related Products
# in the Market

With the increasing demand for 3D food printers and advancement of technology, there is an increase in the number of commercial food printers available for end users to choose from. In this chapter, a few selected commercial food printers are described.

## 2.1 Natural Machines

### 2.1.1 *Company*

Natural Machines was founded by Emilio Sepulveda and Lynette Kucsma in 2012. The company is based in Spain, and the founders started it because they were concerned about people's health. The traditional food supply chain involves manufacturing and distribution. The food factory is typically in a central location, and distribution is required, which increases costs and affects freshness. Hence, the founders wanted to design a 3D food printer as a mini-manufacturing kitchen to bring fresh food to the table. You can find out more about the company on their official website: www.naturalmachines.com.

### 2.1.2 *Product and Its Working Principles*

Foodini is a 3D printer kitchen appliance developed by Natural Machines. The company designed this printer with the objective of speeding up the process of making foods with fresh ingredients. As part of their intention,

51

Figure 2.1.   Foodini from Natural Machines. (Image courtesy of Natural Machines.)

the company aimed to do away with factory-prepared pre-filled food capsules. Hence, the printer comes with empty stainless steel food capsules and the users have the freedom to print fresh ingredients in the shapes and forms they desire.

Figure 2.1 shows an image of the food printer Foodini. The primary working principle of Foodini is extrusion-based, as illustrated in Chapter 1. The printer pushes the food ingredients down the stainless steel capsules, and the food ingredients flow through the nozzle to print the dish. Through layer-by-layer printing, 3D shapes and forms are created.

The technical specifications of the printer can be summarized through the following table.

|  | **Specifications** |
| --- | --- |
| Weight | 20 kg |
| Size of the printer | 45.8 cm × 43 cm × 43 cm |
| Display | 10″ Interactive |
| Auto-exchange of capsules | Yes, and permits up to 5 capsules in the printer |
| Capsule volume | 100 mL per capsule |
| Capsule bay heater | Available, up to 90 °C |
| Base for print | 11 in. |
| Power supply | 110–220 V |
| Printing volume | Maximum height, 110 mm; diameter, 257 mm |
| Internet connection | Yes, WiFi |

(*Continued*)

| Multiple-use profile | Yes |
|---|---|
| Operating system | Android |
| Power consumption | Maximum 324 W |

### 2.1.3 *Data and File Preparation*

The 3D food printer Foodini also comes pre-loaded with its own software, named Foodini Creator. Shape libraries are available in this software and users can utilize them to start designing the dish on a tablet or laptop.

Foodini has avoided the need to convert the design into a machine language such as G-code by allowing users to send their design over the internet to Natural Machines, which will then send the appropriate commands to the 3D food printer to start printing. Programming and files preparation by the users are thus avoided.

### 2.1.4 *Machine Food Safety-related Certifications*

Food safety is a critical component of the food printer, and hence, the parts that are in contact with the food must be food-safe. All the capsules used in the Foodini printer are made from 18/8 stainless steel, also known as 304, which is a food-grade stainless steel. In addition, the company has obtained the *Conformitè Europëenne* (CE) certification and tested the capsule according to the Council of Europe Resolution CM/Res(2013)9 on metals and alloys used in food contact materials.

Another part that is in contact with the food is the silicone mat, and the company has tested it to be compliant with Regulation (EC) No. 1935/2004 and (EC) No. 10/2011 on materials and articles intended to come into contact with food and Resolution ResAP(2004)5 on silicones used for food contact applications. Similarly, the company has certified and tested the rotating glass base to be compliant with (EU) No. 2015/863. The plastic parts used in Foodini are also made of either polycarbonate or polypropylene, both of which are food-grade and BPA-free.

### 2.1.5 *Strengths*

The main advantages of Foodini include the following:

1) *Availability of multi-extrusion nozzles*. The availability of auto-exchange of capsules is a major strength of Foodini. This is an

important advantage as it allows for the design of dishes that consist of multiple ingredients.

2) *Wide range of materials demonstrated.* Foodini has shown to be able to print several different types of food ink, and examples can be seen in the following subsection.

3) *Food-grade certifications.* Certifications for the critical parts of the 3D food printer that come into contact with the food are available, which makes approval of the 3D food printer usage easier.

4) *Availability of capsule bay heater.* Heating of the food ink is required for the printability of some of the food formulations. This increases the choice of ingredients greatly.

5) *Availability of software Foodini Creator.* The availability of this software makes 3D printing easier for non-engineering background users, which is a good thing as the chefs can concentrate on the design of the dish. In addition, the software comes with shape libraries and users can utilize it to kick start the design.

## 2.1.6 *Applications and Case Studies*

Foodini has been used in several applications. For example, the printing of eggs on Carpaccio has been demonstrated by Chef Carles Tejedor of Oil Motion, as shown in Figure 2.2. Without the availability of the 3D food printer, the design of such dishes will not be possible. Foodini extends the

Figure 2.2.   Printing of eggs on Carpaccio by Chef Carles Tejedor of Oil Motion. (Image courtesy of Natural Machines.)

creativity of the chefs greatly and improves the aesthetics and presentation of a dish. It also allows for customization according to the customers.

Other than printing on top of the food that is prepared, the Foodini 3D printer has also been used to print the whole meal for dysphagia patients. People with dysphagia have difficulty swallowing food, so it is important to serve them food that they can swallow easily. With the availability of 3D food printers, there has been an increase in the demand for printing meals for dysphagia patients. Meals for dysphagia patients are usually served as a lump of food, which affects their dignity. Figure 2.3 shows an example of a meal that is printed using Foodini, which allows patients to maintain their dignity.

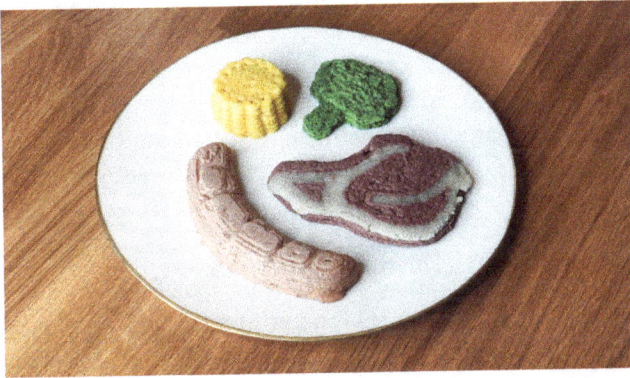

Figure 2.3.    Foodini from Natural Machines. (Image courtesy of Natural Machines.)

3D food printing of fresh vegetables is another example of what has been demonstrated using Foodini. Pant *et al.* [1] used Foodini to design nice dishes using various vegetable inks, and one of the dishes printed is shown in Figure 2.4.

Figure 2.4.    3D food printing of fresh vegetables. Reproduced with permission from [1]. Copyright (2021) from Elsevier.

## 2.2  byFlow

### 2.2.1 *Company*

byFlow is a Dutch company that was started in 2015 as a family business on a high-tech campus in Eindhoven. The company started with designing a 3D printer that could print multiple materials, including plastics, bronze, ceramic clays, silicone rubber, and food. In 2017, to target the food industry, the company decided to dedicate their Focus 3D food printer solely to 3D food printing. To enhance the user experience when using the Focus 3D food printer, byFlow Studio was launched in 2019, which is an online platform for 3D food printing featuring an easy-to-use design tool for personalized shapes, as well as a library of standard designs and recipes. In 2021, they announced a new and patented technology, and their 3D food printer was able to use solid or paste food and process it with temperature control and larger ingredient supply. You can find out more about the company on their official website: https://www.3dbyflow.com/about.

### 2.2.2 *Product and Its Working Principles*

The Focus 3D food printer is a 3D food printer kitchen appliance designed and developed by byFlow. The company designed this printer with the aim of enabling users to create new food designs that are not possible with traditional food preparation. In addition, the design of dishes can be easily customized without the trouble and resources required to create moulds. These will enable users to perform quick experiments to discover new opportunities and products, which include both textures and shapes. This capability will enable users to attract publicity and new customers. Thus, Focus 3D is designed to be dedicated to food printing and is shown in Figure 2.5. The primary working principle of Focus 3D is extrusion-based, and the technique is illustrated in Chapter 1.

One of the distinctive features of Focus 3D is that it is a foldable food printer. The food printer can be folded into a suitcase and safely transported. It is portable and can be stored neatly when not in used.

byFlow has adopted the "open kitchen" concept when designing Focus 3D. Focus 3D comes with 360° visibility, as illustrated in Figure 2.5. It is easy to use and clean. Focus 3D also comes with a touch screen interface to make it easier for users to use. The cartridges are reusable, and users can fill and replace them to experiment with their own ingredients.

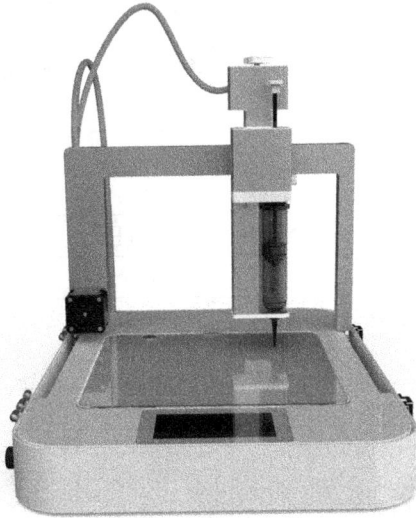

Figure 2.5.   Focus 3D from byFlow. (Image courtesy of byFlow.)

The technical specifications of Focus 3D can be summarized through the following table.

| Printing | |
|---|---|
| Printing technology | FFF |
| Number of extruders | 1 |
| Printing area | 210 × 190 × 135 mm$^3$ |
| Cartridge volume | 30 mL |
| Cartridge temperature control | No |
| Printing platform temp. control | Up to 80° |
| Average printing time | 8–10 min |
| Printing speed | 10–50 mm/s |
| Nozzle size (diameter) | 0.8 mm × 1.2 mm × 1.6 mm |
| Printing materials | Any food in paste consistency |
| | |
| **Mechanics** | |
| Folded | 440 × 325 × 110 mm$^3$ |

*(Continued)*

(*Continued*)

| Unfolded | 440 × 325 × 460 mm³ |
|---|---|
| Weight | 3D Printer, 8 kg; print head, 1 kg |
| Main frame material | Aluminium |
| Case material | Plastic (ABS) |
| Cartridge material | Plastic (PP) |
| Guiding | High-precision linear guide rails |
| | |
| **Software** | |
| File type | G-code |
| Connectivity to 3D printer | WiFi |
| Recommended slicing software | byFlow Studio, Slic3r, Simplify3D |

### 2.2.3 *Data and File Preparation*

Focus 3D comes with a software named byFlow Studio. It is an online platform where users can design their dish and prepare it for 3D printing. It comes with a design tool that allows users to make personalized 3D designs, which include logos, drawings, text, as well as pictures (see Figure 2.6 for an example). The software also includes a collection of

Figure 2.6.   Customized text and image printed on a cake using Focus 3D. (Image courtesy of byFlow.)

designs that users can use, which range from seasonal-related designs to special designs from their ambassadors.

Besides the designs, the software includes a collection of recipes, which is very diverse and was developed in collaboration with chefs, patisseries, and other food professionals. The taste ranges from sweet to savoury. With this software, the user can easily design, and the machine code is automatically generated and sent to the printer to initiate the print.

### 2.2.4 *Machine Food Safety-related Certifications*

byFlow has also ensured that the components that come into contact with food are certified to be food-safe. For example, the barrel, piston, and Luer dispensing tip from Nordson EFD are in compliance with European Directive 2002/95/EC, restriction of hazardous substances. In addition, the components were also sampled and tested to ensure that no detectable levels of silicone mould release agents were in use during production.

### 2.2.5 *Strengths*

The main advantages of Focus 3D include the following:

1) *Compactibility.* Focus 3D is a foldable and compact 3D printer. This saves space, which benefits small shops with limited space.
2) *Wide range of materials demonstrated.* Focus 3D has been shown to be capable of printing several different types of food inks, as can be seen in the following subsection.
3) *Availability of printing platform temperature control.* This additional feature prevents the food ink from suddenly coming into contact with a cold surface. In addition, it can be used to keep the printed dish warm.
4) *Food-grade certifications.* Certifications for the critical parts of the 3D food printer that come into contact with the food are also available, which makes approval of the 3D food printer usage easier.
5) *Availability of collections of recipes and design tools.* The availability of the collection simplifies the design experience for novices.

## 2.2.6 *Applications and Case Studies*

Focus 3D from byFlow has been demonstrated in several applications. For example, personalized printing of text and images on chocolate bars, cookies, and cakes has been demonstrated. Figure 2.6 shows an example of printing customized text and image on a cake. Using this printer, users can create and print personalized shapes such as logos, text, or artistic toppings. The topping shown here includes, but is not limited to, chocolate, ganache, fondant, marzipan, and fruit puree.

Users have also used Focus 3D to print customized 3D models. Figure 2.7 shows an example of a customized 3D model. Such capabilities

Figure 2.7.   Customized text and image printed on a cake using Focus 3D. (Image courtesy of byFlow.)

are suitable for specific occasions and festivals, without the need to create expensive moulds. In addition, it can serve customers who are looking for something exclusive for presents or corporate gifts.

3D printing of unique decorations and garnishes is also possible with Focus 3D from byFlow, as illustrated in Figure 2.8. It can produce numerous customized decorations at the same time. In addition, these decorations can be made from a range of edible food inks.

Figure 2.8. Customized 3D decorations printed using Focus 3D. (Image courtesy of byFlow.)

## 2.3 Shiyintech

### 2.3.1 *Company*

Based in a lab in Zhejiang University, Shiyintech is a Chinese company that was started by a group of faculty and researchers from Zhejiang University. The company specializes in the 3D food printing business and is interested in bringing 3D printing technology into the kitchen. The company has positioned the food printer as a marketing tool for café and pastry shop owners, as well as a fun kitchen appliance to experiment at home. It also aims to provide nursing homes with easy-to-chew foods in appealing shapes. You can find out more about the company on their official website: https://www.3dfoodbot.com/.

### 2.3.2 *Product and Its Working Principles*

FoodBot is a 3D food printer developed by Shiyintech. It is a second version of Createbot that comes in single extruder S2 or dual extruder D2 configuration. Similar to the first two printers, the primary working principle of FoodBot is extrusion-based. Figure 2.9 shows an image of FoodBot.

The technical specifications of FoodBot can be summarized through the following table.

Figure 2.9.    FoodBot from Shiyintech. (Image courtesy of Shiyintech.)

| Power | 50 W |
|---|---|
| Frame | Sheet metal structure |
| Printing support | WiFi/U disk |
| Print size | 150 × 150 × 70 mm |
| Nozzle diameter | 0.4–1.55 mm |
| Printing speed | 15–70 mm/s |
| $x$–$y$-axis positioning accuracy | 0.1/100 mm |
| $z$-axis positioning accuracy | 0.01/100 mm |

*(Continued)*

| Machine dimensions | 42.0 × 38.1 × 40.0 |
|---|---|
| Machine weight | 15 kg |
| Package dimension | 55 × 55 × 56 cm |
| Package weight | 23 kg |

### 2.3.3 *Data and File Preparation*

Users can download models from the internet and save them on a USB drive. The printer comes with a simple interface that allows the users to easily print their design from a USB drive. The operator can pause and restart the print when he wants. In addition, the temperature and speed of the dispenser can be adjusted by the users to achieve a better print.

### 2.3.4 *Machine Food Safety-related Certifications*

The syringe, plunger, and cap are the components that come into contact with food and have been tested according to EN 1186-3:2002 and EN 1186-1:2002. BS EN 1186 basically covers the materials and articles in contact with foodstuffs and was used to demonstrate food safety in FoodBot.

### 2.3.5 *Strengths*

The main advantages of FoodBot include:

1) *Accuracy.* FoodBot has a range of imprinting heads, from 0.3 mm to 1.2 mm. This enables the user to print delicate print jobs, which are essential for intricate decorative prints.
2) *Temperature control.* Another key strength of FoodBot is the availability of adjustable temperature. This is critical for food inks such as chocolate, where heating is often desired for better flow as well as quick solidification during printing. This is also the reason why FoodBot has been regularly used to print chocolate-related products.

3) *Easy operation.* The simple interface allows the user to operate the machine easily and parameters like speed and temperature can be adjusted. This enables users to perform tuning to improve the quality of the print.

4) *Food-grade certifications.* Certifications for the critical parts of the 3D food printer that come into contact with the food are also available, which makes approval of the 3D food printer usage easier.

### 2.3.6  *Applications and Case Studies*

In its various showcases, FoodBot has been largely focusing on chocolate. To demonstrate the capabilities of FoodBot, Shiyintech has demonstrated personalized printing of chocolate. Figure 2.10 shows an example of customized chocolate prints in the shape of a rabbit. Such prints can be served as desserts.

Figure 2.10.   Customized 3D chocolate rabbit made using Foodbot. (Image courtesy of Shiyintech.)

Because of the high accuracy as well as temperature control, 3D chocolate printing with overhang can also be performed as illustrated in Figure 2.11. Figure 2.11 shows a print made of white chocolate that is a structure with overhang and is a good example of edible decoration.

Figure 2.11.   3D chocolate print with overhang made using FoodBot. (Image courtesy of Shiyintech.)

## 2.4 Wiiboox

### 2.4.1 *Company*

Wiiboox is a 3D printing company founded in 2014. With a focus on education, food, scientific research, and manufacturing, the company is committed to becoming a global leader in 3D printing solutions. As a company, Wiiboox has a wide range of printers, which include 3D Food Printer, FDM 3D Printer, SLA 3D Printer, and DLP 3D Printer. The company has the support of technical teams from the University of Southern California and Zhejiang University. You can find out more about the company on their official website: https://www.wiiboox.com.

### 2.4.2 *Product and Its Working Principles*

WiibooxSweetin is a 3D food printer developed by Wiiboox that was launched in 2018. Wiiboox has also developed another printer, named WiibooxSweetin Latte. WiibooxSweetin is a 3D food printer which can print 3D food shapes, while WiibooxSweetin Latte focuses on making patterns on drinks such as coffee.

Figure 2.12 shows an image of the food printer WiibooxSweetin. Similarly, WiibooxSweetin uses a direct extruder to push the food ink out

Figure 2.12.   WiibooxSweetin from Wiiboox.

for 3D printing. The printer comes with a touch screen display, which is the key interface. In addition, it has features like self-calibration through auto-levelling. The features of the printer that come with the printer can be summarized through the following list:

- Touch screen display
- Medical-grade reusable tube
- Auto-levelling
- Real-time temperature control

The specifications of the printer are summarized in the following table:

| Dimension | 192 mm × 380 mm × 420 mm |
|-----------|--------------------------|
| Weight    | 10 kg                    |

*(Continued)*

| Power | 50 W |
|---|---|
| Printing size | 110 mm × 110 mm × 75 mm |
| Positioning accuracy | $XY$: 0.1/100 mm; $Z$: 0.01/100 mm; $E$: 0.01/100 mm |
| Nozzle diameter | 0.4–1.55 mm |
| Nozzle quantity | 1 |

Wiiboox has a second printer, WiibooxSweetin Latte, the focus of which is to make customized patterns on drinks such as coffee. This is done so that each cup of drink can be easily customized and unique, as well as interesting to the customers. Wiiboox has made uploading images easy, and it can be done with a phone. Alternatively, customers can choose the preset patterns. Before giving the signal to start the printer, you can adjust the brightness and contrast of the picture, as well as the diameter of your cup.

### 2.4.3 *Data and File Preparation*

The 3D food printer WiibooxSweetin is easy to operate. Users can choose the desired shape via their own USB thumb drive. In addition, Wiiboox has a touch screen, which makes interfacing easier.

### 2.4.4 *Machine Food Safety-related Certifications*

The plastics, stainless steel, and silica gel used in the 3D printer were tested with FDA 21 CFR 177.1210 with reference to FDA 21 CFR 176.170 (c) by Wiiboox. FDA 21 CFR 177.1520 and FDA 21 CFR 175.300 test methods were also used for these parts to demonstrate their food safety documentation.

### 2.4.5 *Strengths*

The main advantages of WiibooxSweetin include the following:

1) *Availability of auto-levelling*. The availability of auto-levelling allows for self-calibration, which is important if you require fine-resolution 3D prints.

2) *A 0.4 mm minimum diameter of pinhead.* The Wiiboox printer comes with a 0.4 mm minimum diameter of the pinhead, which enables it to print in fine resolution.

3) *Availability of real-time temperature control.* The 3D food printer comes with real-time temperature control. This feature is important to enhance the printing performance. In addition, it can also be used to protect the nutritional ingredients from damage.

### 2.4.6 *Applications and Case Studies*

WiibooxSweetin can be used for several food inks. As illustrated on the company's website, chocolate can be printed using WiibooxSweetin. One reason for this is its real-time temperature control, which makes printing chocolate relatively easier.

The availability of the small pinhead diameter also enables finer-resolution prints. Figure 2.13 shows the 3D printing of gingko nut paste using WiibooxSweetin.

Similarly, Figure 2.14 shows the printing of yam using Wiiboox Sweetin. The formula of this food ink was inspired by a traditional Teochew dessert, *orh nee.*

To increase the amount of protein intake, soy can be mixed with carrot. Figure 2.15 shows a complex 3D shape printed using carrot mixed

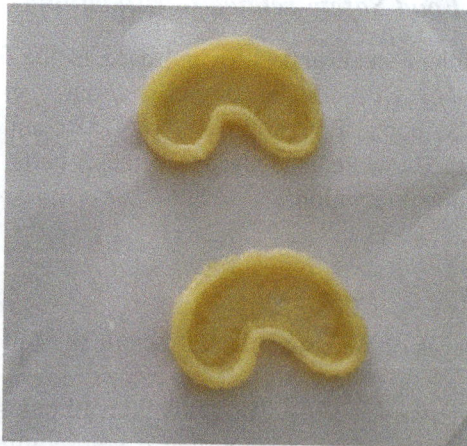

Figure 2.13.    Printing of gingko nut paste using WiibooxSweetin.

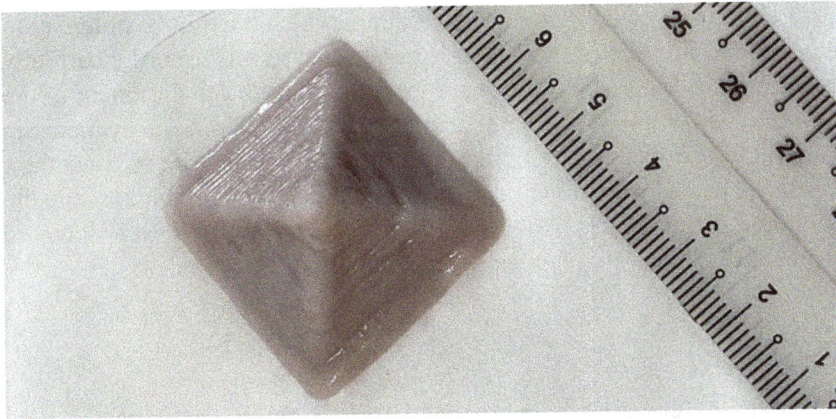

Figure 2.14.   Printing of yam paste using WiibooxSweetin.

Figure 2.15.   Printing of carrot mixed with soy using WiibooxSweetin.

with soy. This print demonstrates the possibility of printing a complex shape as well as food mixing to better meet nutrition requirements.

## 2.5  Conclusion

In this chapter, we made a quick overview of the commercially available 3D food printers. Various research papers have mentioned newer features, such as the use of laser or infraredheating, which have huge potential to be included in these printers in the near future. In addition, the currently commercially available printers are extrusion-based. Advancement of the technologies will result in more techniques like binders to be available in the future, just like how 3D printing has evolved for plastics. The current printing speeds of commercially available 3D food printers are suitable for home use and cafés. However, the speed and volume need to increase for venues like nursing homes, where scalability is important.

## Reference

[1]  Pant, A., Lee, A. Y., Karyappa, R., Lee, C. P., An, J., Hashimoto, M., Tan, U.-X., Wong, G., Chua, C. K. and Zhang, Y. (2021). 3D food printing of fresh vegetables using food hydrocolloids for dysphagic patients, *Food Hydrocolloids*, 114, p. 106546.

## Problems

1. Name some safety standards that are used by companies to illustrate that their printers are food-safe.
2. What are some of the food inks demonstrated by the various commercial 3D food printers?
3. All the commercial 3D food printers mentioned are extrusion-based. Discuss what you think is the reason.
4. What do you think are the features of 3D food printers that were demonstrated in research papers, but have not yet been incorporated into commercial printers?
5. Why do you think that the availability of auto-exchange of capsules is useful? Give examples to demonstrate the benefit.
6. Imagine you are a chef; list down the desired features that you would like to have in your 3D food printer. You may include features that are currently not available in commercial printers.

# Chapter 3

# 3D-Printed Desserts and Snacks: Dairy Products, Chocolate, Sugars, and Doughs

## 3.1 Introduction

The earliest scientific studies in the field of 3D food printing focused on easily extrudable food inks; namely, chocolate, cheese, frosting, and cookie dough [1]. The ability of 3D printing to customize intricate shapes offers unique advantages to the food industry [2]. Complex designs and geometries of food can be prepared in less time, with less labour, and with higher reproducibility [2]. This is especially true for the preparation of desserts. Additionally, the use of 3D printing has enabled the personalization of desserts, which is a novel initiative to gain attention [3]. Among the myriad 3D printing methods, the extrusion method with temperature control is the most appropriate for semi-solid food ingredients, which are heat-sensitive in nature. Examples of such ingredients include eggs and dairy products, which are commonly found in desserts [3]. Laser sintering techniques are limited to ingredients that undergo thermal fusion, such as sugars and fats [4]. Considering the wide range of applicable materials, extrusion-based 3D printing, in particular direct ink writing (DIW) 3D printing, is the current mainstream method of 3D food printing of desserts.

In this chapter, we discuss recent developments in the 3D printing of the four main categories of desserts, namely: (1) dairy products, (2) chocolate, (3) sugars, and (4) doughs. In addition, we also highlighted some works relating to 3D-printed Asian (Singaporean) desserts and snacks such as *chwee kueh, kueh dadar*, and *orh nee*. We also discuss the medical and societal implications of 3D-printed desserts and snacks.

## 3.2  3D-Printed Dairy Products

### 3.2.1  *Introduction*

3D printing offers unconventional routes to prepare conventional dairy products. A *milk cube* is an example of a well-known and delicious snack that is consumed widely around the world [5]. 3D printing can alter or fortify the nutritional contents of foods. For example, milk proteins fortified with whey protein isolate have been printed to meet the requirements of fitness enthusiasts, athletes, and young children, who may require a high protein intake [6]. Other forms of milk, such as heat-desiccated milk powder (HDMP) and heat acid coagulated milk (HACM), can be 3D-printed to produce a variety of dairy sweetmeats [7, 8]. Since the appearance of 3D food printing, there has been a growing interest in evaluating the 3D printability of dairy ingredients with personalized, textural, structural, functional, and sensorial characteristics [8]. This section discusses recent developments in the 3D printing of dairy products in terms of printing methods, materials, properties, and characterization.

### 3.2.2  *3D Printing Methods of Dairy Products*

This section introduces two main methods of 3D printing that have been employed for dairy products: (1) cold extrusion 3D printing, and (2) hot-melt extrusion 3D printing. Extrusion-based printing is the mainstream method to print dairy products due to the nature of the food ink; many dairy products can be prepared from liquid/paste materials, which makes extrusion a suitable candidate for the method of printing. Depending on the chemical and physical properties of the food materials, either cold extrusion or hot-melt extrusion is selected.

#### 3.2.2.1  *Cold extrusion 3D printing of dairy products*

Cold extrusion is a method to pattern extruded food inks without additional temperature control. Methods requiring the use of high temperatures such as hot-melt extrusion and selective laser sintering are not always suitable for creating 3D models comprising temperature-sensitive nutrients. Milk is an example of a food abundant in nutrients such as calcium and protein, which are temperature-sensitive. Such nutrients may degrade and denature at high temperatures. As an alternative, cold extrusion 3D printing allows the printing of food inks at room temperature [9]. Cold extrusion 3D printing is dependent solely on ink rheology. Therefore, parameters such as viscosity, yield stress, and storage modulus of the ink are essential for determining the printability and structural integrity of printed structures [4]. Food additives are often included in the ink to give the desired rheological properties for printing [9, 10]. For instance, Liu *et al.* performed room-temperature printing of the ink comprising milk protein concentrate, whey protein isolate, glycerol, and xanthan gum as additives [6]. The inclusion of multiple additives increases the complexity required for optimizing the concentration of each component to achieve ink printability [4]. To address such problems, Lee *et al.* demonstrated that milk ink can be successfully printed by only changing the concentration of the milk powder [4]. They have also demonstrated the capability of multi-material printing by cold extrusion with milk and other edible inks, including chocolate, coconut, maple syrup, and blueberry inks (Figure 3.1). This method allowed them to fabricate 3D milk models with a rigid enclosure containing different fillings or different layered structures to offer a variety of tastes and textures [4].

### 3.2.2.2 *Hot-melt extrusion 3D printing of dairy products*

This section describes studies that used hot-melt extrusion for the 3D printing of dairy products. Hot-melt extrusion refers to the process of applying heat and pressure (e.g., with a rotating screw) to melt the raw materials and extrude them through a die [11]. In the mid-19th century, hot-melt extrusion was first used in the plastics industry to manufacture polymeric insulation coverings for wires [11]. Today, hot-melt extrusion is commonly used in the pharmaceutical and food industries. 3D printing of dairy products has been mainly studied for the printability of skimmed milk powder (SMP) and semi-skimmed milk powder (SSMP) [4, 12], processed cheese [13], and milk protein concentrate [6, 14] with cold extrusion 3D printing. In contrast, hot-melt extrusion permits the ink to be dispensed at an elevated temperature; the printed inks are solidified immediately after the extrusion owing to thermal gelation and/or phase

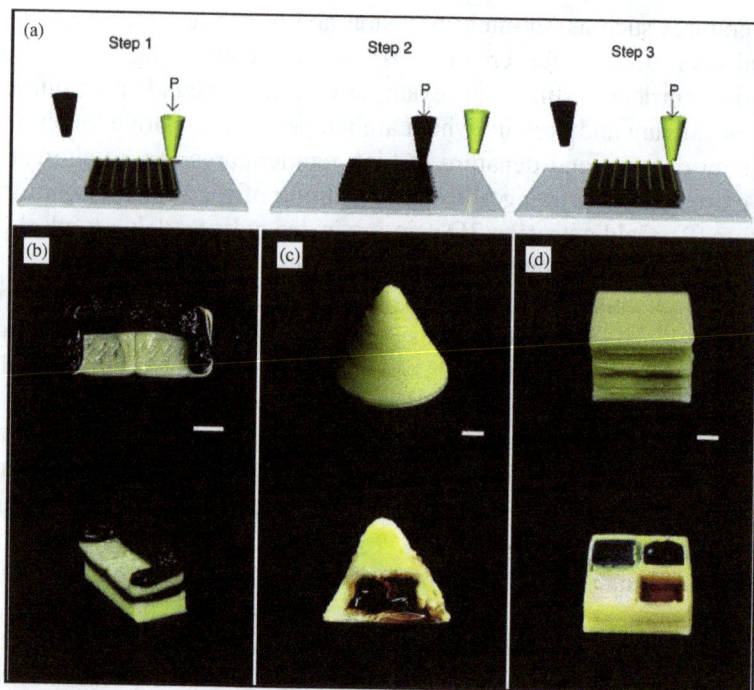

Figure 3.1.    Multi-material 3D-printed food structures. (a) Schematic drawing showing multi-material direct ink writing 3D printing. (b) 3D-printed sofa with different layers of milk and chocolate inks. (c) 3D-printed cone with chocolate syrup as an internal filling. (d) 3D-printed cube having four distinct compartments holding blueberry syrup, chocolate syrup, milk cream, and maple syrup as internal fillings. (Scale bar = 5 mm.) Reproduced with permission from [4]. Copyright (2020) from the Royal Society of Chemistry.

transition [7, 10]. Food inks for hot-melt extrusion must be thermo-responsive and are preferred to be shear-thinned with fast viscosity recovery [14].

Studies have been conducted to determine the effects of temperature on the physical properties of dairy inks, such as their rheological and textural properties. For instance, Joshi *et al.* studied the temperature dependence of the rheological properties of milk inks used in hot-melt extrusion 3D printing [7]. They identified that the formulation consisting of HDMP/SSMP at the ratio of 55/5 (60% w/w) yields the best print fidelity at a 40° C extrusion temperature among the four other ink formulations they investigated. Importantly, because the process involves phase transition of the materials, the print temperature is a key factor that determines the quality of the printed food. For example, Ross *et al.* studied the effects of temperature using hot-melt extrusion on the textures of 3D-printed processed cheese [15]. They discovered that processed cheese samples printed at a higher temperature 65 °C are more resilient, less adhesive, gummier, harder, and springier than those printed at a lower temperature 40 °C [15].

### 3.2.3 *Ink Formulation, Properties, and Characterization*

Studies on the effect of 3D printing methods on the structural properties (such as texture arising from microstructures) of dairy products are relatively new, having started to receive attention only in the late 2010s [13]. In contrast, the properties of the inks are another crucial factor in extrusion-based food printing determining the properties of the printed product. This section describes the rheological and textural properties of the different types of dairy inks studied in recent literature.

#### 3.2.3.1 *Rheological properties of dairy inks*

The rheological characteristics of food inks are essential for satisfactory 3D food printing [16, 17]. A fluid possessing shear-thinning behaviour can be extruded with relative ease at a high shear rate during extrusion through a narrow nozzle [12]. In addition, the yield stress of the food inks allows for the preservation of the printed structures after deposition [12]. Generally, dairy food inks exhibit shear-thinning behaviour and can be printed with extrusion-based 3D printing. The following subsections discuss several ink formulations of various dairy inks and their rheological properties.

- Milk and water inks

Milk inks, aqueous inks containing milk-based materials, have been widely demonstrated for 3D printing. Lee *et al.* studied the rheological properties of milk inks consisting of milk powder and water for cold extrusion 3D printing [4]. Milk inks containing concentrations of 10–75% w/w of commercial milk powder exhibited shear-thinning properties [4]. Both the viscosity and the yield stress (i.e., minimum shear stress required to start flow) of the milk inks increased with the milk powder concentrations; spreading in printed milk inks was observed at milk powder concentrations of <65% w/w due to the low yield stress of the ink. In contrast, the printed ink with the milk powder concentrations of >70% w/w were maintained without spreading [4]. The value of the storage modulus ($G'$) of the milk ink increased with milk powder concentration, suggesting strong bond formation within the colloidal network; this stabilization allowed the printed structure to retain its shape after extrusion [4]. Overall, the milk ink containing 70% w/w milk powder was most suitable for direct 3D printing and maintained the structural integrity of the 3D-printed milk product upon deposition.

- Milk and potato starch inks

Yang *et al.* studied the rheological properties of milk protein composite gel, which comprises potato starch, skimmed milk, cream, and icing sugar [5]. They observed that the milk protein composite gel exhibits shear-thinning behaviour [5]. The value of the composite gel's $G'$ (which represents its mechanical strength to resist deformation) was higher than the value of $G''$. The higher $G'$ value gave tan $\delta$ < 1 (where $\delta = G''/G'$), which indicated that the gel was predominantly elastic and could support its weight during printing.

- Milk and whey protein inks

Liu *et al.* studied the effects of whey protein isolate on the printing performance of milk protein concentrate [6]. Increasing the ratio of whey protein isolate to milk protein concentrate reduced the apparent viscosity of the food ink, facilitating the extrusion of the ink paste. All combinations of inks containing varying ratios of milk protein concentrate and whey protein isolate exhibited shear-thinning properties and were pseudo-plastic fluids [6]. The values of $G'$ of all inks studied were consistently higher than the values of $G''$, regardless of the composition. However, the increase in the whey protein content reduced the values of $G'$ and $G''$ for all inks that were studied. Correspondingly, their loss tangent (i.e., tan $\delta$) increased as the whey protein content increased, suggesting that the protein pastes became liquid-like and flowable. The authors reported that the milk protein concentrate/whey protein isolate at a ratio of 5/2 (35% w/w of the total dry matter content in the ink paste) gave the most desirable combination for 3D printing in terms of its rheological property, geometrical and shape accuracy, and mechanical property to support the weight of its deposited layers.

- Heat-desiccated milk powder ink and semi-skimmed milk powder

Joshi *et al.* studied the rheology of milk composite formulations (containing various amounts of HDMP and SSMP) induced by temperature changes during hot-melt extrusion-based 3D printing [7]. HDMP is prepared from pre-heated and homogenized milk, followed by heat-dehydration, micro-pulverization, and drying. HDMP is high in fat content (33%) and is not natively printable due to its low yield stress [7]. Therefore, corn flour (15% w/w) was included in the ink formulations as a gelling agent to improve their printability. Increasing the proportion of SSMP also increased the apparent viscosity of the inks due to its high hydrophilicity; the high hydrophobicity of the SSMP compelled the molecules to crowd (in the presence of water) and increased its apparent viscosity [7]. Conversely, an increase in the percentage of HDMP reduced the apparent viscosity, possibly due to the lubricating effect from the fat molecules [12]. Increasing the proportion of SSMP while reducing the proportion of HDMP led to higher $G'$ values of the ink, indicative of an increase in the elastic gel-like behaviour. The higher $G'$ was due to the higher hydrophilicity of the SSMP, resulting in lower mobilization of the water molecules within its network [7], which conferred mechanical stability to the printed structure. Similarly, SSMP increases the yield stress while HDMP decreases the yield stress of the inks. The higher yield stress caused by SSMP was attributed to the reduced movement of particles, alluding to its higher viscosity at a higher content of SSMP [6].

As the temperatures increased from 25 °C to 65 °C, the viscosities of all formulations decreased due to the melting of fat and reduced intermolecular bonding within the ink [14, 18]. The yield stress of the ink at 40 °C was lower than that at 25 °C because of a weaker network within the ink. Shear recoverability, the capability of the ink to reconstruct its matrix when a shear force is applied, declined with decreasing HDMP and increasing SSMP concentrations in the ink. Overall, the higher proportion of SSMP and the lower proportion of HDMP yielded less spreading, which improved the resolution, shape fidelity, and structural integrity when printed at 40 °C [7].

- Heat acid coagulated milk ink

Bareen *et al.* printed inks containing HACM semi-solids, which were commonly used in the manufacturing of dairy sweetmeats [8]. Casein, which makes up 80% of milk protein [19], undergoes coagulation when heated to ~80 °C at low pH; this coagulation produces a viscous mass called HACM semi-solids commonly called soft cheese. Soft cheese serves as a base material for producing a variety of dairy products [20].

Bareen *et al.* added maltitol (a moisture-binding and bulking agent used in food processing) to HACM to enhance its rheological property for printing [8]. A previous study has indicated that the use of maltitol improves the hydrophobic interactions within the casein matrix, resulting in lower hardness and improved texture properties of low-fat mozzarella cheese [21]. The inks consisting of HACM, whey protein isolate, maltitol, and water were patterned by extrusion-based 3D printing at room temperature. The spreading of the ink was not prominent for the inks containing high concentrations of whey protein isolate and maltitol. The inks containing whey protein isolate and maltitol in the ratio of 4:2 exhibited complete shape retention of the printed products. The presence of whey protein isolate prevented syneresis, improved the shape-holding ability of the ink, and promoted rapid recovery of mechanical strength to self-support the deposited layers during printing [6]. The addition of whey protein isolate and maltitol also increased the apparent viscosity of the HACM inks. The inks showed a decrease in the apparent viscosity with an increasing shear rate, indicating non-Newtonian shear-thinning behaviour. Additionally, the yield stress of the ink increased with the addition of whey protein isolate and maltitol, which enhanced the self-supporting ability of the printed ink. Inks with yield stresses of 700–1,400 Pa were reported to be suitable for 3D printing [8]. Increasing the maltitol content decreased the shear recoverability of the ink, as maltitol disrupts the network within the aggregated caseinate particles. On the other hand, increasing the whey protein isolate content increased the shear recoverability of the ink due to the high cross-linking density between casein and whey proteins. The higher the shear recoverability of the ink, the faster it could restore its mechanical strength post-extrusion to support the subsequently printed layer. The addition of whey protein isolate and maltitol increased the $G'$, $G''$, and complex modulus of the inks. This change was attributed to the increase in the total solid content within the formulation, particularly the whey protein isolate, which facilitated the formation of a strong network to aggregate particles [22].

- Cheese ink

Ross *et al.* formulated inks containing Cheddar cheese with varying pH and maturity (i.e., consisting of two batches of aged cheese in different percentages — one batch was aged for 1–2 months and the other was aged for 6–8 months) [15]. Emulsifying salts, such as trisodium citrate, were added to the ink to solubilize casein and to produce a homogeneous cheese mixture. They studied the rheology and texture of the mixture.

Inks containing a higher amount of mature Cheddar cheese were softer and more flowable than young Cheddar cheese. During the ageing, Cheddar cheese experienced progressive proteolysis, causing a decrease in the amounts of intact casein, which reduced the number of casein–casein interactions to soften the cheese [15, 23]. On the other hand, the presence of young Cheddar cheese in the ink confers higher stability to the printed structures due to higher structural protein content than that of mature cheese. Among the various formulations tested, the ink with the lowest viscosity exhibited poor printing accuracy due to spreading. Importantly, the ink with the highest viscosity also displayed poor printing accuracy due to gaps in the printed structure from inconsistent extrusion. The inks with viscosities of 7.55–10.94 Pa.s at 65 °C (i.e., printing temperature) were deemed to produce a reasonable print with little spreading [15]. In addition, differences in the tested pH (pH = 5.4, 5.6, and 5.8) did not have any effect on the printing accuracy of the cheese inks.

### 3.2.3.2 *Textural studies on 3D-printed dairy products*

Fan *et al.* evaluated the effects of different infill patterns of a 3D-printed composite gel consisting of milk proteins on their textures; such 3D-printed foods offer a way to control the textures for people with different abilities to chew [5]. In this study, the composite gel comprising potato starch, skimmed milk, cream, and icing sugar was printed using an extrusion-based 3D printer with variations in infill patterns (Hilbert curve, honeycomb, and rectilinear), infill levels (10%, 40%, and 70%), and perimeters (3, 5, and 7). The perimeter represented the number of external layers of the printed structures printed in the horizontal direction; the infill pattern described the type of pattern fabricated by the nozzle at every layer. The infill level referred to the percentage of solid construct that was fabricated at the inner part of the model. Porous structure was not observed in the samples with a 70% infill level, which could be explained by (1) subsidence deformation, and (2) swelling of the composite gel [24]. On the other hand, samples with infill levels lower than 40% experienced an occasional collapse of the inner parts. The hardness and gumminess of the 3D-printed products increased with increasing infill levels and perimeters but did not show any significant differences among the varying infill patterns. The authors noted that the textural properties (i.e., gumminess or hardness) were only measured in the vertical direction by top-down compression, while the textural properties are, in principle, anisotropic. Interestingly, when the infill level of the 3D-printed product was 100%, its firmness, gumminess, and hardness were significantly lower than those made by moulding.

Similarly, Tohic *et al.* reported a lower hardness of 3D-printed cheese than untreated cheese (control) [13]. It was proposed that the shear during extrusion had a greater impact on the textural properties of the cheese than melting alone. On the other hand, adhesiveness was significantly higher for 3D-printed cheese samples than for untreated cheese. Lastly, two different extrusion rates were used for the 3D printing: low-speed printed cheese (LSPC) at 4 mL/min and high-speed printed cheese (HSPC) at 12 mL/min. However, there were no significant differences between HSPC and LSPC for hardness, adhesiveness, springiness, cohesiveness, and resilience. These observations suggested that extrusion speed during printing may not have any impact on the texture of 3D-printed cheese. Overall, 3D printing can make structural changes to 3D-printed dairy products; such changes can be used to augment their textural and sensory properties suited for consumers.

## 3.3  3D Printing of Chocolate

### 3.3.1  *Introduction*

Among various food products, chocolate has been extensively used for 3D food printing [25, 26]. Chocolate is a characteristic ingredient made up of complex crystals, allowing it to be solid at room temperature and melt into a viscous liquid at human oral temperature [27]. Dark chocolate contains many different ingredients such as cocoa, sugar, and emulsifier; some varieties may contain milk powder [28]. This section discusses the recent developments in the 3D printing of chocolate products and their ink formulation, properties, and characterization.

### 3.3.2  *3D Printing Methods of Chocolate*

This section introduces and describes the methods that have been employed for the 3D printing of chocolate, namely (1) cold extrusion, (2) hot-melt extrusion, (3) laser sintering, and (4) inkjet printing.

### 3.3.2.1 *Cold extrusion*

Similar to 3D printing of milk products, cold extrusion is one of the established methods to 3D-print chocolates. Cold extrusion bypasses the need for temperature control in hot-melt extrusion. Karyappa *et al.* demonstrated the direct printing of chocolate-based inks using a direct ink writing 3D printer to fabricate complex 3D shapes without the use of temperature control [9]. The printing of chocolate-based inks was made possible by formulating inks with adequate rheological properties at room temperature. They also extended the application of this method to print chocolate-based models containing different types of chocolate-based inks, with features of a semi-solid enclosure and liquid filling using multiple nozzles. Cold extrusion provides a unique route for 3D food printing, especially when heat-sensitive food ingredients or additives are present in the food ink. Additionally, the use of multiple nozzles allowed for control over the distribution of ink within the printed structure, which could have potential applications in the modification of food textures and controlled release of nutrients [9].

### 3.3.2.2 *Hot-melt extrusion*

Hot-melt extrusion is another popular approach for chocolate 3D printing. Cocoa butter exists in six major crystal polymorphic forms (Form I–Form VI), which can be divided into three main groups: $\alpha$ (alpha), $\beta'$ (beta prime), and $\beta$ (beta). All of them possess different melting temperatures [29]. Form V ($\beta_2$ crystal) has a melting temperature between 33.8 °C and 35 °C, and it is known to be the most important crystal that contributes to a final chocolate product with greater stability, glossy finish, and better texture [30–32]. Therefore, adequate temperature control during printing is essential for the preservation of stable $\beta$ crystals to achieve suitable textural, glossy, and snap properties in the chocolate [27].

Mantihal *et al.* used hot-melt extrusion 3D printing to print chocolate with different support structures [29]. Support structures helped to build complex 3D geometries by stabilizing the overhangs [33]. In their work, they controlled the nozzle temperature at 32 °C to ensure the extrusion of the melted chocolate. The characterization of the flow behaviours of chocolate suggested that the chocolate melted between 28 °C and 30 °C. [29]. The nozzle temperature was kept at 32 °C because it prevented Form V ($\beta_2$ crystal) from melting completely; retaining their crystal seeds would promote stable crystal growth after deposition. Overall, hot-melt extrusion of chocolate provides advantages in terms of accessibility and simplicity. However, attention to temperature is essential due to the nature of chocolate that affects the quality of the printed chocolate. The rheological property of molten chocolate is also highly sensitive to printing temperatures.

### 3.3.2.3 *Laser sintering*

Researchers from TNO, the Netherlands, demonstrated laser sintering of the drinking chocolate powder Nesquik to produce chain links and company logos [1]. Laser sintering is a type of powder-bed 3D printing. For successful implementation of food printing, the food materials must adhere to a specific recipe to allow the agglomeration of powders during printing. Previously, 100% sugar powders were used for laser sintering by the company CandyFab. Chocolate powder Nesquik was successfully printed with a content of 39.3% sugar [34]. Coincidentally, the sugar in Nesquik had the same composition as the sugar present in many white, milk, or dark chocolate formulations [34]. The fat content in Nesquik was 1.7%, which was significantly less than the 26–40% for white, milk, and dark chocolate formulations [34]. The presence of sugars and fats allowed them to be melted, shaped, and fused upon re-solidification. Despite this success, however, there has been very little academic research that involves the use of 3D laser sintering of chocolate products.

### 3.3.2.4 *Inkjet 3D printing*

Inkjet 3D printing has been demonstrated for methods of rapid prototyping. Unlike DIW 3D printing, inkjet 3D printing typically requires inks with a low range of viscosity for printing, which may not be directly applicable to chocolate-based materials. Takagishi *et al.* demonstrated the 3D printing of chocolate using an electrostatic inkjet 3D printer [35]. Electrostatic inkjet printing was created to improve print precision and overcome the challenges that come with printing highly viscous foods such as chocolate. Additionally, electrostatic inkjet 3D printing allowed the printing of chocolate without additives. Additives may undermine the ideal taste of 3D-printed chocolates. A high voltage was applied between the liquid chocolate inside the syringe and the base plate. The electrostatic force pulled the ink to create a small amount of discharge each time. Drawing with the ink was achieved by moving the base plate. A syringe pump was used to control the pressure inside the syringe, making printing feasible.

Three types of discharge were observed (i.e., drop, droplet, and multi-cone). The state where the discharged droplet was larger than the nozzle diameter was defined as a *drop*. The state where the discharged droplet was smaller than the nozzle diameter was defined as a *droplet*. Lastly, the state where multiple cones formed at the nozzle tip and the droplets discharged in many different directions was defined as a *multi-cone*. The *drop* state produced droplets lined up at equal intervals and was the easiest to control of the three states, making it favourable for constructing the outline of the chocolate during 3D printing. The *droplet* state produced prints with a width between 15 µm and 200 µm. This state could be used to manipulate structures to directly influence the human taste receptors (which has a diameter of 50 µm). The printed line width of the printed chocolate decreased as the applied voltage increased from 5 kV to 12 kV, above which *multi-cone* discharge occurred. In this state, the deposition of the chocolate ink was forced in various directions, and the printed line width also increased. Additionally, an increase in the gap distance of >1 mm between the nozzle tip and the print surface also led to an increase in the line width as the electrostatic field was weakened and the ink was deposited in the *drop* state.

### 3.3.3 *Ink Formulation, Properties, and Characterization*

The fluid property of the food ink is another key requirement for the successful 3D printing of food. This section describes the rheological and textural properties of the different types of chocolate inks studied in recent literature.

### 3.3.3.1 *Rheological properties of different chocolate inks*

Rheology is essential for defining the flow properties of chocolate. Rheology can also help predict the different textural qualities of printed chocolate [36]. The several factors influencing the rheological properties of chocolate include (1) fat content, (2) surfactant/emulsifier, and (3) particle size. [37]. A surfactant (for instance, lecithin) homogenizes the hydrophilic domains (i.e., sugar) and the hydrophobic domains (i.e., fats) present in the chocolate, enhancing the flowability of the melted chocolate [37]. Other components present in chocolate, such as cocoa butter, milk fat, and sugar, may affect the flow properties of the chocolate [38, 39]. The following sections describe the rheological properties of various combinations of chocolate inks employed in the 3D printing of chocolate products.

- Cocoa powder in chocolate syrup and paste

Karyappa *et al.* formulated inks containing different concentrations of cocoa powder in chocolate syrup at 10–25% w/w and in the chocolate paste at 5–12% w/w to study their rheological properties for cold extrusion 3D printing [9]. For the range of cocoa powder concentrations investigated, the addition of cocoa powder increased the viscosity of the ink by $10^4$ for chocolate syrups and by $10^2$ for chocolate pastes. The viscosity of all the inks decreased as the shear rate increased, suggesting shear-thinning behaviours. The degree of shear-thinning behaviour, as indicated by the shear-thinning index ($n$), decreased in the syrup and increased in the paste with increasing cocoa concentration. With increasing cocoa concentration, the flow consistency index ($K$) increased in both the syrup and the paste. A larger value of $K$ measures higher mechanical strength to hold the printed structures after they have been dispensed from the nozzle [40]. Additionally, both the storage modulus ($G'$) and the loss modulus ($G''$) of the chocolate syrup and paste increased with cocoa concentration. The solid-like behaviour of all the inks containing cocoa powder was confirmed as $G' > G''$. Yield stresses for the inks also increased for all inks with increasing cocoa concentration. Spreading of the ink is more distinct for the inks with low yield stress and can be prevented by adding larger amounts of cocoa powders to the syrup (i.e., containing 20% w/w cocoa concentration and above) and paste (i.e., 10% w/w cocoa concentration and above). This example highlighted the importance of rheological properties in cold extrusion 3D printing of a chocolate-based ink, which was customized with the addition of cocoa powder.

- Grated commercial dark chocolate and magnesium stearate ink

Mantihal *et al.* used grated dark chocolate with magnesium stearate as the ink for hot-melt extrusion chocolate printing [29]. Grating helps reduce the chocolate particle size and facilitates melting during printing. Additionally, the magnesium stearate improves the flow behaviour of the ink, as it has been used as a lubricant enhancer in tablets for pharmaceutical production [41]. A temperature ramp study was conducted between 25 °C and 32 °C with a constant shear rate of 100 s$^{-1}$ for 10 min to understand the melting temperature and flow of the ink. The apparent viscosity of the ink decreased linearly in the range of 24–29 °C. Above 29 °C, a flattening plateau was formed, demonstrating that the majority of the cocoa butter crystals were melted. The chocolate was fully melted at 32 °C. The decrease in viscosity by heating facilitates extrusion and increases chocolate flowability through the nozzle. The shear-thinning property of the ink was demonstrated with a decrease in apparent viscosity of the chocolate ink with time when sheared at low shear rates of 50 s⁻¹ and 100 s⁻¹ (for 10 min at 32 °C) to mimic the actual extrusion rate.

In another work, Mantihal *et al.* studied the influence of other additives (i.e., magnesium stearate and plant sterol powders) on the melting behaviour as well as rheological and tribological properties of the chocolate ink [42]. Similar to the previous study, magnesium stearate was added to improve the flowability of chocolate ink [29]. Plant sterol has been shown to reduce the risk of coronary disease [43]. Therefore, plant sterol was added to improve the nutritional content and printability of dark chocolate. The addition of additives did not affect the melting temperatures of chocolate ink. Additionally, there was no difference in the melting temperatures of the chocolate ink containing either magnesium stearate or plant sterol. Chocolate inks containing additives also exhibited pseudoplastic shear-thinning behaviour with a decrease in viscosity in response to an increasing shear rate, allowing smooth dispensing through the printer nozzle. The presence of additives did not have an observable effect on the apparent viscosity of chocolate inks at their printing temperature of 32 °C. However, yield stress was significantly higher in chocolate inks containing additives than those without them. Lastly, the flow consistency index ($K$) of chocolate inks also increased with those additives. The introduction of additives increased the number of solid particles within chocolate ink, thereby affecting the flow consistency of the ink. The high values of yield stress and $K$ indicated that a high shear force was required to make chocolate inks flow.

- Chocolate with corn syrup, paracetamol, and ibuprofen inks

A study conducted with 153 paediatric participants revealed that the primary reason for the refusal to take medication was taste aversion [44]. As most active pharmaceutical ingredients have an unpleasant taste, a form of medication with a palatable dosage has been developed to circumvent the problem of taste aversion [45]. 3D printing of chocolate contributed to this effort. Karavasili *et al.* developed paediatric-friendly 3D-printed chocolate-based oral medication containing paracetamol and ibuprofen,

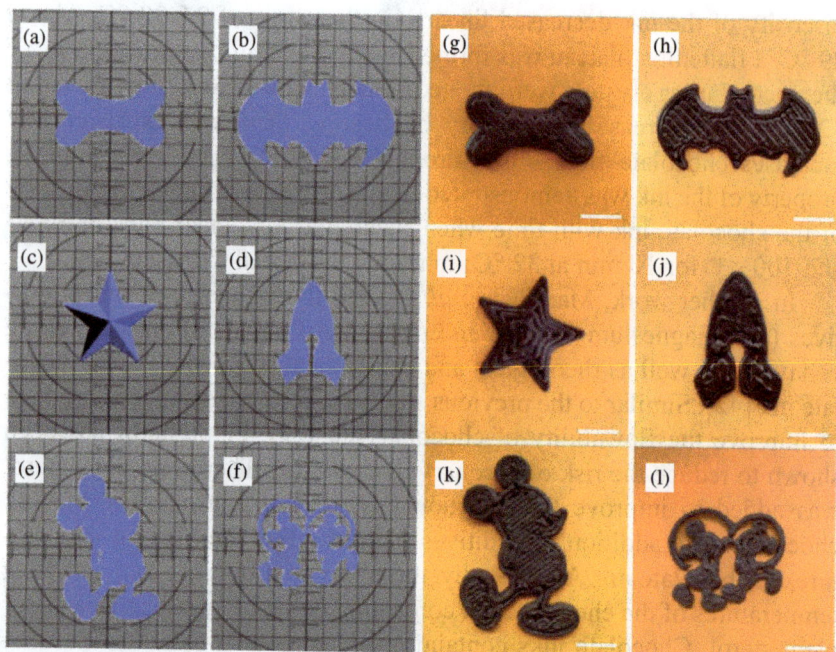

Figure 3.2.    3D printing of chocolate-based dosage forms. (a)–(f) Schematic illustration of the STL files of the printed dosage forms. (g)–(l) Different shapes of the 3D-printed chocolate-based dosage forms. (Scale bar = 20 mm.) Reproduced with permission from [46]. Copyright (2020) from Elsevier.

both of which are widely used within the paediatric population for pain relief and fever reduction [46]. The chocolate-based oral dosage forms were printed in various shapes resembling cartoon characters (Figure 3.2). Inks containing bitter chocolate and corn syrup were prepared at various ratios. An ink with lower ratios of syrup (1:0.7 and below) was non-extrudable. In contrast, high ratios of syrup produced sticky inks that were difficult to handle in 3D printing. Inks containing a 1:1 chocolate-to-syrup ratio enabled smooth extrusion and produced self-standing multi-layered printed structures.

The addition of corn syrup increased the apparent viscosity for both blank and drug-loaded chocolate inks, which can be attributed to the high moisture content of the viscous corn syrup [46, 47]. Additionally, drug-loaded chocolate inks had lower viscosities than corresponding blank chocolate inks, which could be due to the evaporation of water during drug loading at 79 °C [48]. All the chocolate inks demonstrated shear-thinning behaviour.

### 3.3.3.2 *Textural studies on 3D-printed chocolate products*

Textural and physical properties are important traits of any edible material as they suggest the quality of the product. For instance, a common way to ascertain the good quality of a chocolate product is from its glossy appearance and snap (and sound!). A good snap allows the chocolate to be broken apart easily. The ability to snap the 3D-printed chocolate is related to its mechanical properties and could be designed by changing the infill structures within the construct [49].

Infill patterns have been widely used to control the mechanical properties of 3D-printed constructs. In the context of 3D chocolate printing, the effects of the infill patterns (Figure 3.3) and percentages have been investigated [49]. Infill is the structure printed within the 3D construct, where the ink is dispensed in a selected pattern and density as pre-determined by the slicing software. All infill percentages and patterns were able to hold their 3D-printed chocolate shapes. An increase in infill percentage increased the force required to break the 3D-printed samples (i.e., hardness). Additionally, 3D-printed chocolate samples containing flow enhancers (i.e., magnesium stearate and plant sterol) were significantly harder than control samples without flow enhancers, regardless of infill percentage. The presence of additional solids particles was previously shown to influence the chocolate matrix, providing higher mechanical strength in dark chocolate [50, 51]. The addition of flow enhancers may increase the particulate content within chocolate samples, increasing the hardness of the 3D-printed dark chocolate. In addition, moulded samples required a higher force to break than 3D-printed samples at a 100% infill pattern (or regardless of infill pattern). The reduced hardness of 3D-printed

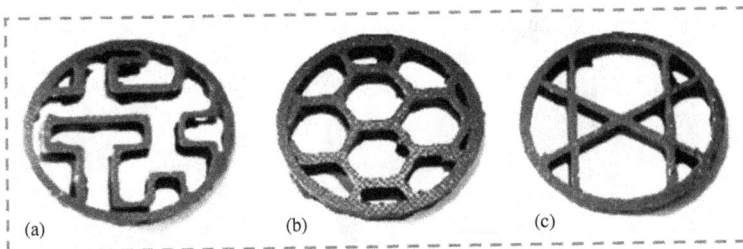

Figure 3.3.   Illustration of the type of infill pattern applied in the 3D printing of chocolate: (a) Hilbert curve, (b) honeycomb, and (c) star. Reproduced with permission from [49]. Copyright (2018) from Elsevier.

chocolate is due to the weakened interactions between the particles arising from the consecutive deposition of layers [49]. This observation is also in line with the previous study of 3D-printed cheese showing lower hardness than casted cheese [13].

3D-printed chocolate with a Hilbert curve infill required a lower force to break than honeycomb and star infills for the infill percentages of 5–60%. The Hilbert curve infill does not have a criss-cross structure, unlike honeycomb and star infills. Criss-cross infill structures produce more mechanically stable models than unintegrated infill patterns, yielding 3D-printed constructs with larger hardness [52, 53]. However, at 100% infill, most of the infill patterns displayed similar line patterns and required a comparable magnitude of forces to break the samples.

In terms of internal supporting structures, Mantihal *et al.* observed that 3D-printed chocolate models with cross support required the highest force to break. 3D models with cross support also possessed a firmer texture than samples with parallel support and no support (Figures 3.4 and 3.5) [29]. These demonstrations highlighted the importance of the support structures that stabilize 3D models and control food texture.

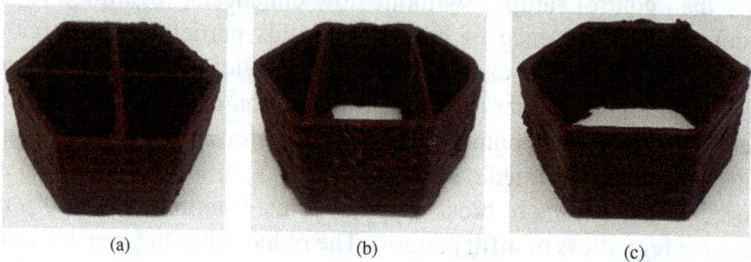

(a)                    (b)                    (c)

Figure 3.4.   3D-printed hexagonal-shaped chocolate models with different support configurations: (a) with cross support, (b) with parallel support, and (c) without support. Reproduced with permission from [29]. Copyright (2017) from Elsevier.

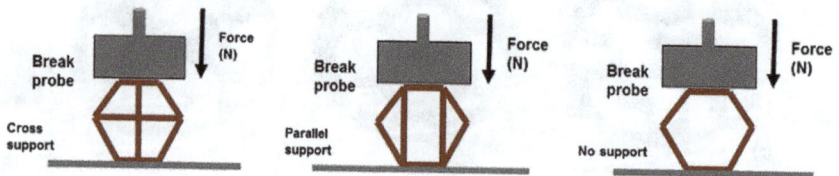

Figure 3.5.   Schematic drawing showing the positioning of the three distinct printed chocolate designs for snap properties analysis using a texture analyser. Reproduced with permission from [29]. Copyright (2017) from Elsevier.

# 3.4 3D-Printed Sugar

## 3.4.1 *Introduction*

Sugars are commonly present in foods and considered to be the primary source of energy for all forms of life. They are also the structural building blocks of deoxyribonucleic acid in all living things and cellulose in plants. Since the industrial revolution, there have been developments in the large-scale manufacture of molten sugar in the candy-making industry. One of the developments has been the extrusion of highly viscous molten sugar close to its glass transition temperature to fabricate linear geometries such as candy sticks [54]. Sugar has also been used traditionally to mimic glass in movies and stage plays because of its low cost, transparency, and ease of preparation.

Oskay and Edman developed one of the earliest 3D printers for sugar in the CandyFab Project [55]. An early do-it-yourself (DIY) 3D printer called CandyFab 4000 was introduced in 2007 based on laser sintering. The subsequent versions in 2008 and 2009 (CandyFab 5000 and 6000) were referred to as SHASAM, an acronym that stands for "selective hot air sintering and melting". In the 3D printing industry, 3D Systems Culinary Lab developed a powder-bed sugar printer called ChefJet Pro as a culinary device [54]. In laser sintering, powdered materials (in this case, sugar) are adhered together to form structures after being briefly melted by a laser beam. Heat sintering, on the other hand, uses either infrared radiation or a heated surface to melt materials before their fusion during solidification. These printers harness a focused heat source moving over a sugar bed, fusing the sugar particles together to form large 3D sugar sculptures.

To date, however, research on 3D printing of sugar has been largely limited. Due to degradable and biocompatible properties, sugars are compelling materials for transient devices such as small-scale robots [56] and biomedical devices such as dissolvable stents [57, 58]. Due to its solubility in water, sugar is also used as a sacrificial material in biomedical engineering [59]. The following section discusses the methods and properties of recent 3D-printed sugar products for food and other applications.

## 3.4.2 *3D Printing Methods of Sugar*

This section briefly introduces and describes the methods that have been employed for 3D printing of sugar, namely (1) hot-melt extrusion, and (2) laser sintering.

### 3.4.2.1 *Hot-melt extrusion 3D printing of sugar*

Leung studied glass 3D printing using sugar as an analogous material [54]. Molten sugar (mixture of table sugar and corn syrup) is optically transparent and has a comparable temperature–viscosity relationship and solidification properties as molten glass, but at a much lower temperature [54]. The lower working temperature of sugar (100–150 °C) decreases the complexity to make molten materials and greatly reduces the dangers of working with high-temperature materials. Leung also fabricated a relatively low-cost (<2,000 USD), desktop-sized, temperature-controlled extrusion-based 3D printer to print molten sugar and corn syrup ink. Cooling fans were added to blow air at the printed object and speed up the hardening of 3D-printed sugar products. However, in humid air, the surfaces of the printed sugar objects became sticky within hours due to the hygroscopic nature of sugar. The relative humidity of the ambient air is suggested to be kept below 30% [54]. A dehumidifier and airtight containers can be used to store 3D-printed sugar products [54].

Farzin *et al.* used extrusion-based 3D printing to fabricate sugar stents of changeable dissolution profiles. The stent can prevent clot formation and hold blood vessels together during vascular anastomosis [57]. Vascular anastomosis is a surgical procedure that connects blood vessels such as a coronary artery bypass. In their work, the sugar-based stent helped to hold the blood vessels together during the operation and was dissolved once the blood flow was restored. Additionally, the use of 3D printing allowed the design of stents with suitable geometry and architecture to overcome differences in blood vessel sizes between patients, improving the success of the surgical procedure.

### 3.4.2.2 *Selective laser sintering 3D printing of sugar*

Gervasoni *et al.* 3D-printed sugar-based magnetic composite structures via selective laser sintering to develop transient sugar-based small-scale robots for minimally invasive procedures (Figure 3.6) [56]. They sugar-based mm-scale helical swimmer structures; the 3D-printed swimmer demonstrated corkscrew motion while being manipulated by magnetic fields, and subsequently dissolved in water. During selective laser sintering, cross-sections produced from the computational 3D model were scanned by the laser; melting and fusing of the sugar powders allowed them to form solid parts. Upon the completion of printing of each layer of the cross-section,

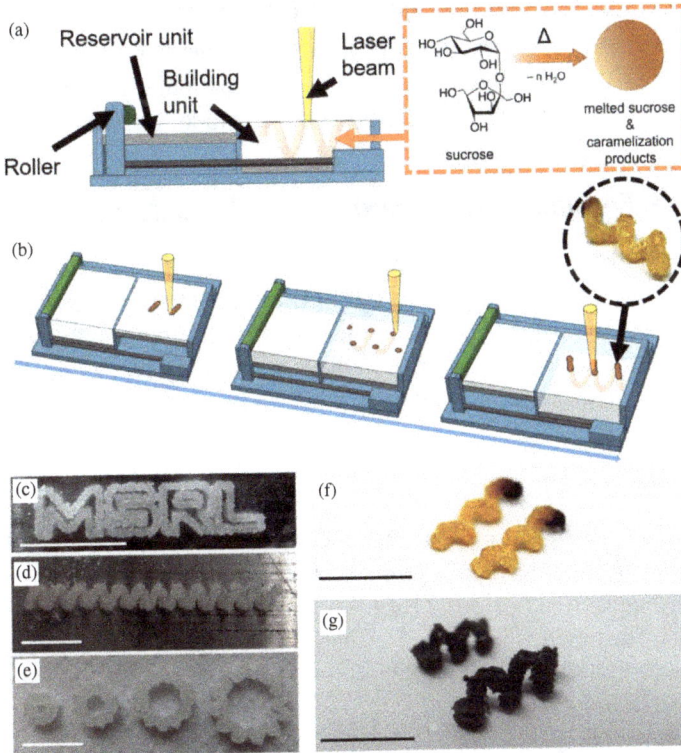

Figure 3.6.  (a) Schematic illustration of the selective laser sintering device used for sugar printing. It demonstrates the fabrication of sucrose-based constructs. The left container is the reservoir unit where sugar is kept. The right container is the building unit where the constructs are fabricated by laser melting. Both containers are raised and lowered to transfer the exact amount of sugar powder into every layer. A roller is used to move the powder from the reservoir unit into the building unit. (b) Schematic drawing displaying the steps used in the printing of a sugar-based helical construct with selective laser sintering. (c)–(g) Microscopy photos showing saccharide-based 3D-printed constructs fabricated with different laser powers and different scan speeds. (c) A glucose-based MSRL logo (scale bar = 20 mm). (d) A glucose-based helical structure (scale bar = 10 mm). (e) A set of sucrose-based gears (scale bar = 10 mm). (f) A pair of sucrose-based caramelized helices (scale bar = 10 mm). (g) A pair of sucrose with barium ferrite composite helices (scale bar = 10 mm). Reproduced with permission from [56]. Copyright (2020) from John Wiley and Sons.

the platform in the building unit moved down while the reservoir unit moved up vertically by one layer of thickness. A new layer of powder was moved from the reservoir unit into the building unit. This process was repeated for each printed layer until the entire model was printed.

### 3.4.3 *Ink Formulation, Properties, and Characterization*

This section describes the physical properties and observations of the different sugar combinations used in 3D printing. Extrusion-based printing requires sugar in a form of liquid solutions (often mixed with corn syrup or other fluids ensuring desired rheological properties), while sintering requires sugar in a form of homogeneous grains.

#### 3.4.3.1 *Sucrose powder*

Gervasoni *et al.* observed that sucrose produced a smoother and more homogeneous finish to its 3D-printed structures than glucose using selective laser sintering 3D printing [56]. Sucrose was used to 3D-print objects that are at least two times smaller than glucose [56]. These phenomena could be attributed to the larger crystal sizes of glucose (~400 μm) compared with sucrose (~40 μm).

The effects of laser power on the physical properties (i.e., stiffness and appearance) of the sintered sucrose were studied. The strength and robustness of the sucrose-sintered slabs were measured by a cantilever beam test. Sintered sucrose structures printed with a weaker laser power yielded higher stiffness than those printed with a stronger laser power. A change in the appearance of the sintered sucrose (from a white or pale-yellow dull colour to an opaque brown colour) occurred when a strong laser beam was used. This change in appearance indicated the formation of caramelized sugar. Caramelization causes the sugars to lose their crystalline form and reduces the mechanical strength of the 3D-printed sucrose structures. However, no insoluble species were produced in the caramelization as all the 3D-printed sucrose samples dissolved in water at around 11 min.

#### 3.4.3.2 *Mixture of glucose, sucrose, dextran, and sodium citrate ink*

Extrusion-based 3D printing of glucose and sucrose was performed to fabricate sugar-based stents. Stents made from only glucose and sucrose were, however, brittle and would break into pieces when exposed to mechanical loads. To improve the mechanical properties of the stent, dextran was added to improve the ductility and compression strength of 3D-printed sugar constructs [57]. In addition, sodium citrate was added to the ink to lower the chances of thrombosis during the surgical procedure. Eventually, 11% w/w glucose, 69% w/w sucrose, 17% w/w dextran (86 kDa), and 3% w/w sodium citrate were used as a formulation of the ink to achieve the desired mechanical, adhesion, ductility, and blood-clotting properties for the vascular stents.

To perform 3D printing, the ink was heated in the nozzle before being dispensed; the viscosity of the ink was reduced to facilitate the extrusion through the nozzle. The temperature range of 85–90 °C, corresponding to an ink viscosity of 50–70 Pa.s, was observed to be ideal for 3D printing using a nozzle gauge number of 18G or 20G [57]. The printed stents were gradually dissolved in transparent tubes perfused with phosphate-buffered saline at body temperature (~37 °C) and a flow rate similar to that of blood. The sugar-based stent also did not exhibit any signs of toxicity towards human umbilical vein endothelial cells. Temperatures were kept below 110 °C for the dissolution of sucrose, glucose, and dextran in water for the experiments to prevent the formation of by-products such as furfural and heterocyclic, which were produced above 150 °C [60].

## 3.5  3D-Printed Doughs

### 3.5.1  *Introduction*

Dough, a mixture of flour and liquid, is one of the traditional edible materials to be used as a base ingredient for 3D food printing. Unlike most of the 3D-printed food products discussed earlier, however, 3D-printed dough often requires post-processing in the form of baking before it can be consumed. Post-processing by baking at high temperatures might result in the deformation of the printed models [61]. A major challenge for formulating dough for 3D printing would be to maintain the structure during and after post-processing. Changes to the printing ink formulation, characteristics of each ingredient, or the post-processing itself are often needed to maintain the 3D-printed structures after post-processing [62]. The following section discusses recent studies involving the methods, formulations, and properties of 3D-printed doughs.

### 3.5.2  *3D Printing Methods of Dough*

This section describes the existing method that has been employed for the 3D printing of doughs, namely cold extrusion 3D printing.

#### 3.5.2.1  *Cold extrusion 3D printing of doughs*

Existing works that demonstrate 3D printing of dough-based inks are done by cold extrusion. 3D-printed doughs are post-processed with heating; temperature control is often decoupled from the process of extrusion printing. For instance, Liu *et al.* printed dough containing flour, freeze-dried mango powder, and olive oil for cold extrusion 3D printing [63]. The addition of mango powder improved the colour, flavour, and nutritional value of the dough. Olive oil gave a smoother surface and improved printability of the dough, which is attributed to the plasticizing and lubrication behaviour of the fat within the ink [64]. The printability of the ink was affected by the amount of water added. An insufficient amount of water would result in an uneven blend of dough, which would cause clogging in extrusion. In contrast, excess water would result in soft dough that is unable to retain its shape after printing.

### 3.5.3  *Ink Formulation, Properties, and Characterization*

#### 3.5.3.1  *Rheological properties of dough inks*

- Dough ink comprising icing sugar (sucrose), butter, gluten flour and egg

Yang *et al.* investigated the effect of the proportions of icing sugar (sucrose), butter, low-gluten flour, and water within the baking dough ink on its rheological and physical properties [17]. The dough ink exhibited pseudoplastic behaviour, and its viscosity decreased with increasing shear rate. The apparent viscosity of the dough ink increased over the same range of shear rate when the amounts of sucrose, butter, and flour were increased in the dough, with sucrose and flour having the least and most effects on the viscosity, respectively. The value of $G'$ for dough ink was higher than its value of $G''$ in the linear viscoelastic region (LVR), indicating that the dough ink has an elastic gel or gel-like structure at room temperature. With increasing amounts of sucrose, butter, or flour added, both the values of $G'$ and $G''$ of the dough ink increased. Sucrose, butter, and flour promoted the stabilization of the network within the printed dough, which was ideal for shape retention after printing. The identified formulation of the dough ink contained icing sugar, butter, low-gluten flour, egg, and water with the weight percentages of 6.6%, 6%, 48%, 10.4%, and 29%, respectively.

- Dough ink comprising flour, fat, sugar, and non-fat milk

Pulatsu *et al.* investigated the different formulations of cookie dough to achieve good printability and post-processing capacity (Figure 3.7) [65]. Additives such as gums, thickeners, and flow enhancers were intentionally left out in their work due to their incompatibility with the consumer trend to eat healthy and natural foods without additives [66]. The presence of additives may result in undesired deformation during baking with an increase in temperature. Among the different ink compositions used, the dough ink containing rice or tapioca flour with reduced sugar and milk content produced adequate yield stress to preserve the 3D-printed shapes during and after printing. The choice of butter and shortening did not affect the yield stress of the ink. Samples exhibiting high yield stress values were selected as ink candidates for their ability to retain their shape during and after printing.

All sample inks of varying compositions displayed solid-like behaviour, with $G' > G''$ in their LVR, which was essential for extrusion 3D printing [67]. However, the higher content of milk and sugar decreased their viscoelastic properties. Butter reduced the viscoelastic properties of the ink compared to shortening; butter contains water that diluted the cookie dough, while shortening was purely hydrogenated vegetable oil without water. A similar study conducted with biscuit dough showed that the presence of hydrogenated fat produced the stiffest dough [68]. Creep-recovery analysis was also performed to determine the ability to regain its structure after shear deformation. Samples containing high levels of milk showed increased deformability and the least elastic recovery, regardless of the type of flour or fat used in the dough formulation. Overall, the dough ink containing 62.5 g shortening, 100 g tapioca flour, 37.5 g sugar, and 32.5 g milk produced the highest yield stress values, lowest

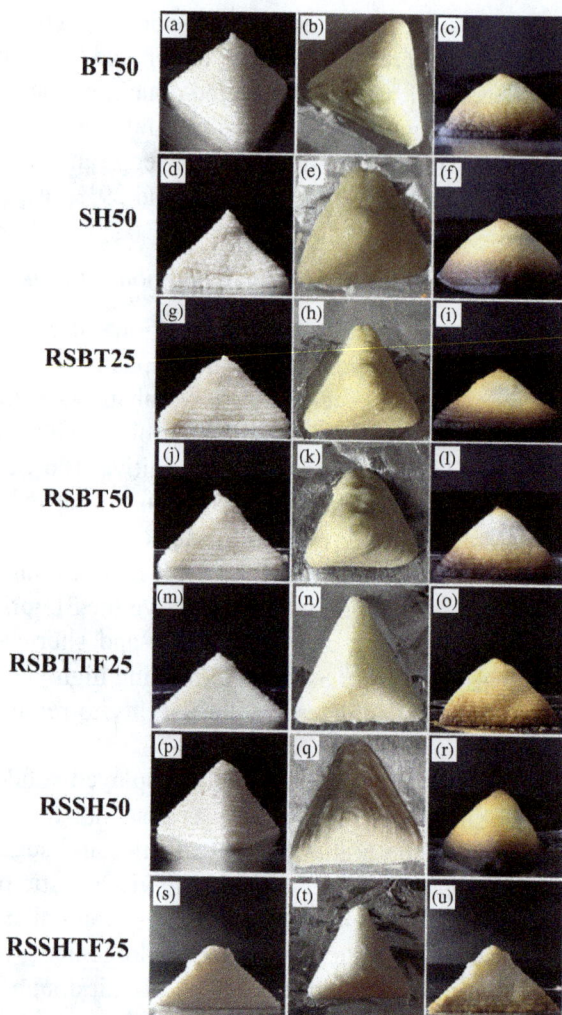

Figure 3.7.   Effects of baking on the structural stability of 3D-printed cookie dough samples. Composition of the printed dough inks are as follows: BT50 (62.5 g butter, 100 g wheat flour, 55 g sugar, 65 g milk), SH50 (62.5 g shortening, 100 g wheat flour, 55 g sugar, 65 g milk), RSBT25 (62.5 g butter, 100 g wheat flour, 37.5 g sugar, 32.5 g milk), RSBT50 (62.5 g butter, 100 g wheat flour, 37.5 g sugar, 65 g milk), RSBTTF25 (62.5 g butter, 100 g tapioca flour, 37.5 g sugar, 32.5 g milk), RSSH50 (62.5 g shortening, 100 g wheat flour, 37.5 g sugar, 65 g milk), RSSHTF25 (62.5 g shortening, 100 g tapioca flour, 37.5 g sugar, 32.5 g milk). Photos of various compositions of 3D-printed pyramidal cookie dough samples in (a), (b), (d), (e), (g), (h), (j), (k), (m), (n), (p), (q), (s) and (t) before baking and (c), (f), (i), (l), (o), (r) and (u) after baking. Reproduced with permission from [65]. Copyright (2020) from Elsevier.

deformation, and highest elastic recovery, thereby making the ideal formulation for 3D printing.

Subsequently, they selected 7 cookie dough samples printed in a 3D pyramidal shape out of 24 cookie dough inks for baking in an oven at approximately 177 °C for 10 min. The rest of the samples did not achieve the desired level of accuracy in their printed structures or could not produce 3D shapes without collapsing. Generally, dough formulations containing reduced sugars shrank after baking except for one, which exhibited spreading due to lower milk content. Additionally, samples containing higher sugar produced a larger collapse in height; during baking, sugar melts and goes through a subsequent transition towards a glassy state [69]. Therefore, the higher amount of sugar resulted in the larger structural deformation. A reduction in sugar in the dough would be essential to produce stable structures post-baking. The type of fat also influenced the shape stability of the 3D-printed dough. It was observed that shortening-containing dough produced a more significant collapse than butter-containing dough when the heat warmed the dough and melted the fats within them. Overall, the printed dough containing 62.5 g shortening, 100 g tapioca flour, 37.5 g sugar, and 32.5 g milk showed no structural deviation upon baking due to its superior rheological properties compared to other samples.

- Dough ink comprising wholegrain rye flour and milk powder

Wholegrain rye flour is an excellent source of dietary fibre with low-gluten content [70]. Gluten is extensible and gives the elastic property to wheat-based doughs, which is, however, undesirable for paste extrusion-based 3D printing. Some 3D printing studies have used low-gluten wheat flour due to its improved printability [16, 17]. Lillie *et al.* investigated the effect of adding wholegrain rye flour to milk powder in the ratios of 1:0, 3:1, 1:1, 1:3, and 0:1, respectively, for printing and post-processing [61]. All the samples displayed elastic-dominating viscoelastic behaviour ($G' > G''$) shortly after preparation. Adding rye flour to milk powder increased the $G'$ of the ink by 50%, whereas adding a higher amount of rye flour resulted in a reduction in $G'$. Yield stress ranged from 10 Pa to 60 Pa for all the samples. Among the samples, the sample containing only rye flour had the lowest yield stress, while the sample containing rye flour and milk powder (1:3 ratio) exhibited the highest yield stress. The increased yield stress stabilized the printed shape of the sample. Dough inks containing higher rye flour amounts exhibited the highest reduction in $G'$, $G''$, and yield stress over time, alluding to the activity of endogenous enzymes (i.e., xylanase and α-amylase) present within the rye flour that solubilized the cell wall and degraded the starch inside [71]. The milk powder-only ink displayed only a small increase in $G'$, $G''$, and yield stress over time, which was likely due to an increase in hydration of the milk powder with time.

### 3.5.3.2 *Textural studies on 3D-printed dough products*

Liu *et al.* studied the effect of olive oil and mango powder on the textures of moulded and 3D-printed dough [63]. The lipid molecules in olive oil disturbed the formation of complexes between the gluten network and starch, causing the dough to reduce its hardness [72]. Additionally, mango powder was also shown to lower the hardness of the dough [73]. The mango powder increased the sugar content in the dough, which led to an increase in hydrogen bonding between the water and the sugar molecules. The increase in hydrogen bonding between the sugar molecules and water increased the osmotic pressure outside the protein chains in the flour, and the amount of moisture entering the spaces within the protein by the osmotic pressure was reduced [63]. The difference in osmotic pressure reduced the degree of gluten formation and lowered the hardness of the dough [74]. Adhesion was slightly reduced with the addition of olive oil but was improved with the addition of mango powder. Comparing moulded dough with 3D-printed dough, the hardness, adhesion, elasticity, and resilience of the 3D-printed dough were significantly lower than the moulded dough. This difference is due to the inherent mechanism of extrusion 3D printing, where the structure is built by stacking printed ink in a layer-by-layer orientation; the adjacent layers of the printed product were only in physical contact with each other and not merged during printing [75]. Since the adjacent layers are not merged, weaker interactions result between the layers. Additionally, as water evaporates during printing, the adhesion, elasticity, and resilience of 3D-printed products are reduced.

Lillie *et al.* studied the effects of milk and rye flour on 3D-printed dough texture post-baking (Figure 3.8) [61]. The presence of milk produced an expansion of 3D-printed dough samples during baking (in

Figure 3.8. Images showing one-layer printed dough samples containing rye flour (R) only, milk powder (M) only, and a combination of R and M in the ratio of 1:1 (a) after printing and (b) after baking. (c) Cross-sectional images of the baked samples at the vertical dotted line in (b) (scale bar = 1 mm) [61]. Reproduced from [61], which is licensed under CC BY 4.0.

the oven at 150 °C), while rye-containing dough samples were flatter. The expansion of milk-containing samples was due to water evaporation. The surface-active proteins in the milk stabilized the gas bubbles produced by water evaporation and increased bubble growth. Rye flour also possesses surface-active proteins [76], but its fibre components obstruct the bubble growth in the same way as in baking bread [77]. This difference resulted in high porosity of the milk-based samples, as shown by X-ray tomography. It also explains why the addition of rye flour enhanced the baking stability of the milk-based system. The fracture force (which refers to the force required to break the 3D-printed samples) was significantly higher in the samples containing both rye flour and milk powder than in the samples containing either rye flour or milk powder only. The presence of milk contributed to greater sample glossiness (due to fats present in the milk), expansion, sweetness, and expansion, whereas rye flour increased the overall hardness, prevented excessive expansion, and gave better shape stability to the 3D-printed dough samples. Samples printed with five layers yielded a higher fracture force than those printed with a single layer. For samples printed with five layers, those containing 100% rye flour and 75% rye flour showed lower fracture force than those containing <50% rye flour. The highest fracture force was recorded for samples containing equal amounts of rye flour and milk powder.

## 3.6  3D-Printed Asian/Singaporean Snacks

Asia is widely known for its foods and delicacies, ranging from snacks to desserts. In Singapore, these snacks are readily available and are commonly consumed daily. These Asian snacks and desserts are traditional and considered popular among the elderly. However, the elderly often experience a decline in swallowing ability. In extreme cases, they present with dysphagia, a clinical condition that impedes swallowing. One of the major contributions of 3D food printing is its ability to alter the texture of printed foods for safe consumption by dysphagia patients. Personalization of food is another contribution of 3D food printing that allows the production of foods with unique shapes, designs, and tastes. In this section, we present some of the pioneering and unique works to recreate some of the popular Asian snacks and delicacies using 3D food printing, namely, *chwee kueh, kueh dadar,* and *orh nee,* for the customization of texture and shape.

### 3.6.1  *Chwee Kueh*

*Chwee kueh* refers to a type of steamed rice cake served together with preserved radish (Figure 3.9). The name *chwee kueh* is a combination of two words: *chwee* from the Hokkien dialect, meaning water, and *kueh* from the Malay language, meaning bite-sized snack usually made from rice. *Chwee kueh* is a popular breakfast dish in many Southeast Asian countries, including Singapore, Malaysia, and Thailand.

The basic ingredients for the preparation of *chwee kueh* are rice flour and water, which are mixed to give a slightly sticky slurry. The mixture is subsequently poured into small saucer-shaped aluminium cups and steamed, giving them a distinctive bowl-like appearance after cooking. These steamed rice cakes are soft and have a pudding-like texture. Additionally, they are usually served together with diced preserved radish and chilli sauce.

Figure 3.9.   Image showing *chwee kueh*s topped with preserved radish.

To recreate *chwee kueh* by 3D printing, the *chwee kueh* ink was formulated with 15% w/w rice flour in water (Figure 3.10). The resulting mix has a thin consistency; it was not sufficiently viscous and inadequate for printing. The mix was therefore double-boiled for ~3 min with constant stirring to increase its viscosity.

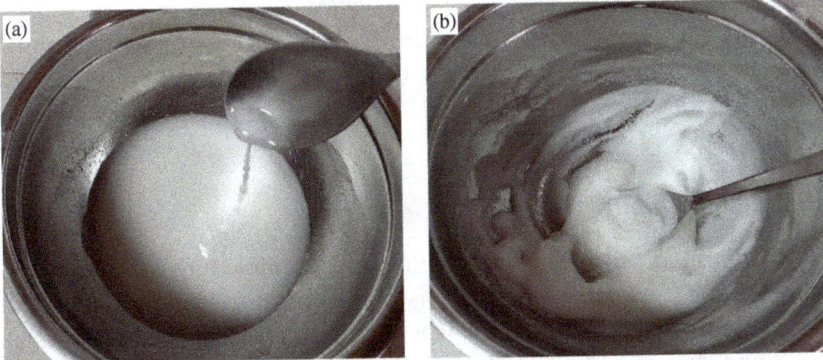

Figure 3.10.   *Chwee kueh* food ink preparation using 15% w/w rice flour in water. (a) *Chwee kueh* food ink was thin and poorly viscous. (b) *Chwee kueh* food ink after being double-boiled for 3 min with constant stirring.

The ink was subsequently cooled and could be printed into different shapes and designs using a 1.5-mm-diameter nozzle (Figure 3.11). The 3D-printed *chwee kueh* constructs were later steamed for ~15 min to produce the taste of soft, cooked rice. Future developments to the taste of the 3D-printed *chwee kueh* may include adding tapioca starch, oil, and salt to enhance its texture and taste.

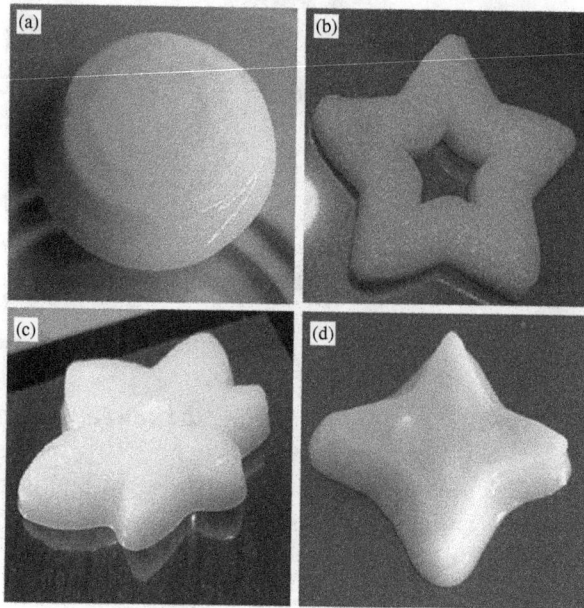

Figure 3.11.   3D-printed *chwee kueh* in different designs and shapes: (a) bowl-shaped, (b) five-pointed star-shaped with hollow centre, (c) flower-shaped, and (d) four-pointed star shaped.

### 3.6.2 *Kueh Dadar*

*Kueh/kuih dadar*, also known by other names, including *dadar gulung*, *kuih ketayap*, and *kuih lenggang*, is a sweet-tasting rolled-up coconut crepe (Figure 3.12). *Kueh dadar* is a popular traditional Southeast Asian snack and delicacy. *Kueh/kuih* is a Malaysian word referring to a bite-sized snack or dessert. Commonly found in Southeast Asia, such foods are prepared from rice or sticky glutinous rice. The green crepe skin takes its

Figure 3.12.    Image of *kueh dadar*.

colour from pandan extract, which is isolated from the leaves of the tropical plant *Pandanus amaryllifolius*, also simply known as pandan. Other key ingredients for making the crepe skin include flour, eggs, and coconut milk. The pandan-flavoured crepe skin is filled and wrapped with grated coconut and sweetened with palm sugar.

To create *kueh dadar* by 3D food printing, Lee *et al.* explored the 3D printability of coconut cream and coconut oil [78], which are key ingredients for the crepe skin of *kueh dadar*. No research had been done previously on the influence of oil content on the printability of food inks. Oil is frequently added to food inks to obtain the desired flavour, texture, and calories. Oils may offer additional functions such as promoting neurological development in infants as they serve as a medium for the absorption of fat-soluble vitamins A, D, E, and K [79]. High oil content may lower the viscosity of the mixture and cause phase separation [80]. Oil separation causes unanticipated changes in the rheological characteristics, making 3D printing of these food inks difficult. Having a stable ink (i.e., where no oil separation happens) is also important to circumvent the ink from becoming rancid, deteriorating the vitamins, and developing harmful compounds [81].

In their work, the extent of phase separation for mixtures containing coconut cream and different weight concentrations of coconut oil was studied. The oil, which had been separated from the mixture upon preparation, was immediately removed by filtering the mixture using a sieve and weighed. The oil separation ratio of each of these mixtures was obtained by taking the ratio of the weight of the separated oil to the total weight of the oil initially added to the mixture. Therefore, as the amount of oil separated increased, the oil separation ratio also increased. With higher concentrations of initially added oil, the oil separation ratio increases. On the other hand, increasing the water content leads to an increase in the amount of dispersed water [82]. The increase in dispersed water provided greater emulsification to establish a continuous interface between the water and oil that prevented phase separation. However, high water content within the ink is not favourable for 3D printing, as this would lower the viscosity of the ink. Overall, three compositions of coconut cream inks were chosen for rheological studies based on their low oil separation ratios, namely, ink A (containing 25% w/w water with 10% w/w oil), ink B (25% w/w water with 12.5% w/w oil), and ink C (contains 33% w/w water with 10% w/w oil). All the inks (A, B, and C) exhibited shear-thinning behaviour: a decrease in viscosity from an order of $10^3$ to $10^2$ Pa·s with a corresponding increase in shear rates between 0.01 and 100 $s^{-1}$. With an increase in water content in the inks (25% to 33% w/w water), a reduction in the viscosity of the ink was observed [78]. Similarly, an increase in oil content (10% to 12.5% w/w water) also led to a decrease in the viscosity of the ink and yield stress of the ink due to weaker internal interactions. Ink C showed the lowest yield stress and $G'$ among the three inks; lateral spreading of ink C was observed in the printed structures, while 3D-printed structures with inks A and B were well maintained. Using ink A, different 3D structures were printed with a direct ink writing 3D printer (Figure 3.13). All the printed structures were self-supporting, and the deposited inks displayed structural integrity. These demonstrations confirmed the printability of coconut cream ink containing 25% w/w water with 10% w/w oil for the creation of 3D-printed food structures, including *kueh dadar*.

To prepare the filling of *kueh dadar*, the ingredients water, *gula melaka* (palm sugar) syrup, and coconut butter were mixed in the ratio of 1:5:10 by weight to form a paste. This paste was termed the *gula melaka* coconut butter ink. The *gula melaka* coconut butter ink was 3D-printed into a waffle-shaped structure to demonstrate its printability (Figure 3.14). Additionally, alternating layers of coconut cream and *gula melaka* coconut butter inks were printed within a single waffle-shaped structure to demonstrate the use of multi-material 3D food printing to create novel

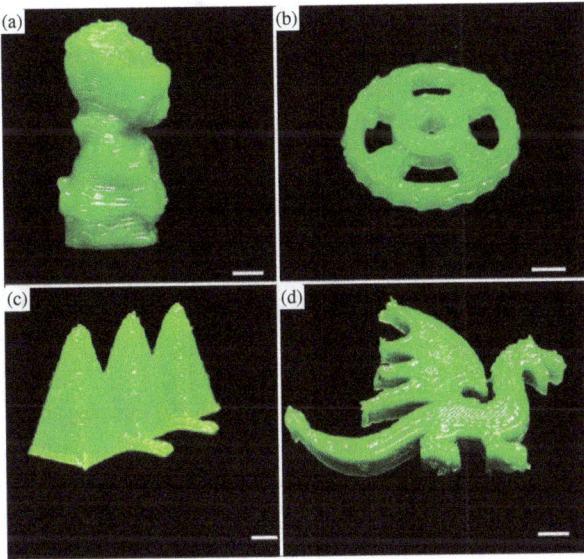

Figure 3.13. 3D-printed structures using coconut cream ink containing 25% (w/w) water with 10% (w/w) oil: (a) humanoid, (b) wheel, (c) pyramids, and (d) dragon (scale bar = 5 mm) [78]. Reproduced from [78], which is licensed under CC BY 4.0.

Figure 3.14. 3D printing of the *gula melaka* coconut butter ink. (a) Top-down view and (b) side view of a ten-layered *gula melaka* cream butter ink printed in a waffle-shaped design. (c) Single alternating layers of coconut cream and *gula melaka* coconut butter inks printed into a single waffle-shaped structure. (d) Multiple alternating layers of coconut cream and *gula melaka* coconut butter inks printed into a single waffle-shaped structure.

shapes for the *kueh dadar* delicacy. Future work will involve the use of both coconut cream and *gula melaka* coconut butter inks to design and fabricate *kueh dadar* and other traditional foods. The designed foods are also intended for individuals suffering from dysphagia.

### 3.6.3 *Orh Nee*

*Orh nee* is a traditional Teochew dessert and delicacy made with yam paste as the primary ingredient, cooked with shallot oil, and served with gingko nuts (Figure 3.15). *Orh nee* has been one of the popular traditional desserts among elderly Singaporeans.

To prepare the yam paste food ink, the yam was first peeled, sliced into thin pieces, and steamed until it became tender. Subsequently, the yam was formed into a paste and then sieved to remove any lumps, which may impede printing. The yam paste was then printed into various shapes and designs. In the same way, gingko nuts were boiled until they became tender, mashed, and blended with water to achieve a paste-like consistency. The paste was sieved again to remove lumps to obtain a paste for 3D printing; the gingko nut ink was used as a decoration as a part of *orh nee* desserts (Figures 3.16 and 3.17).

Figure 3.15.   A picture showing *orh nee* decorated with gingko nuts.

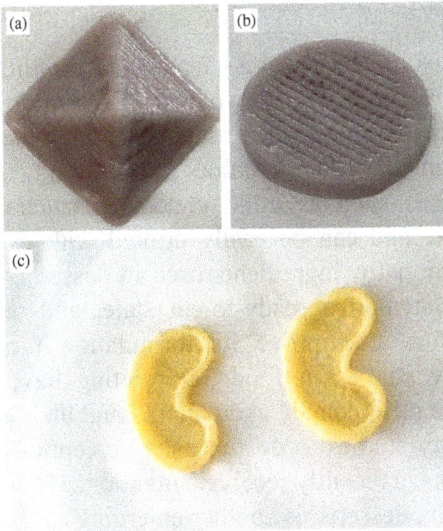

Figure 3.16. 3D printing of yam pastes and gingko nut paste inks. (a) Yam paste ink printed into a pyramid shape, (b) yam paste ink printed into a round shape with inner lattice design, and (c) gingko nut paste ink printed into two halves of a cashew nut shape with hollow centres.

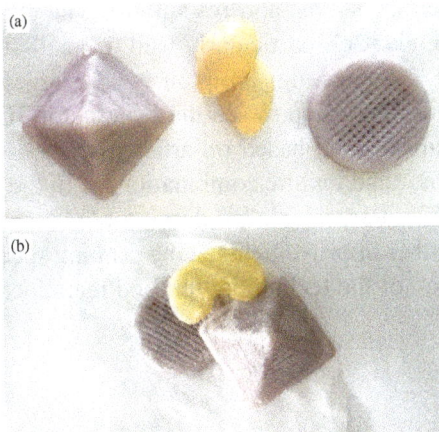

Figure 3.17. 3D-printed *orh nee* dessert consisting of yam paste and gingko nuts. (a) Assembled shapes of 3D-printed yam paste and gingko nuts (boiled). (b) Assembled shapes of 3D-printed yam paste and 3D-printed gingko nuts (in a half cashew nut-shaped design).

## 3.7 Outlook

This chapter discussed the progress in the 3D printing of desserts and snacks. To date, research has shown that 3D printing for desserts and snacks has been favourable because their base ingredients can be easily designed and printed with available 3D printing technologies. For instance, ingredients such as milk, chocolate, sugars, and dough are natively extrudable and can be easily printed with extrusion-based 3D printing [1]. Additionally, ingredients used in desserts, such as milk and chocolate, are printed in a ready-to-eat state, and this would reduce the time and effort to improve its formulation for post-processing. Importantly, many studies in food 3D printing have also considered aspects such as the infill density and patterns and their effects on texture. In addition, sensory studies to evaluate the perception of consumers on 3D-printed foods have recently received attention. 3D printing of regional foods, such as Asian desserts, is another emerging application of 3D food printing. Current research efforts — in terms of methods, materials, and design — will lead to the adoption and use of 3D food printing.

The ability to customize the design and composition of food by 3D printing will benefit dessert and snack creation, as the improved precision for crafting food into unique designs and replicating them with accuracy allows the technology to be adopted by small-scale food production facilities, including bakeries, shops, and restaurants. The current capability of 3D food printing lends itself particularly well to small-scale production of desserts and snacks. Interestingly, Lipton *et al.* suggested that the cake industry may be the first group to see the benefits of 3D food printing as higher premiums are offered based on artistry, whereas other shops may produce novel flavour and texture combinations with the layered manufacturing approach [62]. Large-scale production of 3D-printed foods is the next milestone. Further improvements in the printing speed would pave the way for the adoption of the technology in food manufacturing industries.

## References

[1]  Voon, S. L., An, J., Wong, G., Zhang, Y. and Chua, C. K. (2019). 3D food printing: a categorised review of inks and their development, *Virtual and Physical Prototyping*, 14, pp. 203–218.
[2]  Attalla, R., Ling, C. and Selvaganapathy, P. (2016). Fabrication and characterization of gels with integrated channels using 3D printing with

microfluidic nozzle for tissue engineering applications, *Biomed Microdevices*, 18.

[3] Chow, C. Y., Thybo, C. D., Sager, V. F., Riantiningtyas, R. R., Bredie, W. L. and Ahrné, L. (2021). Printability, stability and sensory properties of protein-enriched 3D-printed lemon mousse for personalised in-between meals, *Food Hydrocolloids*, p. 106943.

[4] Lee, C. P., Karyappa, R. and Hashimoto, M. (2020). 3D printing of milk-based product, *RSC Advances*, 10, pp. 29821–29828.

[5] Yang, F., Cui, Y., Guo, Y., Yang, W., Liu, X. and Liu, X. (2021). Internal structure and textural properties of a milk protein composite gel construct produced by three-dimensional printing, *Journal of Food Science*, 86, pp. 1917–1927.

[6] Liu, Y., Liu, D., Wei, G., Ma, Y., Bhandari, B. and Zhou, P. (2018). 3D printed milk protein food simulant: improving the printing performance of milk protein concentration by incorporating whey protein isolate, *Innovative Food Science & Emerging Technologies*, 49, pp. 116–126.

[7] Joshi, S., Sahu, J. K., Bareen, M. A., Prakash, S., Bhandari, B., Sharma, N. and Naik, S. N. (2021). Assessment of 3D printability of composite dairy matrix by correlating with its rheological properties, *Food Research International*, 141, p. 110111.

[8] Bareen, M. A., Joshi, S., Sahu, J. K., Prakash, S. and Bhandari, B. (2021). Assessment of 3D printability of heat acid coagulated milk semi-solids "soft cheese" by correlating rheological, microstructural, and textural properties, *Journal of Food Engineering*, 300, p. 110506.

[9] Karyappa, R. and Hashimoto, M. (2019). Chocolate-based Ink Three-dimensional Printing (Ci3DP), *Scientific Reports*, 9, p. 14178.

[10] Gholamipour-Shirazi, A., Norton, I. T. and Mills, T. (2019). Designing hydrocolloid based food-ink formulations for extrusion 3D printing, *Food Hydrocolloids*, 95, pp. 161–167.

[11] Crowley, M. M., Zhang, F., Repka, M. A., Thumma, S., Upadhye, S. B., Kumar Battu, S., McGinity, J. W. and Martin, C. (2007). Pharmaceutical applications of hot-melt extrusion: part I, *Drug Development and Industrial Pharmacy*, 33, pp. 909–926.

[12] Lille, M., Nurmela, A., Nordlund, E., Metsä-Kortelainen, S. and Sozer, N. (2018). Applicability of protein and fiber-rich food materials in extrusion-based 3D printing, *Journal of Food Engineering*, 220, pp. 20–27.

[13] Le Tohic, C., O'Sullivan, J. J., Drapala, K. P., Chartrin, V., Chan, T., Morrison, A. P., Kerry, J. P. and Kelly, A. L. (2018). Effect of 3D printing on the structure and textural properties of processed cheese, *Journal of Food Engineering*, 220, pp. 56–64.

[14] Liu, Z., Bhandari, B., Prakash, S., Mantihal, S. and Zhang, M. (2019). Linking rheology and printability of a multicomponent gel system of

carrageenan-xanthan-starch in extrusion based additive manufacturing, *Food Hydrocolloids*, 87, pp. 413–424.

[15]  Ross, M. M., Crowley, S. V., Crotty, S., Oliveira, J., Morrison, A. P. and Kelly, A. L. (2021). Parameters affecting the printability of 3D-printed processed cheese, *Innovative Food Science & Emerging Technologies*, p. 102730.

[16]  Yang, F., Zhang, M., Fang, Z. and Liu, Y. (2019). Impact of processing parameters and post-treatment on the shape accuracy of 3D-printed baking dough, *International Journal of Food Science & Technology*, 54, pp. 68–74.

[17]  Yang, F., Zhang, M., Prakash, S. and Liu, Y. (2018). Physical properties of 3D printed baking dough as affected by different compositions, *Innovative Food Science & Emerging Technologies*, 49, pp. 202–210.

[18]  Ouyang, L., Yao, R., Zhao, Y. and Sun, W. (2016). Effect of bioink properties on printability and cell viability for 3D bioplotting of embryonic stem cells, *Biofabrication*, 8, p. 035020.

[19]  Huppertz, T., Fox, P. and Kelly, A. (2018). *Proteins in Food Processing*, "The caseins: structure, stability, and functionality" (Elsevier), pp. 49–92.

[20]  Kessler, H. (2002). *Food and Bio Process Engineering-Dairy Technology*, "Alternative methods of preservation" (Verlag A. Kessler).

[21]  Li, H., Yu, H., Liu, Y., Wang, Y., Li, H. and Yu, J. (2019). The use of inulin, maltitol and lecithin as fat replacers and plasticizers in a model reduced-fat mozzarella cheese-like product, *Journal of the Science of Food and Agriculture*, 99, pp. 5586–5593.

[22]  Wang, X., He, Z., Zeng, M., Qin, F., Adhikari, B. and Chen, J. (2017). Effects of the size and content of protein aggregates on the rheological and structural properties of soy protein isolate emulsion gels induced by CaSO4, *Food Chemistry*, 221, pp. 130–138.

[23]  Brickley, C., Auty, M., Piraino, P. and McSweeney, P. (2007). The effect of natural Cheddar cheese ripening on the functional and textural properties of the processed cheese manufactured therefrom, *Journal of Food Science*, 72, pp. C483–C490.

[24]  Al-Muslimawi, A., Tamaddon-Jahromi, H. and Webster, M. (2013). Simulation of viscoelastic and viscoelastoplastic die-swell flows, *Journal of Non-Newtonian Fluid Mechanics*, 191, pp. 45–56.

[25]  Hao, L., Mellor, S., Seaman, O., Henderson, J., Sewell, N. and Sloan, M. (2010). Material characterisation and process development for chocolate additive layer manufacturing, *Virtual and Physical Prototyping*, 5, pp. 57–64.

[26]  Lanaro, M., Forrestal, D. P., Scheurer, S., Slinger, D. J., Liao, S., Powell, S. K. and Woodruff, M. A. (2017). 3D printing complex chocolate objects:

platform design, optimization and evaluation, *Journal of Food Engineering*, 215, pp. 13–22.

[27]  Afoakwa, E. O. (2016). *Chocolate Science and Technology* (John Wiley & Sons).

[28]  Afoakwa, E. O., Paterson, A., Fowler, M. and Ryan, A. (2008). Flavor formation and character in cocoa and chocolate: a critical review, *Critical Reviews in Food Science and Nutrition*, 48, pp. 840–857.

[29]  Mantihal, S., Prakash, S., Godoi, F. C. and Bhandari, B. (2017). Optimization of chocolate 3D printing by correlating thermal and flow properties with 3D structure modeling, *Innovative Food Science & Emerging Technologies*, 44, pp. 21–29.

[30]  Afoakwa, E. O., Paterson, A. and Fowler, M. (2007). Factors influencing rheological and textural qualities in chocolate — a review, *Trends in Food Science & Technology*, 18, pp. 290–298.

[31]  Chen, Y. W. and Mackley, M. R. (2006). Flexible chocolate, *Soft Matter*, 2, pp. 304–309.

[32]  El-Kalyoubi, M., Khallaf, M., Abdelrashid, A. and Mostafa, E. M. (2011). Quality characteristics of chocolate — containing some fat replacer, *Annals of Agricultural Sciences*, 56, pp. 89–96.

[33]  Suntornnond, R., An, J. and Chua, C. K. (2017). Roles of support materials in 3D bioprinting — present and future, *International Journal of Bioprinting*, 3.

[34]  Lanaro, M., Desselle, M. R. and Woodruff, M. A. (2019). *Fundamentals of 3D Food Printing and Applications*, "3D printing chocolate: properties of formulations for extrusion, sintering, binding and ink jetting" (Elsevier), pp. 151–173.

[35]  Takagishi, K., Suzuki, Y. and Umezu, S. (2018). The high precision drawing method of chocolate utilizing electrostatic ink-jet printer, *Journal of Food Engineering*, 216, pp. 138–143.

[36]  Masen, M. and Cann, P. (2018). Friction measurements with molten chocolate, *Tribology Letters*, 66, p. 24.

[37]  Afoakwa, E. O., Paterson, A. and Fowler, M. (2008). Effects of particle size distribution and composition on rheological properties of dark chocolate, *European Food Research and Technology*, 226, pp. 1259–1268.

[38]  Lee, S., Heuberger, M., Rousset, P. and Spencer, N. (2002). Chocolate at a sliding interface, *Journal of Food Science*, 67, pp. 2712–2717.

[39]  Servais, C., Ranc, H. and Roberts, I. (2003). Determination of chocolate viscosity, *Journal of Texture Studies*, 34, pp. 467–497.

[40]  Costakis Jr, W. J., Rueschhoff, L. M., Diaz-Cano, A. I., Youngblood, J. P. and Trice, R. W. (2016). Additive manufacturing of boron carbide via continuous filament direct ink writing of aqueous ceramic suspensions, *Journal of the European Ceramic Society*, 36, pp. 3249–3256.

[41] Kikuta, J.-I. and Kitamori, N. (1994). Effect of mixing time on the lubricating properties of magnesium stearate and the final characteristics of the compressed tablets, *Drug Development and Industrial Pharmacy*, 20, pp. 343–355.

[42] Mantihal, S., Prakash, S., Godoi, F. C. and Bhandari, B. (2019). Effect of additives on thermal, rheological and tribological properties of 3D printed dark chocolate, *Food Research International*, 119, pp. 161–169.

[43] AbuMweis, S. S. and Jones, P. J. (2008). Cholesterol-lowering effect of plant sterols, *Current Atherosclerosis Reports*, 10, p. 467.

[44] Mennella, J. A., Roberts, K. M., Mathew, P. S. and Reed, D. R. (2015). Children's perceptions about medicines: individual differences and taste, *BMC Pediatrics*, 15, pp. 1–6.

[45] Walsh, J., Cram, A., Woertz, K., Breitkreutz, J., Winzenburg, G., Turner, R., Tuleu, C. and Initiative, E. F. (2014). Playing hide and seek with poorly tasting paediatric medicines: do not forget the excipients, *Advanced Drug Delivery Reviews*, 73, pp. 14–33.

[46] Karavasili, C., Gkaragkounis, A., Moschakis, T., Ritzoulis, C. and Fatouros, D. G. (2020). Pediatric-friendly chocolate-based dosage forms for the oral administration of both hydrophilic and lipophilic drugs fabricated with extrusion-based 3D printing, *European Journal of Pharmaceutical Sciences*, 147, p. 105291.

[47] Beckett, S. T. (2011). *Industrial Chocolate Manufacture and Use* (John Wiley & Sons).

[48] Aidoo, R. P., Depypere, F., Afoakwa, E. O. and Dewettinck, K. (2013). Industrial manufacture of sugar-free chocolates — applicability of alternative sweeteners and carbohydrate polymers as raw materials in product development, *Trends in Food Science & Technology*, 32, pp. 84–96.

[49] Mantihal, S., Prakash, S. and Bhandari, B. (2019). Textural modification of 3D printed dark chocolate by varying internal infill structure, *Food Research International*, 121, pp. 648–657.

[50] Nedomová, Š., Trnka, J. and Buchar, J. (2013). Tensile strength of dark chocolate, *Acta Technologica Agriculturae*, 16, pp. 69–71.

[51] Svanberg, L., Ahrné, L., Lorén, N. and Windhab, E. (2011). Effect of sugar, cocoa particles and lecithin on cocoa butter crystallisation in seeded and non-seeded chocolate model systems, *Journal of Food Engineering*, 104, pp. 70–80.

[52] Fatimatuzahraa, A., Farahaina, B. and Yusoff, W. (2011). The effect of employing different raster orientations on the mechanical properties and microstructure of Fused Deposition Modeling parts. In *2011 IEEE Symposium on Business, Engineering and Industrial Applications (ISBEIA)*, IEEE, pp. 22–27.

[53] McLouth, T. D., Severino, J. V., Adams, P. M., Patel, D. N. and Zaldivar, R. J. (2017). The impact of print orientation and raster pattern on fracture toughness in additively manufactured ABS, *Additive Manufacturing*, 18, pp. 103–109.

[54] Leung, P. Y. V. (2017). Sugar 3D printing: additive manufacturing with molten sugar for investigating molten material fed printing, *3D Printing and Additive Manufacturing*, 4, pp. 13–18.

[55] Sher, D. and Tutó, X. (2015). Review of 3D food printing, *Temes de disseny*, 31, pp. 104–117.

[56] Gervasoni, S., Terzopoulou, A., Franco, C., Veciana, A., Pedrini, N., Burri, J. T., de Marco, C., Siringil, E. C., Chen, X. Z. and Nelson, B. J. (2020). Candybots: a new generation of 3D-printed sugar-based transient small-scale robots, *Advanced Materials*, 32, p. 2005652.

[57] Farzin, A., Miri, A. K., Sharifi, F., Faramarzi, N., Jaberi, A., Mostafavi, A., Solorzano, R., Zhang, Y. S., Annabi, N. and Khademhosseini, A. (2018). 3D-printed sugar-based stents facilitating vascular anastomosis, *Advanced Healthcare Materials*, 7, p. 1800702.

[58] Wang, X., Yan, Y., Zhang, R., Fan, Y., Cui, F., Feng, Q. and Liang, X. (2004). Anastomosis of small arteries using a soluble stent and bioglue, *Journal of Bioactive and Compatible Polymers*, 19, pp. 409–419.

[59] Goh, W. H. and Hashimoto, M. (2018). Fabrication of 3D microfluidic channels and in-channel features using 3D printed, water-soluble sacrificial mold, *Macromolecular Materials and Engineering*, 303, p. 1700484.

[60] Bergdoll, M. S. and Holmes, E. (1951). The heating of sucrose solutions. I. The relationship of 5-(hydroxymethyl)-furfural to color formation, *Journal of Food Science*, 16, pp. 50–56.

[61] Lille, M., Kortekangas, A., Heiniö, R.-L. and Sozer, N. (2020). Structural and textural characteristics of 3D-printed protein-and dietary fibre-rich snacks made of milk powder and wholegrain rye flour, *Foods*, 9, p. 1527.

[62] Lipton, J. I., Cutler, M., Nigl, F., Cohen, D. and Lipson, H. (2015). Additive manufacturing for the food industry, *Trends in Food Science & Technology*, 43, pp. 114–123.

[63] Liu, Y., Liang, X., Saeed, A., Lan, W. and Qin, W. (2019). Properties of 3D printed dough and optimization of printing parameters, *Innovative Food Science & Emerging Technologies*, 54, pp. 9–18.

[64] Pareyt, B. and Delcour, J. A. (2008). The role of wheat flour constituents, sugar, and fat in low moisture cereal based products: a review on sugar-snap cookies, *Critical Reviews in Food Science and Nutrition*, 48, pp. 824–839.

[65] Pulatsu, E., Su, J.-W., Lin, J. and Lin, M. (2020). Factors affecting 3D printing and post-processing capacity of cookie dough, *Innovative Food Science & Emerging Technologies*, 61, p. 102316.

[66] Portanguen, S., Tournayre, P., Sicard, J., Astruc, T. and Mirade, P.-S. (2019). Toward the design of functional foods and biobased products by 3D printing: a review, *Trends in Food Science & Technology*, 86, pp. 188–198.

[67] Su, J.-W., Gao, W., Trinh, K., Kenderes, S. M., Pulatsu, E. T., Zhang, C., Whittington, A., Lin, M. and Lin, J. (2019). 4D printing of polyurethane paint-based composites, *International Journal of Smart and Nano Materials*, 10, pp. 1–12.

[68] Manohar, R. S., Rao, P. H., Manohar, R. and Rao, P. (1999). Effect of mixing method on the rheological characteristics of biscuit dough and the quality of biscuits, *European Food Research and Technology*, 210, pp. 43–48.

[69] Chevallier, S., Colonna, P., Buleon, A. and Della Valle, G. (2000). Physicochemical behaviors of sugars, lipids, and gluten in short dough and biscuit, *Journal of Agricultural and Food Chemistry*, 48, pp. 1322–1326.

[70] Cardoso, R. V., Fernandes, Â., Heleno, S. A., Rodrigues, P., Gonzaléz-Paramás, A. M., Barros, L. and Ferreira, I. C. (2019). Physicochemical characterization and microbiology of wheat and rye flours, *Food Chemistry*, 280, pp. 123–129.

[71] Fabritius, M., Gates, F., Salovaara, H. and Autio, K. (1997). Structural changes in insoluble cell walls in wholemeal rye doughs, *LWT: Food Science and Technology*, 30, pp. 367–372.

[72] Sudha, M., Srivastava, A., Vetrimani, R. and Leelavathi, K. (2007). Fat replacement in soft dough biscuits: its implications on dough rheology and biscuit quality, *Journal of Food Engineering*, 80, pp. 922–930.

[73] Ajila, C., Leelavathi, K. and Rao, U. P. (2008). Improvement of dietary fiber content and antioxidant properties in soft dough biscuits with the incorporation of mango peel powder, *Journal of Cereal Science*, 48, pp. 319–326.

[74] Maache-Rezzoug, Z., Bouvier, J.-M., Allaf, K. and Patras, C. (1998). Effect of principal ingredients on rheological behaviour of biscuit dough and on quality of biscuits, *Journal of Food Engineering*, 35, pp. 23–42.

[75] Liu, Z., Zhang, M., Bhandari, B. and Wang, Y. (2017). 3D printing: printing precision and application in food sector, *Trends in Food Science & Technology*, 69, pp. 83–94.

[76] Wannerberger, L., Eliasson, A.-C. and Sindberg, A. (1997). Interfacial behaviour of secalin and rye flour-milling streams in comparison with gliadin, *Journal of Cereal Science*, 25, pp. 243–252.

[77] Delcour, J., Vanhamel, S. and Hoseney, R. (1991). Physicochemical and functional properties of rye nonstarch polysaccharides. II. Impact of a fraction containing water-soluble pentosans and proteins on gluten-starch loaf volumes, *Cereal Chemistry*, 68, pp. 72–76.

[78] Lee, C. P., Hoo, J. Y. and Hashimoto, M. (2021). Effect of oil content on the printability of coconut cream, *International Journal of Bioprinting*, 7, p. 354.

[79] Milner, J. A. and Allison, R. G. (1999). The role of dietary fat in child nutrition and development: summary of an ASNS workshop, *The Journal of Nutrition*, 129, pp. 2094–2105.

[80] Zheng, H., Mao, L., Yang, J., Zhang, C., Miao, S. and Gao, Y. (2020). Effect of oil content and emulsifier type on the properties and antioxidant activity of sea buckthorn oil-in-water emulsions, *Journal of Food Quality*, 2020.

[81] Decker, E. A., Chen, B., Panya, A. and Elias, R. J. (2010). Understanding antioxidant mechanisms in preventing oxidation in foods, *Oxidation in Foods and Beverages and Antioxidant Applications*, pp. 225–248.

[82] El-Din, M. N., El-Hamouly, S. H., Mohamed, H., Mishrif, M. R. and Ragab, A. M. (2013). Water-in-diesel fuel nanoemulsions: preparation, stability and physical properties, *Egyptian Journal of Petroleum*, 22, pp. 517–530.

## Problems

1. What are the methods used for the 3D printing of dairy, chocolate, sugars, and dough products?
2. What are the differences among the methods used for 3D printing of dairy, chocolate, sugars, and dough products?
3. What are the ingredients and additives present in the dessert inks? What are their purposes?
4. What are the key requirements to design food inks for extrusion-based food printing?
5. Explain the effects of infill pattern and density on the textural properties of 3D-printed desserts such as milk and chocolate.
6. Describe the challenges involved in the 3D printing of desserts and snacks.
7. What are the desserts and snacks you envision to be 3D-printed in the future?
8. How might 3D printing contribute to maintaining traditional foods in different parts of the world?

# Chapter 4

# 3D Food Printing of Fruits and Vegetables

Fruits and vegetables are important elements for maintaining a well-balanced and nutritious human diet [1, 2]. They contain essential vitamins, dietary fibres, minerals, carbohydrates, antioxidants, bioactive compounds, and phytonutrients to assist the human body to stay healthy [2–4]. The World Health Organization (WHO) and the Food and Agriculture Organization (FAO) also recommend a daily intake of at least 400 g of fruits and vegetables to prevent micronutrient deficiencies and reduce the risk of non-communicable diseases (NCDs), including cardiovascular diseases (CVDs), strokes, diabetes, and certain cancer types [5, 6]. Therefore, it is essential to include the recommended servings of fruits and vegetables in order to maintain a balanced and healthy diet.

Most fresh and unprocessed fruits and vegetables have high water and fibre contents and are not natively extrudable [7]. Because of their low viscosity and high biological variance that is influenced by environmental conditions [3, 4], fresh fruits and vegetables are considered to be the most challenging material to print, among various food inks (e.g., confectionery, dairy, hydrogels, and meats). Furthermore, fresh fruits and vegetables are highly perishable and can degrade easily if not properly stored [4]. Therefore, there is not much research available in the literature on the use of fresh fruits and vegetables for 3D food printing. Researchers have also looked into using fruit- and vegetable-derived ingredients for 3D food printing of fruits and vegetables (for instance, juices and concentrates, dehydrated and freeze-dried fruit, and vegetable powders).

117

This chapter will cover the processing steps for 3D food printing of fruits and vegetables, which include selection of fruits and vegetables for printable food formulations, customization of printable food formulations, preparation of fruits and vegetables for 3D printing, optimization of 3D printing parameters for fabricating 3D-printed edible structures, and extension of the shelf-life of 3D-printed fruits and vegetables. Special considerations for the 3D food printing of fruits and vegetables compared to other food types will be highlighted as well. This chapter also discusses a case study of how fresh fruits and vegetables can be 3D-printed to help dysphagic patients feed better.

## 4.1  Processing Steps for 3D Food Printing of Fruits and Vegetables

Numerous processes are involved in the 3D printing of fruits and vegetables, and they may be broken down into the following major processing steps [4] (see Figure 4.1):

1) Selection of fruits and vegetables for printable food ink formulations.
2) Customization of printable food ink formulations.
3) Preparation of fruits and vegetables for 3D printing.
4) Optimization of 3D printing parameters for fabricating 3D-printed edible structures.
5) Extension of the shelf-life of 3D-printed fruits and vegetables.

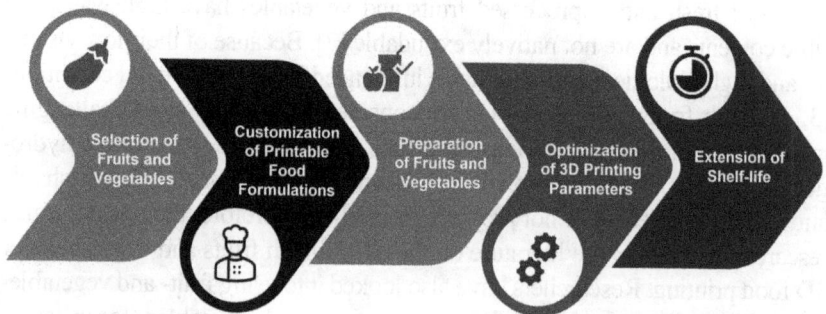

Figure 4.1.   Processing steps for 3D food printing of fruits and vegetables [4].

## 4.2 Selection of Fruits and Vegetables for Printable Food Ink Formulations

Aside from the printability and rheological qualities of the food inks, the final sensory quality and nutritional attributes of the 3D-printed food are also significant factors to consider. Proper selection of the raw materials can allow the customization of nutritional content for various groups of people distinguished by age, gender, and culture [4]. Many different types of fruits and vegetables can be used for 3D food printing, such as fresh fruits and vegetables, frozen fruits and vegetables, juices and concentrates from fruits and vegetables, dehydrated fruit and vegetable powders, and freeze-dried fruit and vegetable powders (see Figure 4.2).

As the nutritional attributes of the same fruits and vegetables will change in different forms, it is vital to select the fruits and vegetables in their most optimized forms for 3D food printing. This section will discuss the advantages and disadvantages of each type of fruits and vegetables to help with a better selection of fruits and vegetables for food ink formulations.

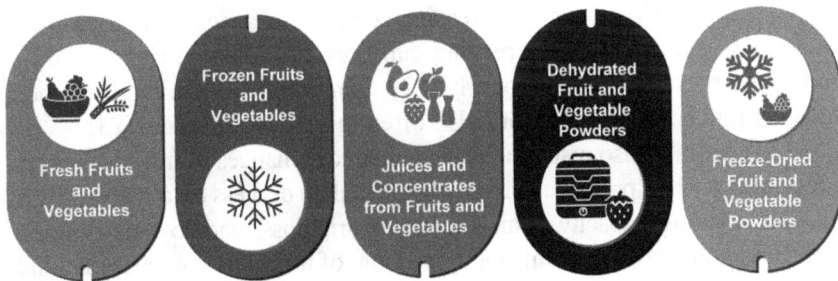

Figure 4.2.   Various types of fruits and vegetables for 3D food printing.

### 4.2.1 *Fresh Fruits and Vegetables*

Fresh fruits and vegetables are vibrant in colour and sometimes aesthetically pleasing. Aside from being widely available in the market, fresh fruits and vegetables also have the best taste and texture when they are gathered in season. Fresh fruits and vegetables often have higher nutritional contents than their processed counterparts because nutrients are lost during processing. Fresh fruits and vegetables, on the other hand, are not inherently extrudable, and additional steps are necessary to make printable food inks. Furthermore, they are perishable food items that oxidize (enzymic browning) and deteriorate rapidly. Thus, it is critical to look into maintaining their stability after harvesting and subsequent storage [8, 9].

To formulate 3D-printable food inks, fresh fruits and vegetables are first cooked and then puréed until they approach a creamy liquid or paste in consistency. If the skins of the fruits and vegetables are removed before puréeing, the purée consistency will be smoother. Fruits and vegetables that have been puréed are easy to swallow and digest and contain more fibres than juices and concentrates. To keep the food colour vibrant, it is best recommended to avoid overcooking of the fruits and vegetables before puréeing [10].

## 4.2.2 *Frozen Fruits and Vegetables*

As far as fruit and vegetable preservation is concerned, freezing is a proven and excellent method for long-term preservation [9] that helps reduce deterioration and spoilage, while increasing consumers' access to non-seasonal fruits and vegetables [11]. Freezing can preserve most nutrients in fruits and vegetables when done properly [12]. Compared to fresh produce, most frozen fruits and vegetables are also often more cost-effective [13].

Freezing may also result in physico-chemical changes in frozen fruits and vegetables, such as protein changes, which result in dryness or toughening, loss of water-binding capacity, which leads to drip loss, and loss in firmness. Depending on the water permeability and rate of heat removal, different kinds of cell damage can also develop during the freezing process [14]. Fruits and vegetables derive their structure and texture from the stiff cell walls that retain water and other chemical substances. Since fresh fruits and vegetables contain very high water content relative to their weight, ice crystals form and thaw more easily than other food types due to the water phase change that occurs during freezing [9]. Hence, fruits and vegetables are frozen by freezing the water in their cells. The expansion of water during the freezing process causes the ice crystals to rupture the cell walls [12]. As a result, thawed fruits and vegetables are considerably softer than raw ones. For instance, frozen tomatoes will turn watery and mushy after thawing. Frozen vegetables have fibrous structures to allow them to retain their structural integrity during thawing. Frozen high starch vegetables, such as lima beans, corns, and peas, are less susceptible to textural changes after thawing [12]. Fruits, on the other hand, have softer structures and are more prone to losing their firmness. Fruits, unlike vegetables, also have cellular structures that make them less resistant to freezing [9, 15, 16]. Consumers typically expect more from frozen fruits, comparing the difference in texture, colour, freshness, and aroma of frozen fruits to unfrozen fruits. Textural changes, on the other hand, are less important for frozen vegetables because they are usually consumed after thawing [9].

### 4.2.3 *Juices and Concentrates from Fruits and Vegetables*

Fluids within fruits and vegetables can be extracted mechanically by either squeezing or hydro-distillation [17]. These extracted fluids derived from fruits and vegetables are often referred to as juices. The yield and quality of juices and concentrates made from fruits and vegetables are essentially determined by their variety and maturity level. In addition, the maturity and variety of the fruits and vegetables can also influence the overall taste, flavour, colour, and total solids of the extracted juices. For instance, the sugar level rises as the fruit matures and can affect the yield of the fruit juice concentrates directly too. Concentrates are juices that have a portion of their naturally occurring water content removed, and they offer numerous advantages over juices [17]. The reduction in the amount of water in juice products can help extend shelf-life and reduce packaging, transportation, and storage costs with better, simplified handling. Some of the common processes for extracting concentrates from juices include freeze-drying, thermal evaporation, direct/reverse osmosis, membrane filtration/distillation, or drying by clathrate hydrates. Non-table-grade fruits and vegetables, which are often small-sized or odd-shaped fruits and vegetables that do not meet retail criteria, are typically used for extracting fruit juices and concentrates and boosting sustainability by minimizing food wastes. Juices and concentrates from various types of fruits and vegetables can also be mixed to provide a well-balanced source of vitamins, minerals, phytonutrients, and other bioactive compounds [18].

Some juices and concentrates may contain added sugars, colouring, and preservatives. Hence, for a healthier option, juices and concentrates without these desired additives should be chosen. Several of the processes for extracting concentrates from juices may also require heating operations, which may result in the destruction of heat-sensitive vitamins, such as vitamins B and C [19].

Furthermore, juices and concentrates may also lack the fibres found in fresh fruits and vegetables.

Yang *et al.* [20] demonstrated the use of lemon juice gel for 3D food printing, in which potato starch was used as a gelling agent for lemon juice. Lemon juice gel was created by combining lemon juice and potato starch, which was then steam-cooked for 20 min. Their formulated lemon juice gel was translucent in appearance, with a flexible and chewy texture. They also looked at how potato starch affected the mechanical and rheological characteristics of lemon juice gels. An optimal lemon juice gel should have good printability as well as adequate gel strength and viscosity and be capable of maintaining the print shape and fusing with the preceding layers. They used an extrusion-based printing process to extrude the lemon juice gel and optimized the printing conditions. They were able to print several 3D constructs under the optimized parameters (see Figure 4.3). In another work by Liu *et al.* [21], a multi-extruder

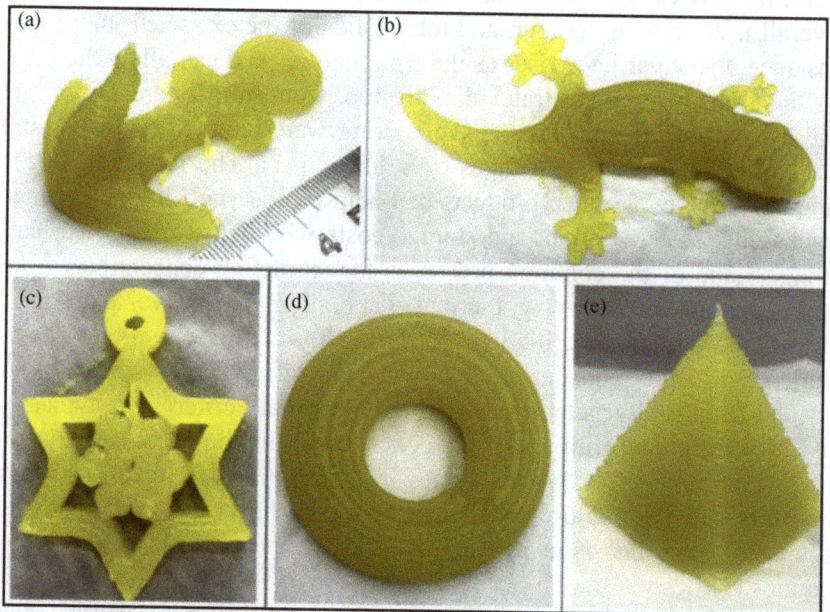

Figure 4.3.   3D-printed lemon juice gel with various designs: (a) anchor, (b) gecko, (c) snowflake, (d) ring, and (e) tetrahedron. Reproduced from [20]. Copyright (2018), with permission from Elsevier.

was used to extrude strawberry juice gel and mashed potatoes to fabricate multi-material constructs. Similarly, strawberry juice gel was made by combining strawberry juice and potato starch, which was then steam-cooked for 20 min. Azam *et al.* [22] also demonstrated that 3D printing of fruit concentrates, with the addition of bioactive compounds and healthy additives, can be an innovative and appealing method of making food (in the form of a cold dish or snack) for people with special requirements.

### 4.2.4 *Dehydrated Fruits and Vegetables*

Fruits and vegetables that have had their water content removed by various means, such as convective drying, microwave drying, spray drying, vacuum drying, osmotic dehydration, and freeze-drying, are known as dehydrated fruits and vegetables [8, 23]. Since freeze-drying does not use heat to induce water to vaporize, it will be covered in more detail in the next section. Fresh fruits and vegetables can contain up to 80% moisture; therefore, drying them can substantially reduce their weight and size as they lose their water content [23]. The majority of dehydrated fruits and vegetables have a long shelf-life and take up minimal storage space. Although dehydrated fruits and vegetables are rich in vitamins, minerals, energy and fibres, nutritional loss can still occur during the dehydration process [23]. For instance, thiamine as well as vitamins A and C are particularly vulnerable to heat and oxidative degradation. In addition, colour loss, lipid oxidation, and enzymatic browning reactions may also occur during the dehydration process. Some of the side effects of enzymatic browning include off-flavour and unpleasant colours. Nevertheless, these dehydrated fruits and vegetables can be blended into powder form and then used as food inks for 3D food printing. Since dehydrated fruit and vegetable powders have higher nutritional stability, they are also ideal for making 3D-printed space food [24, 25].

### 4.2.5 *Freeze-Dried Fruits and Vegetables Powders*

Freeze-drying is one of the most effective techniques for removing water in fruits and vegetables, while preserving their product quality [8]. Freeze-drying comprises three main steps: freezing, primary drying, and secondary drying [26]. To begin the freeze-drying process, fresh fruits and vegetables must first be frozen. The frozen sample is then put under a vacuum and then the ice is sublimated to remove the water content during the primary drying of the freeze-drying process. Any unfrozen water that remains is subsequently desorbed from the drier food matrix for secondary drying. Since freeze-drying operates at high vacuum and low temperatures, this technique is ideal for dehydrating fruits and vegetables that are thermally sensitive and prone to oxidation. In addition, all microbiological and degradation activities would have been interrupted too at this extremely low temperature [8]. Freeze-drying is a time-consuming and costly process compared to other dehydrating processes, and therefore freeze-drying is often used for dehydrating high-value food to retain maximum nutritional values. Similarly, freeze-dried fruits and vegetables can be blended into powder form, and then used as food inks for 3D food printing. Liu *et al.* [27] demonstrated that the addition of freeze-dried mango powder to 3D-printed dough as a flavour additive improves its nutritional value and flavour.

## 4.3  Customization of Printable Food Formulations from Fruits and Vegetables

Food can be easily designed and customized using 3D printing technology in terms of nutritional attributes and sensory qualities for specific consumer groups (children, teenagers, athletics, pregnant women, the elderly, etc.) [28]. Hence, the preferred type, form, and ratios of the various fruits and vegetables must be determined to formulate customized printable fruits and vegetables food inks with the desired nutritional attributes and sensory qualities. The printability and structural stability of these food inks should be considered when designing. In a recent study, Derossi *et al.* [29] customized the ink formulation of a 3D-printed fruit-based snack for 3- to 10-year-old children to provide calcium, iron, and vitamin D in their diet. The ink formulation of this fruit-based snack comprises fresh bananas, canned white beans, dried mushrooms, lemon juice, and dried non-fat milk. Pectin solution was added to help maintain a smooth consistency and prevent phase separation between the fruit tissues and water during ink deposition. They also optimized the 3D paste extruder's flow level and print speed to obtain good ink printability, without compromising the final 3D-printed structures.

# 4.4 Preparation of Various Forms of Fruits and Vegetables for 3D Printing

## 4.4.1 *Formulation of Food Inks with Fresh Fruits and Vegetables*

Since fresh fruits and vegetables are non-natively extrudable in their raw forms, they must be prepared and processed before 3D printing. Fresh fruits and vegetables are often mashed or blended first to make a homogeneous paste and then sieved to remove any coarse particles to prevent nozzle clogging during extrusion. As water directly influences the viscosity of a paste, the water content of fresh fruits and vegetables can affect the paste's printability. Due to the degradation effects and associated costs, it is rather cumbersome to evaporate water from the blended pastes to the desired water content levels [4]. However, a juice processor can be used to separate the juices from the fruit and vegetable pulps, and then reintroduce the juices back to the pulps at a precise concentration to obtain the desired paste viscosity for extrusion [28]. Nevertheless, thickeners are usually added to increase the viscosity of the blended pastes. Fruits and vegetables generally lose their natural colours during the blending and mashing process owing to oxidation, and anti-browning agents, such as acetic acids, ascorbic acids, citric acids, lactic acids, malic acids, and sodium chloride, can be added to the paste to prevent enzymatic browning [4, 29].

## 4.4.2 *Formulation of Food Inks with Juices and Concentrates from Fruits and Vegetables*

Fruit and vegetable juices and concentrates are not 3D-printable on their own, and hence they must be combined with hydrocolloids to form a gel-like structure. As a result, the printability of "juice gel" food ink is influenced by the ink's rheological and viscosity characteristics, which are determined by the hydrocolloids' specific gelation mechanisms, glass transition temperature, and melting point [4]. Common hydrocolloids that are used in the food industry as thickening and gelling agents include gelatine, alginates, agar, cellulose derivatives, microcrystalline cellulose, gellan gum, carrageenan, xanthan gum, potato starch, and pectin. Yang *et al.* [20] used lemon juice gel as a food material for 3D food printing in their study. To create their 3D printable lemon juice gel, they first used a food mixer to thoroughly combine lemon juice and potato starch and then steam-cooked it for 20 min. As the mixture cooled to ambient temperature, it formed a weak gel-like structure. The lemon juice gel was stored at 4 °C before 3D printing.

### 4.4.3 *Formulation of Food Inks with Fruit and Vegetable Powders*

Fruit and vegetable powders are made by crushing dried fruits and vegetables into fine powders. These fruit and vegetable powders can be used to formulate food inks by simply dispersing them in a hydrocolloid matrix, in which the hydrocolloids function as gelling or thickening agents for the powders [30]. These fruit and vegetable powders should be sieved to remove any coarse particles before dispersing them in the hydrocolloid matrix. For instance, Kim *et al.* [30] sieved freeze-dried powders with mesh sizes of 100 and 150 μm in their study. They also prepared hydrocolloid mixtures at 10% concentration and then kept them in the fridge at 4 °C for 1 day to facilitate hydration. The sieved freeze-dried powder was then mixed into the hydrocolloid mixtures continuously to achieve a homogeneous mixture. Since these fruit and vegetable powders have little moisture, they can promote uniformity of the raw ingredients while allowing better printability and storage. This is because it will be easier to manage the water content of food inks while also adjusting the inks' rheological properties. Some of the parameters that also directly impact the rheological characteristics of the fruit and vegetable powders food inks include particle size distribution, powder source, powder volume, and particle swelling. The nutrition value of 3D-printed foods may also be enhanced by increasing the amount of food powder utilized [30].

## 4.5 Optimization of 3D Printing Parameters for Fabricating 3D-Printed Edible Structures

The optimization of the 3D printing parameters has a major influence on the final 3D-printed edible structures' quality and long-term structural integrity. Hence, the 3D printing parameters must be optimized to the food inks used to obtain a quality 3D structure. The several critical 3D printing parameters include print speed, extrusion rate, nozzle size, layer height, and infill density [4].

(1) *Print speed.* Print speed can be defined as the nozzle velocity travelling in the $x/y$ direction [31]. The dimensions and microstructures of the extruded food ink will also be affected by both print speed and extrusion rate. Hence, the print speed must be appropriately optimized in conjunction with the extrusion rate, so that the 3D printer can give good print quality (e.g., printed samples with stable dimensions).

(2) *Extrusion rate.* Extrusion rate is defined as the volumetric rate of material extruded from the print nozzle. A simple mathematical equation [32] that can be used to express the extrusion rate ($ER$) in terms of nozzle diameter ($D_n$), layer height ($h$), and print speed ($V_n$) is shown below:

$$ER\ (mm^3/s) = D_n(mm) * h\ (mm) * V_n(mm/s) \qquad (4.1)$$

However, Equation 4.1 can only help to some extent in coupling the appropriate print speed to the required extrusion rate during the printing process. This is because the wettability, density, viscosity, and rheological properties of food inks can ultimately affect the final extrusion rate too [32].

(3) *Nozzle size.* The diameter of the nozzle, which directly determines the diameter of the extrusion lines, is referred to as nozzle size [33]. The size of the nozzle influences the dimensional parameters of the 3D-printed samples since it directly determines the line width of the extruded material. A smaller nozzle typically provides greater performance on the dimensional parameters of 3D-printed samples, but at the expense of a longer printing time.

(4) *Layer height.* The distance between the last deposited layer of the material and the nozzle tip is defined as the layer height [4]. Smaller layer heights can typically give a better surface finish, but more layers are necessary to print the entire structure.

(5) *Infill density.* The filling percentage of the internal part of an object is termed as the infill density, and it is expressed as a percentage between 0 and 100% [4, 33]; 100% infill density means that the sample is fully solid, whereas 0% infill density means that the sample is hollow. The infill density can be pre-defined by the slicing software by using various print paths. Infill density can also affect the material consumption, printing time, and weight of the printed sample. According to a study by Huang *et al.* [33], infill density has no significant influence on the dimensional qualities of the printed samples, but it can have a considerable effect on their textural properties.

## 4.6  Extension of the Shelf-Life of 3D-Printed Fruits and Vegetables

Given the highly perishable nature of fruits and vegetables, the shelf-life and storage of the 3D-printed fruits and vegetables should also be considered [7]. These 3D-printed fruits and vegetables should also maintain their appearance, taste, and nutritional properties during storage. Hence, it is extremely important to ensure that these 3D-printed fruits and vegetables have a prolonged shelf-life so that they can be pushed to the market for commercialization [4].

Preventing microbial spoilage by fungi or bacteria [34] is one way to improve the shelf-life of 3D-printed fruits and vegetables, as well as ensure their safety and quality. This may be accomplished by thoroughly cleaning or sanitizing any equipment (including nozzles, tubes, piston containers, extruders, etc.) that comes into contact with the 3D-printed fruits and vegetables during the 3D printing process. In addition to sanitizing the equipment, raw fruits and vegetables can also be cleaned in chlorinated water [4]. Other ways to improve the shelf-life of 3D-printed fruits and vegetables include the optimization of the storage conditions and temperatures and the use of appropriate stabilization treatments (such as microwave or ultraviolet C [UV-C]) [4].

## 4.7  Case Study: 3D Printing of Fresh Vegetables for Dysphagic Patients

Dysphagia is a medical term for a swallowing disorder, characterized by difficulty in transporting food or liquids from the mouth to the stomach during a normal feeding process [35–37]. Dysphagia is caused by any condition that impairs the nerves and muscles necessary for swallowing. This disorder is common among the elderly and individuals suffering from debilitating illnesses, such as dementia, stroke, amyotrophic lateral sclerosis, and Parkinson's disease [35]. Oropharyngeal dysphagia (OD) may also cause difficulty in initiating swallowing, coughing, and sometimes even choking, due to the food residues that remain in the oral cavity. As a result, dysphagia patients' reduced food intake often leads to unfavourable conditions such as dehydration, malnutrition, and weight loss. Hence, the food prepared should be soft and safe enough for dysphagia patients to chew and swallow [36].

The textural properties of food may be altered to make it soft and safe (for instance, food can be puréed, mashed, or minced to make it softer for easier swallowing) [36]. Fluids may also be thickened for enhanced absorption by altering their viscoelastic characteristics [36]. Puréed, mashed, and minced foods can be quite unattractive aesthetically and appear unappetizing to dysphagia patients. Hence, there should be additional efforts to assist dysphagia patients in having dignified feeding by allowing them to eat meals that appear, feel, and taste like regular food. It has been demonstrated that similarity to everyday meals increases their willingness to consume meals [38]. Casting and moulding are typically applied to shape puréed, mashed, and minced foods to make them aesthetically enjoyable. However, the casting and moulding methods are technically demanding and require specially skilled manpower. The 3D food printing technology can provide an innovative alternative for automating, standardizing, and shaping texturally modified food so that it is aesthetically appealing, repeatable, and consistent [10, 39]. The food inks used in 3D food printing can be modified texturally by adding food thickeners, stabilizers, additives, and viscoelasticity modifiers, such as pectin, agar, kappa carrageenan, locust bean gum, gellan gum, and xanthan gum [36, 40]. This section discusses a case study by Pant *et al.* on the 3D printing of fresh vegetables using food hydrocolloids for dysphagic patients [36].

### 4.7.1 *Classifications of Vegetables and Formulation of Food Inks*

Pant *et al.* [36] generally classified vegetables into three main categories (namely low starch and high water percentage, high starch and low water percentage, and medium starch and medium water percentage), each of which requires a distinct treatment to print. For demonstrative purposes, they used one representative vegetable from each category for this study (for instance, bok choy, garden peas, and a carrot to represent low starch and high water percentage, high starch and low water percentage, and medium starch and medium water percentage vegetables, respectively) (see Figure 4.4). Hydrocolloids, such as kappa carrageenan (KC), locust bean gum (LBG), and xanthan gum (XG), were used to modify the rheological characteristics and textures of these food inks. These food inks were then characterized according to their rheological, textural, microstructural, and printability characteristics [36].

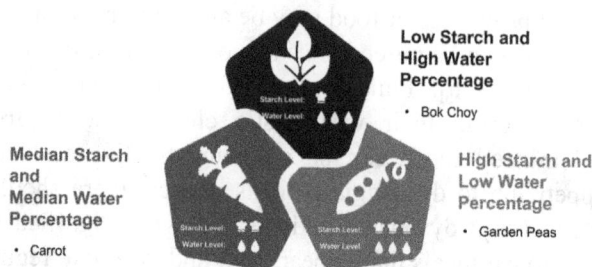

Figure 4.4.   Classifications of vegetables into three main categories: low starch and high water percentage, high starch and low water percentage, and medium starch and medium water percentage [36].

## 4.7.2   *Formulations and 3D Printing of Food Inks*

### 4.7.2.1   *Formulation of food inks*

The vegetables (bok choy, carrot, and garden peas) were manually prepared, including peeling, washing and dicing, before being processed. The bok choy was steamed for 15 min, while the carrot and garden peas were boiled for 15 min. After draining the water, each vegetable was blended individually in a food processor for 5–10 min to a purée-like consistency. The puréed vegetables were sieved to prevent any nozzle clogging during printing [36].

The initial water content percentage (WC %) of each type of puréed vegetables was determined by measuring the weight difference of a sample before and after the drying process [36]. The initial WC % for each vegetable was standardized to a final WC % based on the vegetable type by adding water to that particular batch accordingly. To create distinct food inks for experimental setup (see Table 4.1), different concentrations of food-grade hydrocolloids (KC, LBG, and XG) were added to puréed vegetables of varying WC %.

Table 4.1.   Food ink formulations of bok choy, a carrot, and garden peas in their water content (wt%), XG (wt%), KC (wt%), and LBG (wt%) food hydrocolloids content [36].

| Food Inks | Ink 1 $H_2$0-XG-KC-LBG | Ink 2 $H_2$0-XG-KC-LBG | Ink 3 $H_2$0-XG-KC-LBG | Ink 4 $H_2$0-XG-KC-LBG | Ink 5 $H_2$0-XG-KC-LBG |
|---|---|---|---|---|---|
| Bok Choy | 96-0-0-0 | 96-1.0-0-0 | 96-0.7-0-0.5 | 96-1.0-0-1.0 | 96-1.0-0-2.0 |
| Carrot | 90-0-0-0 | 90-0.3-0-0 | 90-0.3-0.3-0 | 90-0.3-0.5-0 | 90-0.3-0.7-0 |
| Garden Peas | 80-0-0-0 | 85-0-0-0 | 90-0-0-0 | 80-0.1-0.1-0 | 80-0.3-0.3-0 |

### 4.7.2.2 *3D printing of food inks*

Foodini's extrusion-based 3D food printer with a 1.5 mm nozzle was used to 3D-print samples of the various food inks [36]. A 3D hexagonal prism structure with eight layers was printed on a dish to individually test the printability of the food inks (see Figure 4.5a), in which the food inks were considered printable if the printed structures could be maintained minimally for 15 min. Optimizations of the printing parameters such as extrusion rate and print speed were done during the pre-tests. Photographs were taken immediately after the printing of the 3D hexagonal prism structures to assess the shape stability and shape fidelity on a scale of 1 (very poor) to 5 (excellent) (see Figure 4.5a). Pant *et al.* [36] used Wiibox's Sweetin chocolate printer to 3D-print a painting with carrot, corn, and garden peas food inks (see Figure 4.5b).

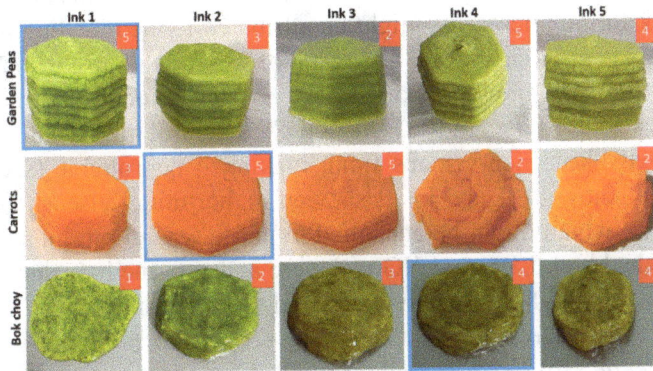

(a)

(b)

Figure 4.5. (a) 3D-printed hexagonal prism structures with garden peas, carrot, and bok choy food inks. Print scores on the top-right corner assessed the shape stability and shape fidelity of the 3D-printed structures. (b) 3D-printed painting with carrot, corn, and garden peas food inks. Reprinted from [36]. Copyright (2021), with permission from Elsevier.

### 4.7.3 *Rheological Properties of Food Inks*

Pant *et al.* [36] evaluated the shear-thinning property, yield stress, and recovery behaviour of the various food inks formulations, which are listed in Table 4.1. The rheological properties of the various food ink formulations can be modified by adjusting the amount of water content and hydrocolloids present. Shear-thinning pseudoplastic behaviour was observed in all food ink formulations, in which their viscosities decreased with increasing shear rates in the range of $0.001–1,000$ s$^{-1}$ (see Figure 4.6a).

They observed that the ink viscosity of the garden peas food ink formulations generally decreased with increasing WC % (see Figure 4.6a). The addition of XG and KC caused garden peas ink 4 and ink 5 to have higher viscosity. The WC % of all five carrot food ink formulations was kept constant at 90%, and they all displayed shear-thinning behaviour and had comparable initial viscosities at low shear rates. They also noticed that the ink viscosity decreased substantially when XG was added to the carrot ink formulation. This might be attributed to the repelling nature of forces between starch and XG as a result of surface charges' incompatibility [41]. Similarly, the WC % of all bok choy carrot food ink formulations was also kept constant at 96%. The bok choy control ink formulation (ink 1) behaved like a non-viscous fluid, making it unprintable (see Figure 4.5a). Bok choy ink 1 and ink 2 showed higher initial viscosities at low shear rates, which might be attributed to phase separation. Inks 3, 4, and 5, which had both LBG and XG and were added to the bok choy ink formulations, had higher viscosity compared to inks 1 and 2 when the shear rate approached the crossover point [36].

Pant *et al.* [36] also assessed the yield stress of the various ink formulations by performing oscillatory amplitude sweep tests (see Figure 4.5b). The mechanical strength of the inks is determined by the storage modulus (G'), a factor that represents the elastic response. The loss modulus (G'') represents the viscous response. The microstructures of the inks can collapse with increasing shear stress, causing liquid-like flow. Ink formulations with higher yield stress are preferred as they may hold up the printed shapes better. Hence, to obtain reasonable mechanical stability, the yield stress of the ink formulations must be properly optimized without negatively affecting the extrusion process. The yield stress of garden peas ink formulations decreases considerably with increasing WC %. Garden peas ink 5 had 0.3% w/w XG and 0.3% w/w KC added to it and was shown to have a substantially higher yield stress than the control garden peas ink 1. For carrot ink formation 2, the sole addition of XG caused the yield stress

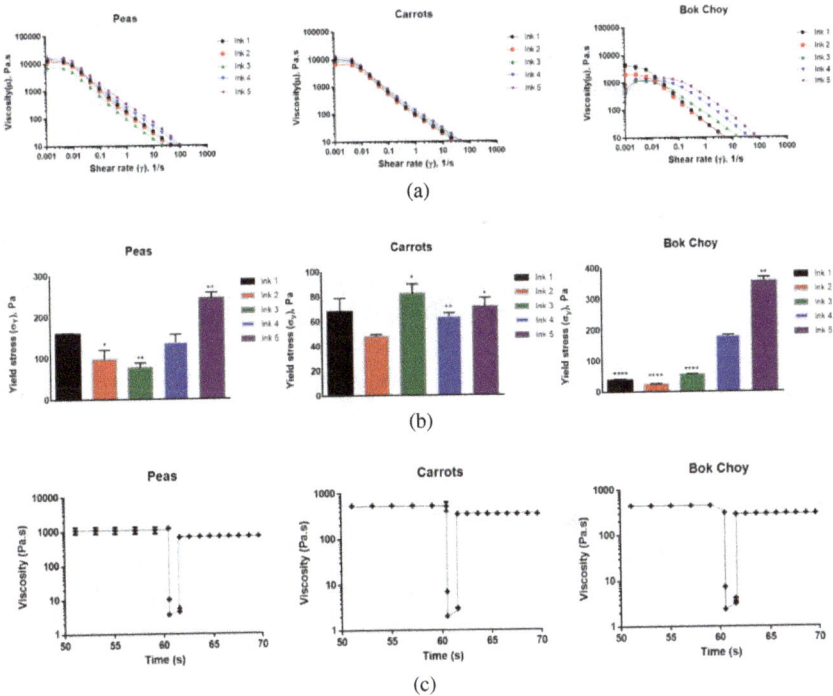

Figure 4.6. Various rheological characterizations of the garden peas, carrot, and bok choy food inks: (a) viscosity vs. shear rate, (b) yield stress of the food inks, and (c) recovery of the food inks. Reprinted from [36]. Copyright (2021), with permission from Elsevier.

and viscosity to decrease. However, the yield stress increased substantially with the addition of both XG and KC, as shown in carrot ink formulations 3, 4, and 5. Similarly, the addition of XG alone reduced the yield stress and viscosity of bok choy ink, while a combination of both XG and LBG can substantially increase the yield stress of bok choy inks. This phenomenon might be attributed to the interactions between LBG and XG that can result in the formation of soft elastic gels [42].

The reversibility of viscosity is another desired characteristic of food ink. They conducted experiments with the most optimized ink formulations for each type of vegetable, namely garden peas ink 1, carrot ink 2, and bok choy ink 4. These three inks were subjected to three different stress levels that reflect the various stages of extrusion-based printing, particularly inks stored without pressure in a syringe, ink extrusion through the nozzle when shear stress is applied, and ink restructuring after deposition when extrusion pressure is withdrawn. All three inks

134 Digital Gastronomy: From 3D Food Printing to Personalized Nutrition

exhibited good reversibility of viscosity. The ink viscosity recovered to its previous level once the shear stress was removed, demonstrating that the food ink formulations could preserve their structural integrity after deposition [36].

### 4.7.4 3D-Printed Structures

Factors such as shape fidelity, syneresis, and structural integrity were used to evaluate the printability of the food ink formulations listed in Table 4.1 with a modified print scoring system. The selected optimized ink formulations should display low syneresis, good shape fidelity, and good structural integrity. Elderly patients had a common perception of food hydrocolloids imparting an artificial flavour to food, which might hamper their intake of these 3D-printed foods [36]. As a result, among the food inks with the same print scores, the ink formulations with the least amount of hydrocolloids were chosen as the best-optimized formulations. Garden peas ink 1, carrot ink 2, and bok choy ink 4 were the selected optimized ink formulations.

A pleasurable, safe, and nutritious meal is most critical to improving the quality of life for dysphagic individuals. Aesthetically attractive meals can lead to a higher intake of food for dysphagic individuals, and 3D printing is an effective tool to provide considerably better presentation than the commonly used silicone moulds. Pant *et al.* [36] demonstrated that innovative, sophisticated, and visually attractive patterns may be created using various food inks in a 3D artwork form with the 3D food printing technology (see Figure 4.5b).

### 4.7.5 Syneresis of Food Inks

Syneresis is the unwanted leakage of water from foods that results in an unappealing aesthetic appearance [43]. Water spreading jeopardizes the overall integrity of the 3D-printed food structures, causing structural instability. Pant *et al.* [36] measured the wetting area on the Whatman filter paper to quantitatively determine the amount of water leakage from the 3D-printed food. Garden peas with greater WC % (garden peas inks 2 and 3) exhibited considerably higher syneresis than the control garden peas ink (garden peas ink 1). Syneresis was observed to decrease significantly when KC and XG were added to the garden peas ink formulations (garden peas inks 4 and 5). Compared to the other four carrot inks, carrot ink with XG gave lower syneresis (carrot ink 2), as XG functions as a weak gelling agent and thickener. The syneresis was considerably increased when a combination of XG and KC hydrocolloids were added into the carrot inks (carrot inks 3, 4 and 5). This is due to the addition of KC inhibiting XG's water swelling capability. With higher hydrocolloids concentrations, bok choy inks 2 to 5 showed reduced fluid seepage. The combination of XG and LBG hydrocolloids in bok choy ink 5 results in a more synergistic gelling effect, decreasing syneresis [42].

### 4.7.6 *Microstructure of Food Inks*

Pant *et al.* [36] also investigated the impact of adding hydrocolloids to food ink formulations by examining the inks' scanning electron microscope (SEM) images (see Figure 4.7). Figure 4.7a(i, ii) and Figure 4.7a(iii, iv) correspond to the garden peas inks without hydrocolloids (garden peas ink 1) and with the addition of the XG and KC hydrocolloids (garden peas ink 5), respectively. Garden peas ink 5 exhibit mesh-like structures that are typically observed in hydrogels (see Figure 4.7a(iii, iv)). Because of the starch content in garden peas inks 1 and 5, the formation of the clump structures aided in the structural integrity of the 3D-printed structures and good printing quality.

Figure 4.7b(i, ii) and Figure 4.7b(iii, iv) correspond to the carrot inks without hydrocolloids (carrot ink 1) and with the addition of the XG hydrocolloids (carrot ink 2), respectively. Fibres, pores, and limited interconnections between the layers were observed in carrot ink 1 (see Figure 4.7b(i, ii)). The addition of 0.3% XG in carrot ink 2 allowed the formation of weak gel-like structures and interconnections of fibres, resulting in better structural integrity of the printed ink (see Figure 4.7b(iii, iv)). In comparison with carrot ink, bok choy ink without any hydrocolloids (bok choy ink 1) showed larger pores in the microstructures (see Figure 4.7c(i, ii)). The microstructure's porosity influences the dimensional stability and extrudability of the printing [41]. The addition of the LBG and XG hydrocolloids in bok choy ink (bok choy ink 4) helped to maintain the pore density while reducing pore size and enhancing the interconnections between fibres and sheets (see Figure 4.7c(iii, iv)). The ink's mechanical stability was enhanced due to the development of a prominent gel-like matrix, as evidenced by an increase in yield stress of bok choy ink 4 compared to bok choy ink 1. Similar to how increasing the cross-linkage

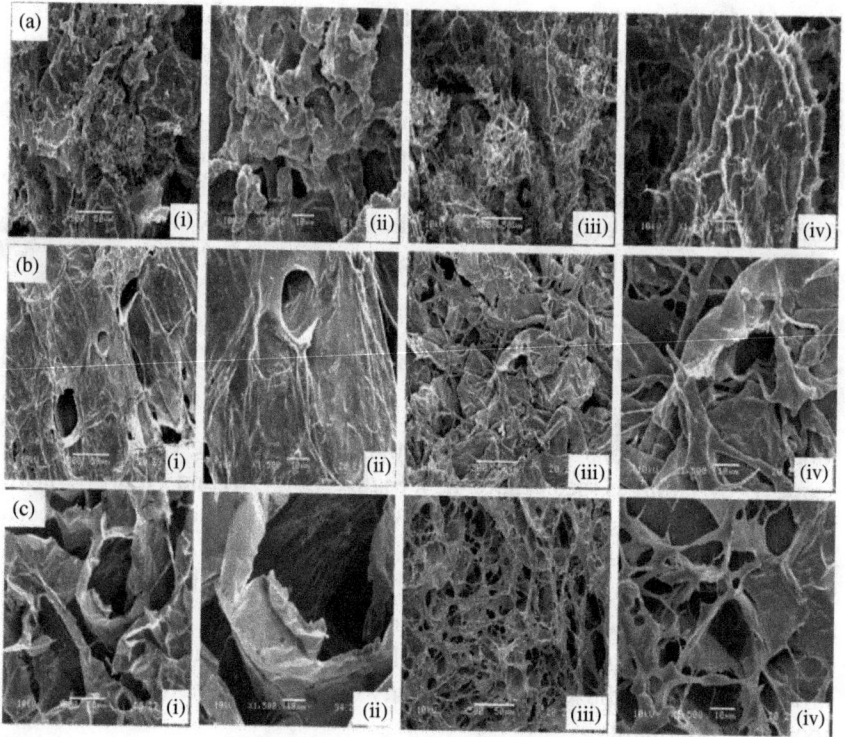

Figure 4.7.  Microstructures of the various food ink formulations: (a(i, ii)) garden peas ink 1; (a(iii, iv)) garden peas ink 5; (b(i, ii)) carrot ink 1; (b(iii, iv)) carrot ink 2; (c(i, ii)) bok choy ink 1; and (c(iii, iv)) bok choy ink 4. Reprinted from [36]. Copyright (2021), with permission from Elsevier.

density of hydrogel may lead to enhanced hydrogel strength, the addition of hydrocolloids in food inks may enhance the inks' mechanical strength due to the increased connections [44].

## 4.7.7 *Textural Properties of Food Inks*

Pant *et al.* [36] also characterized their 3D-printed foods with International Dysphagia Diet Standardization Initiative (IDDSI) fork pressure and spoon tilt tests, and the results are shown in Figure 4.8. The spoon tilt test was used to assess the cohesiveness and adhesiveness of the food. Indent patterns were clearly visible on the various optimized food inks (garden

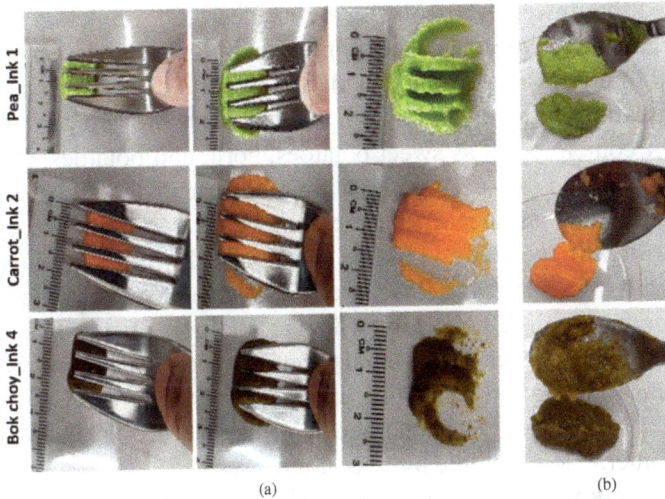

Figure 4.8. IDDSI (a) fork pressure test, and (b) spoon tilt test on optimized food inks (garden peas ink 1, carrot ink 2, and bok choy ink 4). Reprinted from [36]. Copyright (2021), with permission from Elsevier.

peas ink 1, carrot ink 2, and bok choy ink 4) for the fork pressure tests (see Figure 4.8a). From Figure 4.8b, it was observed that a significant amount of the optimized food inks (garden peas ink 1, carrot ink 2, and bok choy ink 4) could slide off the spoon when tilted and flicked once, thus implying that the optimized food ink formulations would not stick to the oral cavity for easy swallowing. These optimized food ink formulations were deemed transitional food based on the fork pressure and spoon tilt test results. The 3D-printed food structures could be easily flattened or disintegrated when pressure was applied. When the printed food comes into contact with water or saliva, it may melt and transform.

## 4.7.8 *Discussions and Conclusion on 3D Printing of Fresh Vegetables*

3D food printing offers an alternate and comparatively better option to using moulds to shape puréed foods, as it can help reduce manual handling and manpower costs [45]. Pant *et al.* [36] established a novel method of classifying various types of vegetables in three main categories, each with a distinctive starch and water content, to make them 3D-printable. They concluded that fewer hydrocolloids are needed for formulating food inks when vegetables have lower water content and higher starch content. Their approach to processing and modifying the textural characteristics of food ink formulations may be used on other vegetables with comparable starch and water content, such as sweet potatoes, corn, and potatoes, which can be treated like garden peas without the addition of food hydrocolloids. Similarly, turnips and beets with medium starch content may be treated like a carrot with the addition of one hydrocolloid, while green leafy vegetables with high water and low starch content, such as kale and spinach, can be processed like bok choy with the addition of two hydrocolloid combinations. However, the optimal formulation for each food ink would necessitate optimization depending on the individual food type, hydrocolloid, and printer used, as each food may have a distinct chemical and physical interaction with the hydrocolloids used.

Their work was particularly noteworthy since they 3D-printed visually appealing and palatable food while retaining flavours and nutrients by using undehydrated vegetables with minimal addition of food hydrocolloids. Food usually tastes better when there are fewer food hydrocolloids present, because high concentrations of food hydrocolloids can decrease taste and aroma perceptions [46].

# References

[1]  Li, D., Li, B., Ma, Y., Sun, X., Lin, Y. and Meng, X. (2017). Polyphenols, anthocyanins, and flavonoids contents and the antioxidant capacity of various cultivars of highbush and half-high blueberries, *Journal of Food Composition and Analysis*, 62, pp. 84–93.

[2]  Wang, M., Li, D., Zang, Z., Sun, X., Tan, H., Si, X., Tian, J., Teng, W., Wang, J., Liang, Q., Bao, Y., Li, B. and Liu, R. (2021). 3D food printing: applications of plant-based materials in extrusion-based food printing, *Critical Reviews in Food Science and Nutrition* 10.1080/10408398.2021.1911929, pp. 1–15.

[3]  Çakmak, H. and Gümüş, C. E. 3D food printing with improved functional properties: a review, *International Journal of 3D Printing Technologies and Digital Industry*, 4, pp. 178–192.

[4]  Ricci, I., Derossi, A. and Severini, C. (2019). *Fundamentals of 3D Food Printing and Applications*, "3D printed food from fruits and vegetables" (Elsevier), pp. 117–149.

[5]  World Health Organization (WHO). (2020). Healthy diet. Retrieved from https://www.who.int/news-room/fact-sheets/detail/healthy-diet

[6] World Health Organisation (WHO). (n.d.). Increasing fruit and vegetable consumption to reduce the risk of noncommunicable diseases. Retrieved from https://www.who.int/elena/titles/fruit_vegetables_ncds/en/

[7] Voon, S. L., An, J., Wong, G., Zhang, Y. and Chua, C. K. (2019). 3D food printing: a categorised review of inks and their development, *Virtual and Physical Prototyping*, 14, pp. 203–218.

[8] Sagar, V. R. and Suresh Kumar, P. (2010). Recent advances in drying and dehydration of fruits and vegetables: a review, *Journal of Food Science and Technology*, 47, pp. 15–26.

[9] Silva, C. L. M., Gonçalves, E. M. and Brandão, T. R. S. (2008). *Frozen Food Science and Technology*, "Freezing of fruits and vegetables" (Wiley), pp. 165–183.

[10] Kouzani, A. Z., Adams, S., Whyte, D. J., Oliver, R., Hemsley, B., Palmer, S. and Balandin, S. (Year). 3D printing of food for people with swallowing difficulties. In *DesTech 2016: Proceedings of the International Conference on Design and Technology* (Knowledge E), pp. 23–29.

[11] Li, L., Pegg, R. B., Eitenmiller, R. R., Chun, J.-Y. and Kerrihard, A. L. (2017). Selected nutrient analyses of fresh, fresh-stored, and frozen fruits and vegetables, *Journal of Food Composition and Analysis*, 59, pp. 8–17.

[12] Wolf, I. D. and Munson, S. T. (1978). *Freezing Fruits and Vegetables* (Revised 1978).

[13] Miller, S. R. and Knudson, W. A. (2014). Nutrition and cost comparisons of select canned, frozen, and fresh fruits and vegetables, *American Journal of Lifestyle Medicine*, 8, pp. 430–437.

[14] Evans, J. A. (2009). *Frozen Food Science and Technology* (John Wiley & Sons).

[15] Sterling, C. (1968). Effect of low temperature on structure and firmness of apple tissue, *Journal of Food Science*, 33, pp. 577–580.

[16] Alonso, J., Canet, W. and Rodriguez, T. (1997). Thermal and calcium pre-treatment affects texture, pectinesterase and pectic substances of frozen sweet cherries, *Journal of Food Science*, 62, pp. 511–515.

[17] Adnan, A., Mushtaq, M. and Islam, T. u. (2018). *Fruit Juices*, eds. Rajauria, G. and Tiwari, B. K., "Fruit juice concentrates" (Academic Press), pp. 217–240.

[18] Lorenzoni, G., Minto, C., Vecchio, M. G., Zec, S., Paolin, I., Lamprecht, M., Mestroni, L. and Gregori, D. (2019). Fruit and vegetable concentrate supplementation and cardiovascular health: a systematic review from a public health perspective, *Journal of Clinical Medicine*, 8, p. 1914.

[19] Darvishi, H., Mohammadi, P., Fadavi, A., Koushesh Saba, M. and Behroozi-Khazaei, N. (2019). Quality preservation of orange concentrate by using hybrid ohmic — vacuum heating, *Food Chemistry*, 289, pp. 292–298.

[20] Yang, F., Zhang, M., Bhandari, B. and Liu, Y. (2018). Investigation on lemon juice gel as food material for 3D printing and optimization of printing parameters, *LWT*, 87, pp. 67–76.

[21] Liu, Z., Zhang, M. and Yang, C.-h. (2018). Dual extrusion 3D printing of mashed potatoes/strawberry juice gel, *LWT*, 96, pp. 589–596.

[22] Azam, S. M. R., Zhang, M., Mujumdar, A. S. and Yang, C. (2018). Study on 3D printing of orange concentrate and material characteristics, *Journal of Food Process Engineering*, 41, p. e12689.

[23] Sablani, S. S. (2006). Drying of fruits and vegetables: retention of nutritional/functional quality, *Drying Technology*, 24, pp. 123–135.

[24] Zhao, L., Zhang, M., Chitrakar, B. and Adhikari, B. (2021). Recent advances in functional 3D printing of foods: a review of functions of ingredients and internal structures, *Critical Reviews in Food Science and Nutrition*, 61, pp. 3489–3503.

[25] Terfansky, M. L. and Thangavelu, M. (2013). *AIAA SPACE 2013 Conference and Exposition*, "3D printing of food for space missions" (American Institute of Aeronautics and Astronautics).

[26] Bhatta, S., Stevanovic Janezic, T. and Ratti, C. (2020). Freeze-drying of plant-based foods, *Foods*, 9, p. 87.

[27] Liu, Y., Liang, X., Saeed, A., Lan, W. and Qin, W. (2019). Properties of 3D printed dough and optimization of printing parameters, *Innovative Food Science & Emerging Technologies*, 54, pp. 9–18.

[28] Severini, C., Derossi, A., Ricci, I., Caporizzi, R. and Fiore, A. (2018). Printing a blend of fruit and vegetables. New advances on critical variables and shelf life of 3D edible objects, *Journal of Food Engineering*, 220, pp. 89–100.

[29] Derossi, A., Caporizzi, R., Azzollini, D. and Severini, C. (2018). Application of 3D printing for customized food. A case on the development of a fruit-based snack for children, *Journal of Food Engineering*, 220, pp. 65–75.

[30] Kim, H. W., Lee, J. H., Park, S. M., Lee, M. H., Lee, I. W., Doh, H. S. and Park, H. J. (2018). Effect of hydrocolloids on rheological properties and printability of vegetable inks for 3D food printing, *Journal of Food Science*, 83, pp. 2923–2932.

[31] Geng, P., Zhao, J., Wu, W., Ye, W., Wang, Y., Wang, S. and Zhang, S. (2019). Effects of extrusion speed and printing speed on the 3D printing stability of extruded PEEK filament, *Journal of Manufacturing Processes*, 37, pp. 266–273.

[32] Derossi, A., Paolillo, M., Caporizzi, R. and Severini, C. (2020). Extending the 3D food printing tests at high speed. Material deposition and effect of non-printing movements on the final quality of printed structures, *Journal of Food Engineering*, 275, p. 109865.

[33] Huang, M.-s., Zhang, M. and Bhandari, B. (2019). Assessing the 3D printing precision and texture properties of brown rice induced by infill levels and printing variables, *Food and Bioprocess Technology*, 12, pp. 1185–1196.

[34] Tomašević, I., Putnik, P., Valjak, F., Pavlić, B., Šojić, B., Bebek Markovinović, A. and Bursać Kovačević, D. (2021). 3D printing as novel tool for fruit-based functional food production, *Current Opinion in Food Science*, 41, pp. 138–145.

[35] Khoo Teck Puat Hospital. Dysphagia. Retrieved from https://www.health whub.sg/a-z/diseases-and-conditions/601/dysphagia

[36] Pant, A., Lee, A. Y., Karyappa, R., Lee, C. P., An, J., Hashimoto, M., Tan, U. X., Wong, G., Chua, C. K. and Zhang, Y. (2021). 3D food printing of fresh vegetables using food hydrocolloids for dysphagic patients, *Food Hydrocolloids*, 114, p. 106546.

[37] Alagiakrishnan, K., Bhanji, R. A. and Kurian, M. (2013). Evaluation and management of oropharyngeal dysphagia in different types of dementia: a systematic review, *Archives of Gerontology and Geriatrics*, 56, pp. 1–9.

[38] Aguilera, J. M. and Park, D. J. (2016). Texture-modified foods for the elderly: Status, technology and opportunities, *Trends in Food Science & Technology*, 57, pp. 156–164.

[39] Dick, A., Bhandari, B., Dong, X. and Prakash, S. (2020). Feasibility study of hydrocolloid incorporated 3D printed pork as dysphagia food, *Food Hydrocolloids*, 107, p. 105940.

[40] Cichero, J. (2015). *Modifying Food Texture, Vol., 2: Sensory Analysis, Consumer Requirements and Preferences*, eds. Rosenthal, A. J. and Chen, J., "Adjustment of food textural properties for elderly patients: food texture properties suitable for the elderly" (Woodhead Publishing).

[41] Liu, Z., Zhang, M. and Bhandari, B. (2018). Effect of gums on the rheological, microstructural and extrusion printing characteristics of mashed potatoes, *International Journal of Biological Macromolecules*, 117, pp. 1179–1187.

[42] Saha, D. and Bhattacharya, S. (2010). Hydrocolloids as thickening and gelling agents in food: a critical review, *Journal of Food Science and Technology*, 47, pp. 587–597.

[43] Mizrahi, S. (2010). *Chemical Deterioration and Physical Instability of Food and Beverages*, eds. Skibsted, L. H., Risbo, J. and Andersen, M. L., "Syneresis in food gels and its implications for food quality" (Woodhead Publishing), pp. 324–348.

[44] Alkayyali, L. B., Abu-Diak, O. A., Andrews, G. P. and Jones, D. S. (2012). Hydrogels as drug-delivery platforms: physicochemical barriers and solutions, *Therapeutic Delivery*, 3, pp. 775–786.

[45] Awwad, S. (2019). *An Exploration of Residents' and Care Partners' Perspectives on 3D Printed Pureed Food in Long-Term Care Homes in Ontario* (University of Waterloo).

[46]  Tournier, C., Sulmont-Rossé, C. and Guichard, E. (2007). *Flavour Perception: Aroma, Taste and Texture Interactions* (Global Science Books).

# Problems

1.  List the processing steps involved in the 3D printing of fruits and vegetables.
2.  What are some of the critical 3D printing parameters that influence the final 3D-printed edible structures' quality and long-term structural integrity? Elaborate.
3.  List the various forms of fruits and vegetables for 3D food printing.
4.  Describe the differences between dehydrated food and freeze-dried food.
5.  Which forms of fruits and vegetables do you think are most suitable for 3D food printing?
6.  Do you think 3D printing is a suitable technique for shaping puréed, mashed, and minced foods? Why?
7.  Discuss the advantages and disadvantages of frozen fruits and vegetables.
8.  Describe how food inks can be formulated with fruit and vegetable powders.
9.  Describe how the shelf-life of 3D-printed fruits and vegetables can potentially be extended.

# Chapter 5

# 3D-Printed Meat, Plant Proteins, and Alternative Proteins

## 5.1 Introduction

Global meat production has grown steadily over the past five decades with the increasing demand for meat consumption [1]. This growth is considered to be unsustainable, as conventional meat production is laden with issues of food security, public health, loss of biodiversity, and animal welfare from livestock production [2].

According to the Food and Agriculture Organization (FAO), the livestock sector makes up 14.5% of the total greenhouse gas emissions resulting from human activities and accounts for 8% of the global freshwater use [3]. Most of the livestock includes chicken, sheep, pigs, goats, and cattle [4, 5]. The contribution of greenhouse gas emissions by global livestock leading to climate change has been a cause for concern as raised by many around the world. Signs of climate change include rising temperatures, rainfall, and drought frequencies leading to flooding and other extreme weather events. The speed and extent of climate change are generally influenced by greenhouse gases emitted from energy production (i.e., combustion of fossil fuel) and non-energy production (i.e., livestock, agriculture, and land-use changes) [6].

Among the meats produced by livestock, beef poses an undesirable requirement for large areas of land and resources [4, 7]. Extensive areas of forests and grasslands were cleared to raise livestock [8]. Intensive factory farming and poor animal welfare conditions in conventional meat production have led to health considerations including nutrition-associated and food-borne diseases [9]. Examples of these diseases include swine flu and bovine spongiform encephalopathy [10]. Frequent administration of antibiotics in animal farming has resulted in the creation of antimicrobial-resistant pathogens [10]. Ethical concerns for farming and slaughtering animals for their meat have also added to the pressure of seeking alternatives to livestock meats. Animal welfare is frequently disregarded in animal farming to ensure production efficiency [11]. Moreover, the animals are force-fed frequently to increase their weight or size before slaughtering [11]. Lastly, global pollution is mainly contributing to waste production in animal farming, primarily due to the preference of the consumers for specific meat cuts [11]. For instance, ~30% of the whole chicken is discarded; the percentage for cattle is higher [12]. These add to the large volumes of food waste generated every year. These unsustainable practices in conventional meat production have demanded us to seek different methods to deliver alternative protein sources to feed the future population [13].

Over the years, multiple potential replacements for livestock meats have been suggested. One of the strategies involves substitution of meat with plant-based foods including peas and soy [14]. The production of plant-based foods could improve food security [15]. Plant-based proteins are generally affordable, highly nutritious, and non-toxic; such proteins can be consumed as food sources and supplements [16]. Other types of meat substitutes include meat analogues based on insects and mycoproteins [14]. Yet another strategy is cultured meats — artificial meats produced from lab-grown animal cells [17].

With the introduction of 3D food printing technology, sustainable meat substitutes can be printed to give meats of desired shapes, flavours, and textures. 3D printing allows the adopting of different food sources obtained from lab-cultured meats, plant-based proteins, and insect-based proteins. The use of various alternative protein sources for 3D food printing could potentially reduce waste production and greenhouse gas emissions.

This chapter discusses some works involving the 3D printing of meats, which include real meats, cultured (or cell-based) meats, plant-based meat alternatives, and other examples.

## 5.2  3D Printing of Real Meat

### 5.2.1  *Introduction of 3D Meat Printing*

High-value steaks only account for 7.2% of the weight of a cattle carcass [18]; the rest of the meat cuts are either sold as low-valued by-products or discarded [19]. Additionally, livestock by-products, including their skin, intestines, fat, and feet, account for 66% and 52% of the live weight of pigs and cattle, respectively [20]. Some by-products such as kidneys and livers are rich sources of carbohydrates, amino acids, fatty acids, and vitamins. The ears, feet, and skin of the animals are made of collagen, and their protein quantities are akin to lean meat [20]. Some of the by-products are consumed as food, while more than half of the live weights of the animals are considered unsuitable for direct consumption. Researchers and engineers have sought new technologies to put these meat by-products to better use [21].

3D food printing provides an excellent opportunity to salvage meat by-products by turning them into customized meat products; such printed meats can exhibit similar, or even improved, nutritional content, texture, and taste. In a sense, 3D food printing can be used to de-animalize animal by-products. De-animalization is defined as the alteration of food structures through methods such as the removal of skin and bones, thereby diminishing disgust by consumers towards the food [22]. Using different food materials such as meat paste and fat slurry, recombined meats can be created by 3D printing to achieve the appearance, nutritional value, and flavour of a steak. In summary, 3D food printing can be applied in meat manufacturing to improve nutrition and to reduce food wastage [23].

Besides its implication in upcycling meat wastes, 3D food printing allows preparing customized 3D-printed meals for individuals with chewing and swallowing difficulties (i.e., dysphagia) [24]. With 3D food printing, custom-made meals could be produced to resemble the original shape of the food while retaining the textural characteristics suitable for consumption by dysphagic individuals. Considering this, texture-modified meat analogues are created via extrusion 3D printing (Figure 5.1).

Among the different methods for 3D food printing, extrusion-based 3D printing is the most frequently employed method for meat printing due to its suitability with meats containing rheology-modified food inks. The printability of the ink is the key consideration in extrusion-based printing for any food ingredients; printability of a food ink refers to its capacity to be extruded by a 3D printer into a free form model and to retain the printed shape [25]. Printability is thus influenced by the operating condition of 3D printing and the rheological properties of the ink [26]. Fibrous ingredients of such meat and its by-products are non-printable by nature [27]; the addition of flow enhancers such as hydrocolloids and fats is necessary to modify their rheological properties so as to yield an extrudable paste-like ink. Examples of hydrocolloids include guar gum (GG), xanthan gum

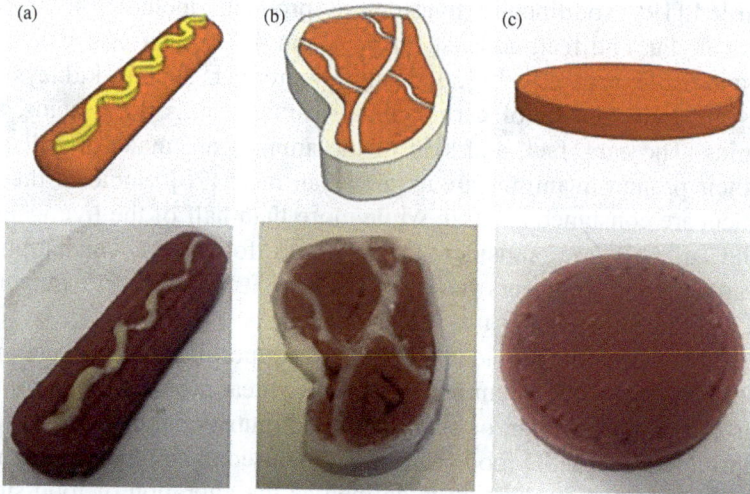

Figure 5.1.   Schematic designs and their corresponding 3D-printed meat types: (a) sausage, (b) steak "'recombined meat'", and (c) patty. Reproduced with permission from [19]. Copyright (2019) from Elsevier.

(XG), and gum tragacanth, all of which serve as viscosity enhancers by thickening the meat paste for extrusion and shape retention [28].

To perform extrusion-based meat printing, raw meats are first minced into a paste for extruding through the nozzle. Mincing the meat is crucial to reduce the particle size of the meat paste to make it smaller than the inner diameter of the nozzle (several hundreds of μm) to avoid clogging [19]. Moreover, by mincing the meat, myofibrillar proteins can be extracted to form stable meat emulsions within the paste [19]. Food additives are added to improve the extrudability of the paste and to ensure the adhesion of adjacent filaments upon printing. For example, the printability of meat pastes containing chicken, pork, and fish can be improved by adding gelatin [27]. Similarly, the enzyme transglutaminase can be added to turkey and scallop pastes before printing, as a binder, to maintain the printed meat structures [29]. Binders can also be added to enhance the mechanical stability of the 3D-printed meat paste after deposition.

Examples of binders that produce heat-resistant gels are (1) transglutaminase, (2) alginate hydrogels, and (3) plasma proteins (such as fibrinogen and thrombin). Transglutaminase is an enzyme derived from animal blood and bacterial fermentation; it catalyzes cross-linking of proteins within the meats and is generally known as meat glue [19]. The binding by transglutaminase can be improved in the presence of sodium caseinate, soy proteins, myosin, gelatin, salt, and phosphates [30]. Transglutaminase is added to the meat paste immediately before printing to avoid the occlusion of the nozzle. Alginate hydrogels consist of a source of calcium ions and sodium alginate; an acidifier (such as lactic acid) is also added to regulate the release of calcium ions [19]. When meat products are in contact with the alginate hydrogels, an irreversible gel is gradually formed. Fibrinogen/thrombin are plasma proteins involved in blood clotting where fibrinogen is enzymatically converted to fibrin by the enzymatic activity of thrombin. Both fibrinogen and thrombin are separately packed and are mixed with the meat paste. The hardness of the meat emulsion is dependent on its contact time with fibrinogen and thrombin, as well as its concentrations.

The following section discusses recent examples of 3D meat printing. The meat inks are classified into poultry, fish, beef, and pork. The ink formulations, properties, and characterizations are discussed.

## 5.2.2 *Ink Formulation, Properties, and Characterization*

### 5.2.2.1 *Poultry*

- 3D printed chicken nuggets

Chicken is often known for its high biological protein content [31]. Chicken muscle proteins contribute to the gelation and emulsification of the meats, which enhance their appearance, texture, and mouthfeel [32]. Wilson *et al.* studied the printability of chicken meat to produce 3D-printed chicken nuggets (Figure 5.2) [32]. The raw chicken meat was first minced to form ground chicken (GC) in order to extract the myofibrillar proteins. The extraction of those proteins helped to produce a stable emulsion through complex interactions with other muscle constituents in the paste. GC pastes offered a smooth texture and flowable consistency, but the gelling strength of GC was not sufficiently high due to the lack of aggregation of the myosin fibres and the poor disulphide bond arrangement [33]. Therefore, refined wheat flour (RWF) was added to GC to increase the mechanical strength of the 3D-printed product.

In the study, three inks containing different ratios of RWF to GC (1:1, 1:2, and 1:3 w/w) were prepared and their rheological and textural

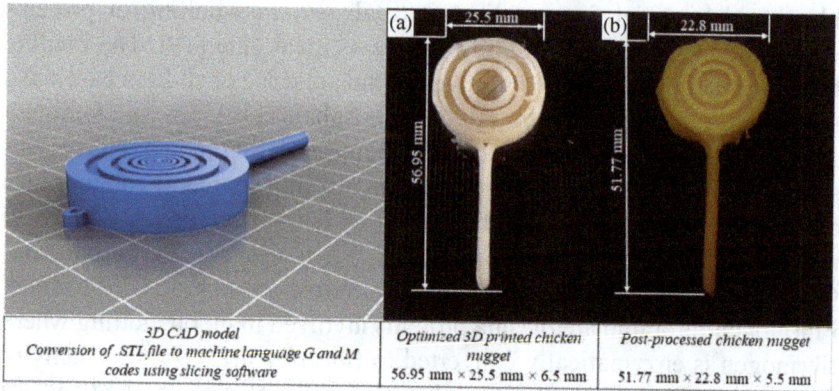

Figure 5.2.   3D CAD design model and the corresponding 3D-printed model of chicken meat. (a) Raw 3D-printed chicken meat. (b) Post-processed (by deep-frying) 3D-printed chicken nugget. Reproduced with permission from [32]. Copyright (2020) from Springer Nature.

properties were investigated. Shear-thinning behaviour, characterized by $n < 1$ ($n$, flow behaviour index), was observed for all three samples. A decrease in apparent viscosity with increasing shear rate was observed. However, the viscosity of the paste generally decreased with an increase in the proportion of GC in the paste (characterized by the decrease in $K$, flow consistency index). On the other hand, a higher viscosity was observed with pastes containing a higher proportion of RWF. The increase in the viscosity with the addition of RWF is due to the molecular junctions formed by the overlapping protein chains in the RWF, which restrict the stretching and arrangement of the polymer chains [34].

The value of $G'$ was higher than the value of $G''$ for all the samples tested, indicating elastic gel-like property at the linear viscoelastic region [35]. $G' > G''$ is also crucial for keeping the structure stable post-printing [36]. The value of $G'$ increased with an increase in the proportion of RWF in the paste, owing to its increased water absorption capacity, which increased the stability of the 3D-printed construct [35]. The formation of the gluten network increased the flexibility of the pastes for continuous extrusion [37]. Increasing the amount of GC decreased the loss tangent, tan $\delta$ (i.e., tan $\delta = G'' / G'$). Note that tan $\delta$ measures the nature of the viscoelastic property exhibited by the sample, with tan $\delta < 1$ for elastic behaviour and tan $\delta > 1$ for viscous behaviour [38]. The reduction in loss tangent can be explained by the increased binding of starch and protein molecules, resulting in more elastic and solid-like behaviour of the 3D-printed product. Lastly, the paste with GC:RWF 2:1 produced the highest yield stress (28.3 Pa) among all the other samples tested.

In terms of texture, GC:RWF 1:1 produced the hardest printed product (with hardness = 72.65 ± 3.58 g) due to the increased protein–protein interactions from RWF with the myofibrillar fibres from the GC [39]. The increase in the ratio of GC resulted in decreased hardness and increased adhesiveness of the printed product. The highest cohesiveness was observed for the ink with GC:RWF = 2:1. There was no significant difference in springiness and resilience among all the samples tested.

Subsequently, the 3D-printed products (GC:RWF = 2:1) were dried using hot air at 58 ± 2 °C for ~10 min. The printed product was then deep-fried at 170–180 °C for ~1 min till a golden-brown colour was observed, indicating that the nuggets were cooked. Initial results showed that hot air drying removed the water from the printed product and helped in retaining the 3D structure during deep-frying. Hot air drying before frying also reduced the intake of oil by the product [40]. This study demonstrated that meat trimmings (and potentially meat by-products) can be used to create shape-customizable chicken meat-based products via 3D food printing.

### 5.2.2.2 *Fish/surimi*

*Surimi* is a type of seafood analogue made from minced and deboned fish meat [41]. To create *surimi*, fish meats are blended with various ingredients (such as enzymes, salt, fat, and polysaccharides) to form a paste; subsequently, the paste is heat-treated [42]. Examples of *surimi* include imitation crab meat, fish tofu, and kamaboko [43–45]. Because of its high nutritional content and functional proteins, *surimi* has received much attention among food products [43, 46]. The increasing demand for *surimi* is due to its relatively low amounts of saturated fats and cholesterol, and the variety of nutrients it offers. Additionally, *surimi* is a viscous gel that can be used as an edible gel ink for the fabrication of different 3D structures [47]. The following subsections describe the recent development of *surimi* food inks for 3D food printing.

- 3D-printed *surimi* with sodium chloride

Wang *et al.* studied the effects of concentrations of sodium chloride (0, 0.5, 1.0, and 1.5 g/100 g) on the rheological and gel properties of the 3D-printed *surimi* constructs (Figure 5.3) [45]. Sodium chloride dissolves

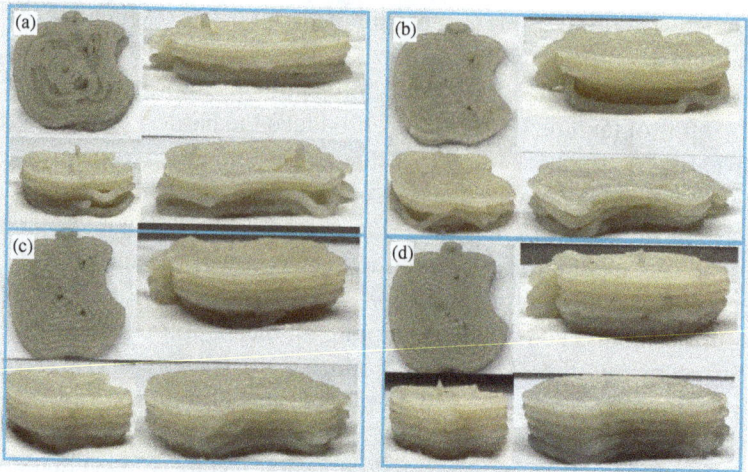

Figure 5.3.   Structure of 3D-printed *surimi* gel samples by varying the sodium chloride concentrations: (a) 0 g/100 g (control), (b) 0.5 g/100 g, (c) 1 g/100 g, and (d) 1.5 g/100 g. Reproduced with permission from [45]. Copyright (2018) from Elsevier.

the myofibrillar proteins and causes unfolding of the proteins. A previous study indicated that *surimi* gels containing different amounts of sodium chloride displayed distinct rheological properties [48]. In this study, all formulated *surimi* gels (with and without sodium chloride) were pseudo-plastic and shear-thinning. The increase in sodium chloride concentration resulted in a corresponding decrease in the viscosity of the *surimi* gels. $G' > G''$ was observed within the linear viscoelastic region for all *surimi* gels, suggesting their ability to form a stable elastic gel-like structure after extrusion. Both $G'$ and $G''$ decreased with increasing sodium chloride concentration, alluding to the denaturation of the myosin tail (light mero-myosin) [45]. In contrast, increasing the sodium chloride concentration led to an increase in *surimi* gel strength due to the swelling and enhanced interactions among the myofibrillar proteins [49]. Gel strength indicates the initial force needed to disrupt the gels. The added gel strength is advantageous as it could support its weight during 3D printing to produce stable constructs. At the same time, increasing the sodium chloride con-centration increased the water-holding capacity of the *surimi* gels. The increased water-holding capacity of *surimi* gels could be explained by the binding of the chloride ions to the myofibrillar proteins; this binding increased the affinity to water by increasing the electrostatic repulsion between the proteins [50]. Low-frequency nuclear magnetic resonance (NMR) relaxation measurements revealed a marked increase in the amount of immobilized water and the disappearance of free water with the addition of sodium chloride, which resulted in the swelling of the myofibrillar proteins. Micrographs revealed that the addition of sodium chloride (up to 1.0 g/100 g) caused cross-linking of the myofibrillar protein due to bonding between free amino acids. *Surimi* gel without sodium chloride was highly viscous and exhibited solid-like behaviours [51], which was not extrudable. The addition of sodium chloride allowed uniform extrusion of the *surimi* slurry, with the maximum sodium chloride concentration (1.5 g/100 g), giving the smoothest surface texture of the 3D-printed object.

- Coaxial extrusion 3D printing of *surimi* and potato starch

Kim *et al.* investigated the use of coaxial nozzle extrusion-based printing to produce potato-starch-coated *surimi* filaments and created fibres consisting of two food materials [42]. Fabrication of 3D-printed fibrous structures containing potato starch-coated *surimi* filaments was demonstrated in one of the three infill patterns: (1) aligned-rectilinear, (2) concentric, and (3) rectilinear (Figure 5.4). Potato starch has been conventionally combined with *surimi* to improve its gel strength, alter its texture, and enhance stability during storage [52–54]. Additionally, potato starch is a good gelling agent because it can bind to water and swell [55, 56], making it appropriate for enhancing the quality of *surimi*-based products. In this work, potato starch was dispensed as the shell ink through the outer part of the coaxial nozzle while *surimi* was dispensed as the core ink through the inner part of the coaxial nozzle.

In the study, the control samples were printed entirely with *surimi* (in both core and shell inks) without the potato starch. All the tested inks exhibited pseudoplastic and shear-thinning behaviour and possessed a gel-like viscoelastic property within the frequency range tested. The control (*surimi* only) sample showed significantly larger values of $G'$ and $G''$ than the potato-starch-containing *surimi* samples. The increase in the potato starch concentrations would enhance the mechanical strength of the 3D-printed structures. The increase in mechanical strength is due to the formation of a dense network from the absorption of water and the swelling of the potato starch granules during heating [57]. While the mechanical strength of the gel can be enhanced, increasing the potato starch

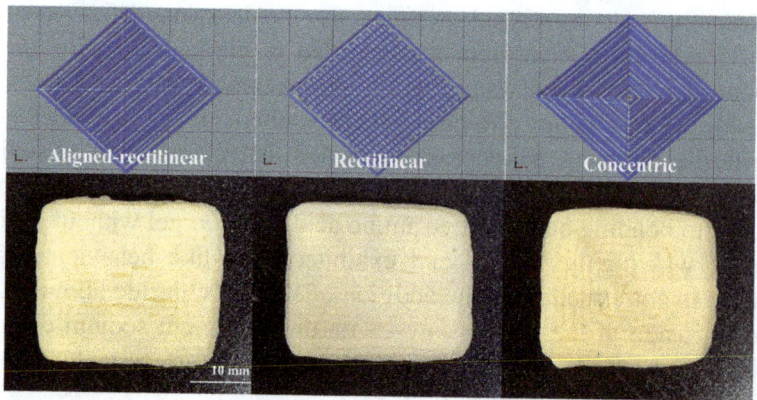

Figure 5.4.    Images showing the design and the appearance of the 3D-printed imitation crab meat with varying infill patterns. Reproduced with permission from [42]. Copyright (2021) from Elsevier.

concentration may result in chain entanglement within the paste, which interrupts the extrusion of the ink [58]. Increasing the potato starch concentration (<12% w/w) also decreased cooking loss (i.e., loss in weight due to water loss from cooking). The reduced cooking loss is due to the enhanced absorption of water by the starch during the paste preparation and its gelatinization during heating [59]. Similarly, an increase in potato starch concentration (<12% w/w) led to an increase in the water-holding capacity of the *surimi* gels, suggesting an increased capacity of the *surimi*-starch gel network to retain water [45, 60]. In total, 12% w/w potato starch with *surimi* yielded the least deformed and most stable 3D-printed sample among all samples tested.

Textural properties of the 3D-printed samples were also studied. In total, 12% w/w potato starch with *surimi* was used to print structures of aligned-rectilinear, rectilinear, and concentric infill patterns at 83% infill density. The control samples exhibited the highest values of hardness and chewiness among all samples (including commercial crab meat and 3D-printed potato-starch-containing *surimi* samples). Concentric and rectilinear infill patterns exhibited greater hardness than aligned-rectilinear infill patterns. These observations were in agreement with previous findings where interlayer bonding of the rectilinear and concentric designs was impervious to deformation because of the alignment of fibres [61]. Infill patterns did not affect the springiness, cohesiveness, chewiness, and resilience of the 3D-printed samples, though commercial crab meat was less springy and chewy than 3D-printed samples. The higher chewiness of the 3D-printed samples was attributed to the presence of potato starch between the layers of *surimi*, which acted as a filler to increase its chewiness [62].

Overall, *surimi*-based meat can be successfully created with coaxial nozzle 3D printing. The 3D-printed product has good printability and similar textural properties to commercially available *surimi*-based imitation crab meat. They showed that coaxial extrusion food printing can be used to produce textured foods through an outer coating and by varying infill patterns. Additionally, their work can also be employed in other 3D-printed foods containing filaments and fibres such as meat analogues.

- *Surimi* and microbial transglutaminase

Dong *et al.* evaluated the influence of microbial transglutaminase concentrations (0.1–0.4%) on the rheological properties, printability, and texture of *surimi* gels and their 3D-printed products (Figure 5.5) [47]. Transglutaminase is an acyl transferase enzyme that facilitates the formation of covalent bonds between the ε-amino groups on the lysine residues and the γ-carboxyamide groups of the glutamine on the myosin heavy chain [63]. Exogenous transglutaminase, such as microbial transglutaminase, is commonly added to *surimi* to improve its textural properties [64].

A decrease in the viscosity of the *surimi* gel with an increase in shear rate was observed, which suggested that the *surimi* gel is a pseudoplastic fluid with shear-thinning properties [45]. The *surimi* gel containing microbial transglutaminase displayed low viscosity; an increase in the microbial transglutaminase concentration increased their viscosities due to an increased degree of cross-linking [65]. Both $G'$ and $G''$ of all the samples (i.e., *surimi* gels with or without microbial transglutaminase) increased with an increase in the angular frequency while maintaining $G' > G''$ throughout their entire linear viscoelastic region. The loss tangent (tan δ) of all the samples was less than 1 for the range of angular frequencies investigated. Note that tan δ gradually decreased as the microbial transglutaminase concentration increased, suggesting that microbial transglutaminase supports solid-like behaviour in the *surimi* gels [52]. The *surimi* gels of low viscosity (with a low concentration of microbial transglutaminase) exhibited poor printing accuracy. However, *surimi* gels of high viscosity (containing 0.4% w/w microbial transglutaminase) resulted

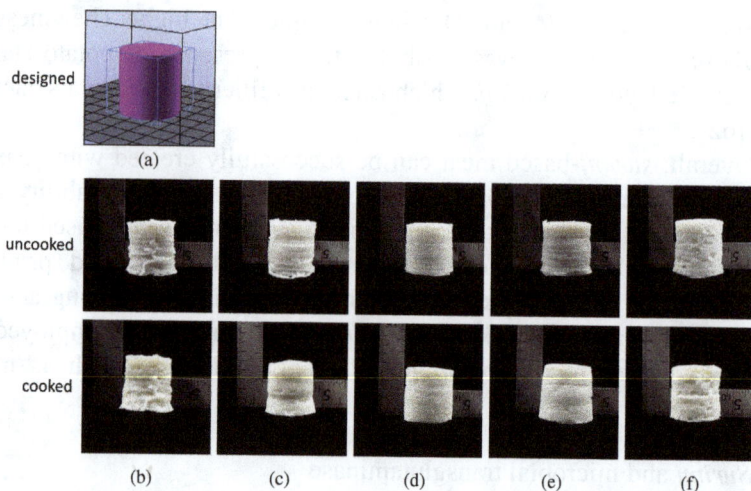

Figure 5.5.   Optical images of 3D-printed *surimi* gels in uncooked and cooked states with different concentrations of added microbial transglutaminase: (a) targeted printing geometry, (b) 0% microbial transglutaminase, (c) 0.1% microbial transglutaminase, (d) 0.2% microbial transglutaminase, (e) 0.3% microbial transglutaminase, and (f) 0.4% microbial transglutaminase. Here, 0–0.4% refers to the grams of microbial transglutaminase enzyme added per 100 g of *surimi*. Reproduced with permission from [47]. Copyright (2020) from Elsevier.

in uneven extrusion. Overall, *surimi* containing 0.2% and 0.3% w/w microbial transglutaminase gave the best 3D-printed shapes, possessing a smooth exterior and showing good printing accuracy matching with the intended computer-aided design model.

From the puncture test, it was found that the gel strength of the 3D-printed and cooked *surimi* increased with increasing microbial transglutaminase concentration up to 0.3% w/w microbial transglutaminase. Increased gel strength enhances the *surimi* structure to support its weight after printing. Increasing the microbial transglutaminase concentration resulted in a gradual increase in the hardness, cohesiveness, chewiness, and resilience of the resulting 3D-printed and cooked *surimi* gels, though these textural properties were not statistically significant for samples containing more than 0.1% w/w microbial transglutaminase. Improvements to the *surimi* textures with the addition of microbial transglutaminase were due to heat-induced formation of disulphide bonds and ε-(γ-glutamyl)lysine bonds in and between protein molecules during cooking, which increased the mechanical strength [66]. On the other hand, there was no significant difference in the springiness of 3D-printed samples regardless of the concentrations of microbial transglutaminase.

Cooking loss is the percentage change in the weight of the food before and after cooking, measured for estimating the loss of protein, water, and other substances during the heating of meat [67]. The cooking loss increased in *surimi* samples with increasing concentrations of microbial transglutaminase. The increase in bonding between proteins facilitated by microbial transglutaminase decreased the interaction between proteins and water, leading to a highly firm and brittle gel that loses water easily [63]. An increase in the proportions of free and unbound water in the *surimi* gels with increasing microbial transglutaminase concentration was observed with low-field NMR. Overall, this work studied the effect of varying microbial transglutaminase concentrations on the 3D printability and physical properties of *S. niphonius surimi*, such as texture and gel strength. The findings from this work expand the knowledge and understanding for producing 3D printable *surimi*.

### 5.2.2.3 *Beef*

- 3D-printed composite beef and lard

Dick *et al.* used dual extrusion-based 3D printing to fabricate composite foods made from lean beef paste with varying infill densities (50%, 75%, and 100%) and numbers of layers (1, 2, and 3 layers) and tested the feasibility of post-processing (i.e., cooking) (Figure 5.6) [68]. Cooking is an example of a post-processing method that causes chemical and physical changes in the food. These changes need to be considered to obtain the desired structure and texture of the 3D-printed food product. Infill density is an essential parameter for 3D food printing because it influences the stability and mechanical strength of the printed structure. Changing the amount of substance deposited within its internal structure affects the overall texture of the printed food. The fat (lard) amount affects the texture of the printed food due to fat displacement after heating; this change may allow the customization of the fat amount and the distribution of 3D-printed meat products. Other relevant variables include moisture and fat retention, cooking loss, shrinkage, and texture of the products, which were studied.

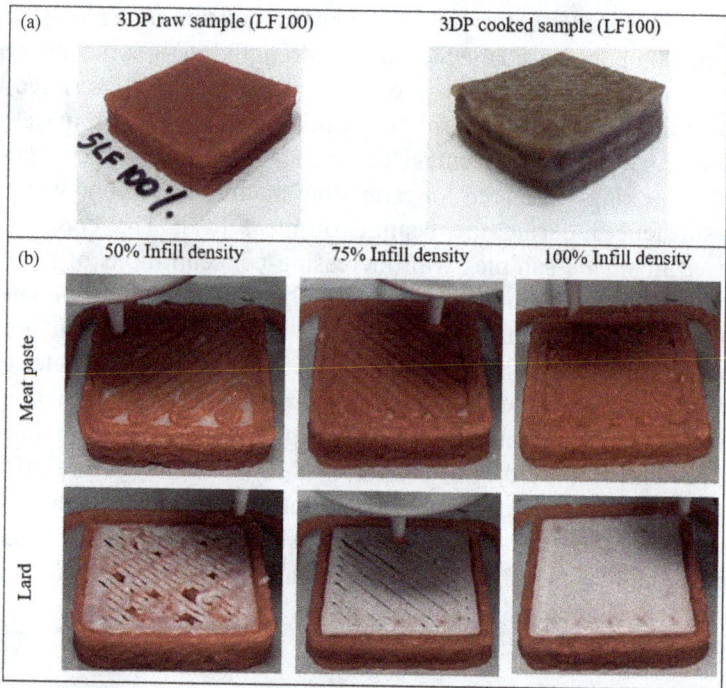

Figure 5.6.   3D-printed multi-layered beef with lard composite models: (a) 3D-printed raw and cooked samples containing one layer of lard and 100% beef infill density, (b) filament stream printed at varying infill densities (i.e., 50%, 75%, and 100%) for meat paste dispensed from a 2 mm diameter nozzle and lard dispensed from a 1 mm diameter nozzle. Reproduced with permission from [68]. Copyright (2019) from Elsevier.

The printing temperature of the raw meat paste (before cooking) was set below 27.9 °C because it exhibits greater solid-like behaviour and can hold its printed shape. $G'$ of the meat paste decreased as the temperature was increased due to the melting of the crystalline fat within the meat paste [69]. Below 27.9 °C, the $G'$ of the meat paste remained higher than $G''$, indicating that the meat paste displayed more elastic and solid-like behaviour than viscous and fluid-like behaviour. Above 27.9 °C, however, the meat paste displayed fluid-like behaviour, as characterized by $G'' > G'$. A controlled ambient temperature (23 ± 1 °C) was therefore chosen to print the meat paste for maintaining the printed structure.

The 3D-printed samples of multi-layered composite beef and lard samples were vacuum-packed and cooked by submerging the samples into an uncirculated water bath controlled at 75 °C for 30 min. Shrinkage and cooking loss are observed in post-processed meat products because of heat denaturation and contraction of the meat proteins [70]. The loss of water and fats are the primary causes of the cooking losses in meat products. Heat generates a contraction pressure within the meat network, which expulses the bulk water, normally held by capillary forces within the meat, resulting in water release and shrinkage of proteins. The effects of fats on cooking loss and shrinkage of the 3D-printed samples were also investigated. Generally, the higher the fat content, the larger the cooking loss and shrinkage. During cooking, the printed fats within the structure melted and displaced from the inner to the outer parts of the structure through the pores, causing the structure to shrink [68].

The puncture test measured the hardness of the 3D-printed meat, which revealed that increasing layers of fat caused a reduction in hardness. The change in hardness was attributed to the increased degree of shrinkage and cooking loss in the samples containing two and three layers of lard. Hardness was greater at the infill density of 75% than at 50% due to higher cooking loss. The cooking loss was mainly due to the loss of moisture, which affected the mechanical stability of the food. The puncture test also allowed the hardness of the crust in grilled meats to be studied [71]. The hardness, chewiness, and cohesiveness of the cooked samples were evaluated with a texture profile analyser (TPA). The hardness and chewiness of the samples increased with the infill density. On the other hand, the hardness and chewiness of the samples decreased with the increase in the number of lard layers. A higher amount of fat reduced the bulk density of the meat since fat is less dense than lean tissue, which confers the ease of bite to the meat [68]. Cohesiveness was higher for the samples with 50% infill density than for samples with 75% and 100% infill densities, while the samples containing three layers of lard showed the highest cohesiveness among all the samples containing lard.

- 3D-printed beef for dysphagic individuals

Dick *et al.* investigated the effects of hydrocolloids on the printability and textural properties of cooked beef pastes (Figure 5.7) [72]. XG, GG, κ-carrageenan (KC), and locust bean (LB) gum were added to the meat. The samples were represented in the format XX/YY for the samples with two different hydrocolloids, and XX for the samples with one

Figure 5.7.    Images of 3D-printed, and 3D-printed and heated (at 120 °C for 15 min in a conventional oven) beef paste containing different percentages and types of hydrocolloid blends. (XG = xanthan gum, GG = guar gum, LB = locust bean (gum), and KC = κ-carrageenan) [72]. Reproduced from [72], which is licensed under CC BY 4.0.

hydrocolloid, followed by (05) and (1) showing their total final concentrations of 0.5% and 1%, respectively [72]. Phase separation during printing was observed in pastes containing no hydrocolloids and only KC; phase separation is attributed to inadequate binding of the food materials to water. When extruding such inks, only water was released, and the aggregation of the meat clogged the nozzle. In contrast, inks with XG or GG (0.5% or above) were successfully 3D-printed.

All formulations (with or without hydrocolloids) had a flow behaviour index ($n$) of less than 0.4, indicating shear-thinning behaviour. A small flow behaviour index favours extrusion by lowering the required extrusion force as the viscosity decreases [73]. Inks containing only KC exhibited the highest apparent viscosity among all formulations, which is alluded to the swelling of KC particles due to the strong ionic and hydrogen bonding with water [74]. The paste containing no hydrocolloids displayed the lowest viscosity among all formulations, while adding hydrocolloids increased the viscosity of the paste.

For all the formulations with or without hydrocolloids, $G' > G''$ was maintained over the frequency sweep tested with a phase angle of less than 45°. This measurement indicated the inks were viscoelastic and predominantly solid-like. Similar to the trend observed for the viscosity, the ink containing only KC showed the highest value of $G'$, while the paste without any hydrocolloids showed the lowest value of $G'$.

The 3D-printed beef samples containing GG/LB(05) and LB(1) exhibited a large deviation from the intended model. These samples showed an increase in tan δ for the tested frequency range, suggesting an increase in viscous characteristics and a decrease in the ability to retain printed shapes [75, 76]. Consequently, the samples showed signs of spreading during 3D printing. Samples containing XG/GG(05) and GG/LB(1) showed a slight dimensional deviation from the intended model.

Textural profile analysis was done on the 3D-printed samples after heating to 120 °C and cooling to 36 ± 1 °C. Slight dimensional changes were observed in the sample with XG after heating due to reduced viscosity and increased flowability. The shape alteration was reduced with short heating times; other heating techniques such as steaming or microwaving may be suitable to maintain the printed shape [72]. Superficial fissures of varying extents were observed in the post-heated 3D-printed samples, resulting from the gel network produced after cooling, which caused a fracture in the superficial layer.

Samples containing KC exhibited greater hardness than others, while samples containing both KC and LB exhibited the highest hardness values. KC and LB behave like water binders during heating; KC produces brittle and hard gels, and LB behaves like a thickener. The blends containing KC and LB offered synergistic effects where the gel was strengthened from brittle (with KC alone) to elastic (with KC and LB) [77]. The highest value for gumminess was also observed with samples containing both KC and LB. However, hardness and gumminess were significantly reduced when KC was paired with other gums such as XG and GG. No significant differences were observed across the different samples for adhesiveness. Samples containing XG alone or when XG was mixed with GG showed lower cohesiveness values, with no significant differences among other samples. It was proposed that XG may impede the formation of a gel network in myofibrillar systems owing to its large molecular size and great hydration properties, leading to weaker gel constructs [78]. This observation is aligned with their findings of lower hardness, gumminess, and cohesiveness.

Two temperature ramps (4–75 °C and 75–23 °C) were performed to understand the textural changes that happened during heating and cooling.

For the sample without hydrocolloids, a decrease in $G'$ was generally observed between 4 °C and 30 °C due to the thawing and melting of fat in the meat paste happening between 20 °C and 34 °C [79]. The values of $G'$ increased with increased heating, exhibiting solid-like properties at 75 °C.

The increase in $G'$ is caused by the continued protein-protein interactions and cross-linking [80], and the shrinking of connective tissue and meat fibres [72]. Upon heating and cooling, the phase angle at 23 °C became greater than its original value, indicating a more rigid structure after heating. The increase in rigidity is caused by the release of water from its denatured protein and by the change in their protein-protein hydrogen interaction, increasing the stability and elastic behaviour of the actin-myosin matrix in the gel [81]. Variations in $G'$ were lesser for samples containing hydrocolloids than those samples without hydrocolloids, suggesting the heat stability conferred by the hydrocolloids. Among the samples containing hydrocolloids, greater visible changes were observed during cooling for KC-containing samples than others.

All textural properties of the 3D-printed beef were comparable to the puréed meats for dysphagic patients [82]. However, based on the International Dysphagia Diet Standardization Initiative (IDDSI) framework, none of the 3D-printed samples met the conditions for puréed foods (level 4). In the IDDSI framework, foods can be classified as liquidized (level 3), puréed (level 4), minced and moist (level 5), soft and bite-sized (level 6), and easy to chew/regular (level 7) [72]. Samples containing XG(1) and XG/GG(1) did not qualify for any of the levels. Samples containing XG/GG(05), LB(1), and GG/LB(1) were classified under IDDSI level 5; samples containing GG(1), GG/KC(1), XG/KC(1), GG/LB(05), and KC/LB(1) were classified under IDDSI levels 6 and 7.

• Post-processing of 3D-printed beef products

Post-processing alters the textural properties of the foods and their suitability for dysphasic patients. Dick *et al.* also investigated the impact of reheating the beef pastes on the texture and physical properties of 3D-printed beef classified under level 4 (puréed) of the IDDSI framework [83]. Pastes containing XG and GG were 3D-printed and reheated using a microwave, convection oven, and steam convection oven to determine the methods suitable for post-processing of the printed beef product (Figure 5.8).

A microwave oven uses electromagnetic waves (radiation) to heat food. High internal rates of heating and uneven heat dissemination

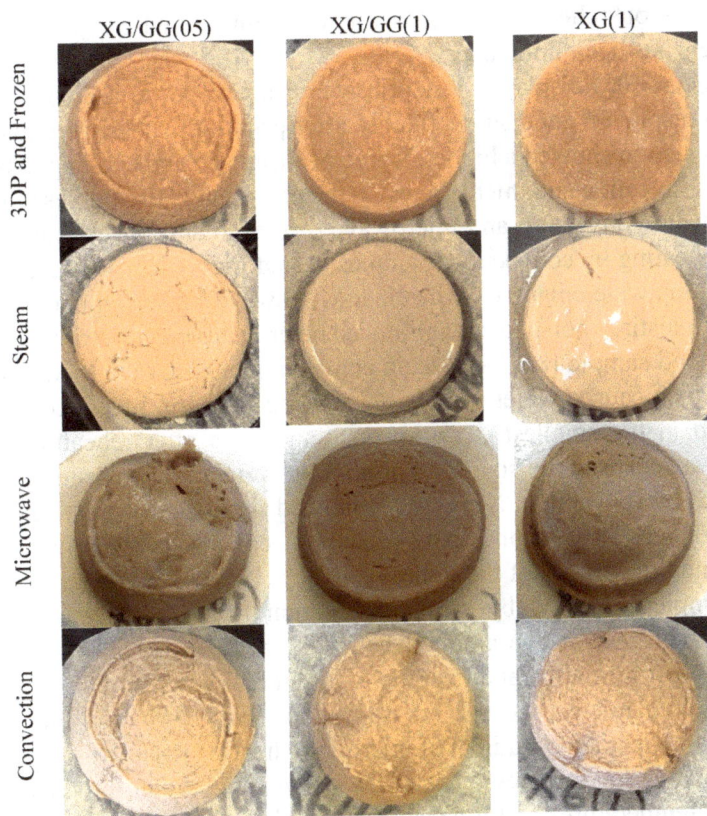

Figure 5.8. Images showing 3D-printed cooked beef pastes in their frozen conditions after steaming, microwave heating, and convection heating. Formulations XG/GG(05) and XG/GG(1) have a 1:1 blend of XG and GG at concentrations of 0.5 g/100 g and 1 g/100 g, respectively. XG(1) = 1 g/100 g of added XG. The photos are not to scale, but have been shown here for the visualization of the overall shape variation. Reproduced with permission from [83]. Copyright (2021) from Elsevier.

prevent crust formation and produce soggy surfaces [84]. On the other hand, a convection oven uses circulating hot air to reduce cooking times [85]. A steam convection oven uses both steam and circulating hot air to cook food and reduce cooking times while regulating the relative humidity and browning. For the cooking of meats, controlled humidity prevents dehydration and produces juicier meat products [86]. Texture changes in meat products have been studied with these methods [87], but the number of studies on reheating methods for texture-modified 3D-printed meat products has been limited [83].

XG/GG(05) displayed low printing accuracy, while XG(1) and XG/GG(1) displayed good shape fidelity after printing. All samples reheated by convection showed a higher degree of shrinkage and cooking loss than steam convection and microwave. On the other hand, all samples reheated with steam convection and microwave displayed smaller cooking losses than reheating by convection. Samples heated with microwave and steam convection were enlarged by heating. Microwave caused an uneven heating with distinct hotspots that produced heterogeneously distributed pores in the food sample [88, 89]. Steam convection exposed the food to moisture, which caused water uptake and volume increase. Steam convection resulted in the smallest deviation in dimensions after reheating, while microwave caused the smallest variation in weight after reheating. When reheated by convection, XG/GG(05) showed significantly greater cooking losses than XG/GG(1) and XG(1) due to the collapse of the printed structures and the outflow of the liquids.

Samples reheated by steam convection became softer than those treated with microwave and convection. Reheating under moisture prevented crust formation and kept the food samples soft. Similarly, samples treated with microwave did not show any crust formation due to the inverted temperature gradient of microwave heating [84]. Crust formation was obvious for samples reheated by convection, which produced high surface temperatures and a negative internal gradient of the temperature [84]. In such cases, the formation of a dry crust increased the hardness of the samples. The formation of a dry crust occurred for all samples reheated by convection, and these samples were excluded from the IDDSI classification. All samples heated by steam convection were classified as IDDSI level 4, while the samples heated by microwave and containing XG(1) or XG/GG(05) were classified as IDDSI level 4.

Overall, the textural properties of 3D-printed and cooked beef samples were dependent on (1) the methods of heating, and (2) the composition of additives. This work also demonstrated that heating by convection resulted in crust formation, which was not a desirable post-processing method to prepare foods for dysphagic patients. Microwave heating offered convenience and yielded a potentially compliant texture, but its internal steam pressure caused structural expansion in the 3D-printed beef samples. Steaming was the most appropriate reheating method, which retained the shape and texture of the 3D-printed beef products.

## 5.2.2.4 *Pork*

- 3D-printed pork for dysphagic patients

Pork is another type of meat that was demonstrated for 3D printing. Dick *et al.* investigated the effect of hydrocolloids on the rheological, textural, and microstructural characteristics of 3D-printed cooked pork paste [24]. The mixtures of XG and GG (at the ratio of XG/GG = 0.3/0.7, 0.5/0.5, and 0.7/0.3 with the total percentage of 0.36%) were used to formulate the pork pastes for 3D printing.

All pork pastes exhibited shear-thinning property; the increasing shear rate allowed extrusion of the pork pastes. The control paste without hydrocolloids was much less viscous than those containing hydrocolloids. Galactose in GG interacted with water molecules, contributing to increasing the viscosity [90]. XG permitted intermolecular interactions to form a frail network of entangled rod-like molecules, blocking hydrolysis that would otherwise reduce the viscosity [24]. As such, both hydrocolloids contributed to increasing the overall viscosity of the pastes. $G' > G''$ was observed for all the pork pastes tested, indicating that the pastes were solid-like and able to self-support the printed structures. On the other hand, yield stress did not vary with the concentrations and the proportion of hydrocolloids (i.e., XG/GG).

Texture profile analysis was conducted on 3D-printed pork samples heated (post-processed) at $36 \pm 1$ °C; the temperature simulated the physiological condition inside a mouth. Samples without hydrocolloids presented the largest hardness value, followed by the sample containing only GG (0.36%); no significant differences in hardness were observed for other formulations. The sample without hydrocolloids also displayed the largest cohesiveness and chewiness; the value of chewiness was 3–8 times larger than for samples containing hydrocolloids. No significant differences were observed in adhesiveness for all the paste formulations tested in this study.

In comparison with a previous study [82], 3D-printed pork products had similar values of cohesiveness, chewiness, and adhesiveness. The previous study conducted the texture analysis at the serving temperature (65 °C), while the samples in this study were tested at the physiological temperature ($36 \pm 1$ °C). A pronounced decrease in the loss tangent (tan δ) was observed for the pork paste without hydrocolloids when the temperature was ramped from 23 °C to 75 °C. The decrease in tan δ is indicative of an increasing elastic behaviour due to increased molecular interactions and cross-linking [80]. The shrinkage of the connective tissue from water loss increased its overall density and toughness [91]. However, tan δ of the sample without hydrocolloids did not recover back to its value at 23 °C after cooling. Samples containing both XG and GG did not show a major reduction in tan δ during either heating (23–75 °C) or cooling (75–35 °C), which confirmed that the viscous property of the samples was retained during post-processing. Cryo-SEM revealed the presence of a network of heterogeneous cavities in samples containing hydrocolloids, indicating greater water retention and more viscous-like behaviour than in the control samples without hydrocolloids after post-processing.

The tested pork pastes possessed textural values close to those of texture-modified meat served to dysphagia patients [82]. The IDDSI test was performed on the 3D-printed pork product printed with the paste containing XG/GG = 0.3/0.7 (Figure 5.9). The 3D-printed pork samples complied with level 6 of the IDDSI framework.

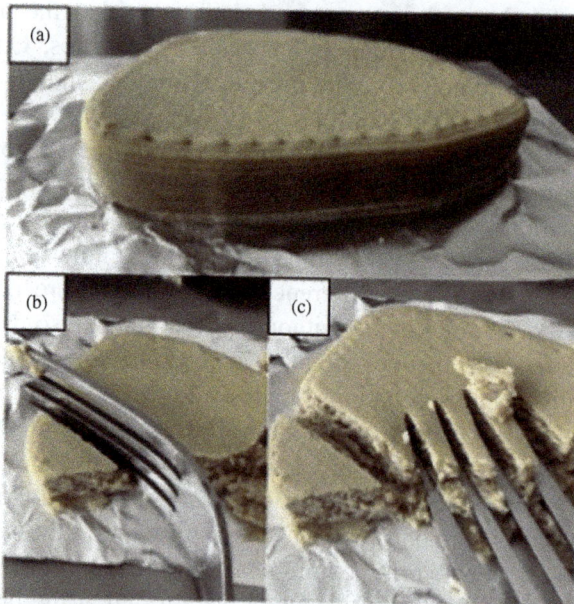

Figure 5.9.  3D-printed pork fillet and demonstration of the IDDSI fork test. (a) 3D-printed pork fillet, (b) IDDSI fork pressure test to cut the 3D-printed pork fillet, and (c) IDDSI fork pressure test to break the printed sample. Reproduced with permission from [24]. Copyright (2020) from Elsevier.

## 5.3 3D Printing of Cultured Meat

### 5.3.1 *Introduction*

Cultured meat has been regarded as a source of alternative protein. The use of healthy animal muscle cells for food production neither compromises its nutritional content nor does it involve the slaughter of animals [92]. Over the years, cultured meat has been called *clean meat* or *cell-based meat*, etc. [93].

A recent survey revealed that 37% of the respondents were willing to pay the same price of regular meat for cultured meat, while 58% of the respondents would consider paying a premium for cultured meat [94]. Another report showed that labels and descriptions of cultured meats affect the level of acceptance among consumers [95]. A better understanding of the technology has boosted the acceptability of cultured meat. The personal benefits of cultured meat have been accepted over quality and flavour [11].

The complexity of the constituents in the skeletal muscle (meat) defines the quality of the entire meat [96]. Adipocytes (fat cells found in the meat) contribute to the flavour and taste of the meats during cooking. The stroma, which is represented by the connective tissue, influences the meat texture. The meat texture, in turn, is one of the most essential considerations for consumers. The quality of the meat is determined by the biochemical processes occurring before and after slaughter, depending on the type of muscle.

Myogenesis (a process of muscle development) is crucial for meat quality and meat-derived foods [97]. Myogenesis begins during the embryonic stage, continues during foetal life, and is nearly completed at birth [98]. Additional episodes of myogenesis happen during muscle regeneration following injury or because of workload adaptation [99]. Hormones, nervous system, growth factor, and cytokine activity regulate muscle growth and development. There is constant crosstalk between different types of cells (including muscle cells, adipocytes, endothelial cells, fibroblasts, and leukocytes), rendering muscle a highly interconnected organ [100, 101]. The outcomes of these interactions can be trophic (stimulating muscle growth) or atrophic (inhibiting muscle growth).

Muscle cells can be grown with relative ease on plastics as a routine *in vitro* cell culture. This method has been used for decades in biology, and it is a well-established model to study the molecular mechanisms of muscle development and decay [96]. This approach has been recently viewed as an appealing alternative to animal meat production. However, a single layer of cultured muscle cells differs significantly from a muscle tissue, which is a highly complex organ. The skeletal muscle contains adiposes, connective tissues, and a vascular bed [96]. Each component contributes differently to the flavour, aroma, and nutritional value of the meat.

3D printing of cultured meat is more advanced than culturing of meat alone because it prints muscle cells, extracellular matrix (ECM) support-ive cells, and fats within the scaffold that promotes cell growth [11]. Bioink, which consists of cells and biomaterials, is an essential part of the printing process because it creates scaffold structures where muscle fibres are produced. The muscle fibres eventually become meat [102]. Following the printing, the meat will be matured further in bioreactors that supply them with nutrients [103]. Cultured meat (without 3D print-ing), on the other hand, is created by actively dividing muscle cells attached to a scaffold or carrier before transporting them into a suitable bioreactor with growth media [10]. However, simple cell culture does not produce highly structured meats such as steaks; the scope of this technol-ogy remains limited at present [104]. As such, 3D printing would be a desirable strategy to achieve complex and structured meat as well as large-scale manufacturing of cultured meat [105].

3D printing of cultured meat has the potential to help improve human health and the environment by reducing the need for livestock farming and removing the requirement for traditional meat manufacturing [9, 106]. Although evidence for sustainability improvements is still scarce [92], 3D printing of cultured meat adds versatility to meat products by allowing them to contain designed nutritional values, shapes, and textures [9, 107]. 3D-printed cultured meat can be utilized to enhance food intake in spe-cific groups of people with medical conditions [108]. For example, 3D printing can provide textured meat to the elderly or patients who have difficulties in swallowing and chewing during their food consumption [109]. There has been a trend of children and adolescents not taking suf-ficient vitamins and minerals. The cultured meat can be created to include healthy content [9, 109]. Furthermore, people with allergies might benefit from 3D printing of cultured meat if meat could be customized to be free of specific allergens [11]. On a global scale, 3D-printed cultured meat may contribute to addressing worldwide issues such as food shortage and famine by using the available food resources effectively [109].

## 5.3.2 *3D Printing and Scaffold*

### 5.3.2.1 *Considerations for 3D printing of cultured meat*

- Introduction to 3D bioprinting for cultured meat

Bioprinting is a process to 3D-print cell-laden structures, and it is widely applicable for 3D meat printing. In bioprinting, 3D structures are built using bioinks, comprising cells, biomaterials, and other molecules [110]. The number of cells present within the bioinks usually ranges from 10,000 to 30,000 per droplet (corresponding to a volume of 10–20 µL). Polymer crosslinkers such as calcium chloride, thrombin, sodium chloride, gelatin, and fibrinogen are found in the cell medium, which can be thermally, photochemically, or enzymatically activated [111]. Biomaterials, such as decellularized extracellular matrix (dECM), melt-cure polymers, and hydrogels, are used to create a conducive microenvironment for cell adhesion, migration, and differentiation [110].

Three primary methods employed for bioprinting include extrusion printing, inkjet printing, and light-assisted printing [112]. Similarly to food printing, extrusion-based printing is most commonly used for bioprinting. Bioink is deposited by layer-by-layer extrusion via a nozzle. Direct printing, indirect printing, and hybrid printing are the three main manufacturing concepts to create spatial patterns of living cells [113]. Inks containing cells (such as cell-laden hydrogels) are used in direct printing. Alternatively, the printing of cell-laden hydrogel layers onto a cell-free sacrificial component or mould is referred to as indirect printing, typically giving rise to the vascular construct after removal of the sacrificial structures. Hybrid printing uses a combination of direct and indirect printing methods [113, 114]. While various tissue constructs — muscle, bone, tendon, cartilage, and skin — have been 3D-printed in the field of tissue engineering, only skeletal muscles are considered to produce meat.

Skeletal muscles are primarily made of muscle fibres, connective tissues, and intramuscular adipose tissues [115]. Recently, there has been interest in the development of human muscle tissues that mimic the native human muscle tissues [116]. In this regard, 3D printing is used to create muscles that are structurally and functionally similar to native muscle tissues [116]. In one work, skeletal muscle tissues have been 3D-printed with three major constituents: (a) human muscle progenitor cell (hMPC) hydrogel bioink, (b) sacrificial acellular gelatin hydrogel bioink, and (c) poly (ε-caprolactone) (PCL) [116]. 3D printing of vascularized tissues [114] and multi-material bioprinting [117] have been demonstrated to create vascularized human-scale tissues. Although these methods have been employed to print human muscles, they can also be used to create animal muscles. Farm animals and humans are similar in scale, and their adipose tissue and skeletal muscle development are also similar to that of humans [115].

- Cell types used for 3D printing of cultured meat

Cells with self-renewing capabilities are the best starting materials for the 3D bioprinting of cultured meat. Examples of these cells include endothelial cells, fibroblasts, myofibres, chondrocytes, and adipocytes [118]. Stem cells, either embryonic or adult, are considered the best sources that meet the technical requirements for cultured meat production. Embryonic stem (ES) cells can proliferate and differentiate into all the cell types needed for cultured meat production [119]. However, developing ES cell lines from farmed animals is a tedious task due to the high risk of cell contamination with non-ES cells. Low efficiency of replated ES cells is another obstacle, as undifferentiated pluripotent stem cells can cause teratomas [120]. Alternatively, adult stem cells such as adipose-derived stem cells (ADSCs) and satellite stem cells are potential sources for developing cultured meat [121].

Satellite stem cells (or myosatellite cells) are mature muscle tissue stem cells that are activated to repair, regenerate, and recover damaged tissues within the mammalian body. These satellite cells isolated from the biopsy are purified using selected cell surface markers, which transform myoblasts into myocytes, then myotubes, and finally myofibres [122]. However, satellite stem cells were found to be incapable of transforming into adipocytes. Similarly, ADSCs derived from bone marrow or adipose tissues possess the capability to differentiate into adipogenic, chondrogenic, myogenic, or osteogenic cell lines. Furthermore, because ADSCs are isolated using minimally invasive liposuction methods, they are suitable for cultured meat production. Liposuction and resection are also shown to produce large numbers of viable cells [123].

- Cell growth media

Growth media are nutrient-containing liquids that have been designed to fulfil the physicochemical cues for cultivating cells [11]. Growth media are essential for cell proliferation and differentiation, which eventually lead to tissue regeneration and maturation. Growth factors and serum are crucial ingredients in the growth medium to achieve tissue maturation. Culturing myosatellite cells requires the use of foetal bovine serum (FBS), which is extracted from either the adult, new-born, or foetus of an animal [124]. However, large-scale use of such a medium is prohibitively expensive and unsustainable. Furthermore, due to its *in vivo* origins, it possesses varied compositions and may potentially carry pathogens [125]. Furthermore, acquiring FBS raises ethical considerations, and its acceptance among vegan consumers is dubious. Crucially, the usage of animal-derived supplementation for cell growth undercuts many anticipated benefits of cultured meat due to its underlying environmental and ethical implications. The identification of culture media without animal-origin constituents has the potential to address all of the aforementioned issues, including cost, contamination, and ethics [126].

Currently, research efforts are being made to identify alternatives to animal-based serum and antibiotics for the production of cultured meat [127, 128]. Examples include the use of synthetic substitutes (i.e., a blend of substances that mimic animal-based serum activity) and natural plant products possessing the same properties as the serum [96]. In earlier work, McFarland *et al.* developed a serum-free media to culture myosatellite cells obtained from turkey *in vitro* [129]. Following that, a plant-based growth medium based on the *maitake* mushroom was developed and found to be comparable to animal serum. This mushroom extract was found to be effective in promoting the expansion of the surface area of fish explants [130]. While the mushroom extracts are applicable for cell culture, potentially allergenic plant-based proteins may pose another problem [13].

Different satellite cells require varying compositions of growth medium due to species-specific responses [131]. For instance, avian cells were grown and differentiated in high glucose Dulbecco's Modified Eagle Media (DMEM-Glutamax I) supplemented with 10% horse serum and 4% chick embryo extract, whereas equine and bovine cells were cultured in DMEM-Glutamax I supplemented with 20% foetal calf serum [132]. Similarly, serum-free growth media for myosatellite from various species, such as sheep [133] and pig [134], were developed.

To date, the only commercially available serum-free growth medium is Ultroser G, an FBS replacer that comprises all the nutrients required for eukaryotic cell growth [135]. However, just like FBS, the usage of Ultroser G has been hindered by its cost [136]. Additionally, the exact formulation of the serum-free media is copyright protected. Alternatively,

Cyanobacteria, a rapidly growing photosynthetic bacteria, have been proposed as a possible source of cell growth supplements for cultured meat. The protein content in Cyanobacteria is up to 70% of their dry matter and these are easily cultured for biomass [137]. Nonetheless, a thorough understanding of metabolic simulation and flux balance analysis to establish the metabolic and functional state of the cell is required for comprehending the best nutritional requirements for *in vitro* cell culture [138].

## 5.3.2.2 *Scaffolds*

Cells used for the development of cultured meat (such as myosatellite) are adherent. Scaffolds provide a framework for cells to attach, proliferate, and mature (often by imitating native three-dimensional tissue). Structured meats contain a distribution of muscle, fat and vasculature located in fixed compartments. These meats — such as steak and meat cuts — can be realized with appropriate scaffolding. These scaffolds must be: (1) biologically active, (2) flexible, (3) porous to allow diffusion, (4) edible, (5) non-toxic/non-allergic, and (6) removable/degradable [13]. Smart and/or dynamic scaffolds have been widely explored in research. For example, mechanical stretching in myoblasts can be facilitated by porous biomaterials (such as alginate, cellulose, chitosan, or collagen) which undergo surface extension and contraction with respect to changes in pH and temperature [139]. Additionally, electrically-stimulated differentiation of muscle cells to create long myotubes has been reported using conductive fibres to culture myoblasts [140].

The fibrous nature of the scaffold is also essential to improve the organoleptic profile of printed meat products. Gelatin and soy protein have been shown to meet the fibrous requirements of meat [141, 142]. For instance, Abbasi *et al.* developed a novel scaffold from textured soy protein, obtained as an easily available by-product of soy (Glycine max). Such plant-based scaffolds may offer affordable routes to realize cultured meats [143]. To remove confluent cultured muscle cell sheets, two methods have been traditionally used — enzymatic and mechanical — but both are destructive to the cultured muscle cells [144]. To address this limitation, a non-destructive detachment method has been developed using a thermo-responsive coating to release cultured cells with their extracellular matrix as an intact sheet [145]. Alternatively, the biodegradation of the laminin (a selective attachment protein) was used to detach the cultured muscle cells as a confluent sheet from a non-adhesive micropatterned surface [146].

The use of conventional 3D cell culture in conjunction with scaffolds does not allow for customization and regulation of cell positioning, resulting in poor tissue maturation. Poor tissue maturation would result in undesired nutritional and organoleptic properties in the cultured meat. On the other hand, 3D bioprinting should permit precise control over cell positioning, cell-cell ratio, and densities of cell distribution. A biomimetic composition of edible materials, capable of supporting cell viability and proliferation, is required to create 3D-bioprinted cultured meat. The composition of edible materials is identified to formulate meat inks. The rheological characteristics of meat inks and the post-printing and post-processing (cooking) stability of the 3D-printed scaffold are essential considerations for successful 3D printing of cultured meats.

### 5.3.3 *Ink Formulation, Properties, and Characterization*

A bioink is a printable medium containing living cells and growth factors, which is widely used for bioprinting and food printing. To print meat analogues, bioink matrices and scaffolds must be edible [141]. Natural polymers such as agarose, alginate, chitosan, and gelatin are used for printing cultured meat [147]. Other natural polymers, such as collagen and fibrinogen, have also been used as scaffolds to enhance the physiological resemblance to skeletal muscles *in vivo* [148]. However, structures printed with collagen tended to deform due to dissolution or swelling, and other biodegradable materials were required to enhance mechanical properties [111]. Although the use of hydrogels enhances the stability of the printed structures, they may compromise cell viability [113]. Overall, bioinks for 3D meat printing must be balanced carefully to meet the requirements for rheological and biological properties.

#### 5.3.3.1 *Contractile 3D bovine muscle tissue for cultured steak*

In this section, we discuss a recently reported method for culturing 3D bovine muscle tissue to produce a cultured steak. Earlier studies have tried culturing myocytes-containing hydrogel around an agarose column to facilitate the formation of ring-shaped muscle tissue [149] and demonstrated the formation of fibre-shaped tissue by bridging myocyte-containing hydrogels between anchors [150]. However, these methods have produced only small-sized muscle tissues not suitable to produce large-scale cultured meat. The development of realistic cultured steak, consisting of adipose cells and aligned muscle cells, remains challenging

[138, 142, 151]. Large muscle tissues with tightly packed, unidirectionally aligned and matured myotubes are needed to produce cultured steak meats with realistic texture [152]. To produce large bovine muscle tissue, block-shaped tissue formation from myocyte culture in porous scaffolds has been demonstrated [141, 142]. These methods, however, cause myotubes to be distributed in an isotropic manner, which resulted in the loss of contractility. Due to this limitation, these methods cannot be directly used for culturing bovine muscle tissues for steak meat [153]. In light of these limitations, Furuhashi *et al.* developed a culturing technique to produce 3D-cultured bovine muscle tissue with unidirectionally aligned myotubes, possessing adequate maturity to display contraction with electrical stimulation [153]. The researchers also produced millimetre-thick bovine muscle tissues composed of highly aligned myotubes by stacking up bovine myoblast-laden hydrogel modules [153].

The culture device comprised anchors with pillars made via stereolithography 3D printing and a substrate made of PDMS. The culture device was made by placing two anchors to the substrate 7 mm apart. The pillars were then arranged to protrude from the surface of the substrate (Figure 5.10). Using this device, the researchers compared the contractility of the cultured bovine muscle tissues consisting of different hydrogels (collagen or fibrin-Matrigel). The collagen or fibrin-Matrigel hydrogel solution containing bovine myocytes was applied to the device, cultured for 2 days, and subsequently allowed to differentiate for 12 days to produce fibre-shaped bovine muscle tissues with a length of 7 mm.

Out of the bovine muscle tissue cultured in fibrin-Matrigel, 100% contracted with applied electrical pulses, while only 56% of those cultured in fibrin-Matrigel contracted without applied electrical pulses. On the other hand, only 33% of the bovine muscle tissue cultured within collagen contracted when electrical pulses were applied while no contractions were

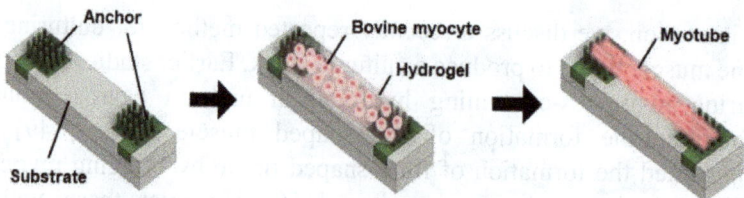

Figure 5.10.  Schematic showing the construction of the contractile bovine muscle tissue [153]. Reproduced from [153], which is licensed under CC BY 4.0.

observed without electrical pulses. The distance of contraction with electrical stimulation was larger in fibrin-Matrigel than in collagen. Additionally, immunostaining with α-actinin demonstrated that the myotubes were aligned to the long axis of the tissue, regardless of the hydrogel contents and the electrical stimulation. Overall, these findings suggested that using fibrin-Matrigel culture under electrical stimulation aided the maturation of contractile bovine muscle tissue with aligned myotubes.

Subsequently, Furuhashi *et al.* assembled myocyte-containing modules to create a millimetre-thick bovine muscle tissue containing highly aligned myotubes. The modules were made of collagen solution containing myocytes moulded on a PDMS template; each module has holes at both ends to facilitate the insertion of pillars fabricated from 3D printing. Additionally, each module has several slits to create striped structures, which assist in the alignment of the myocytes along the long axis (Figure 5.11) [154, 155].

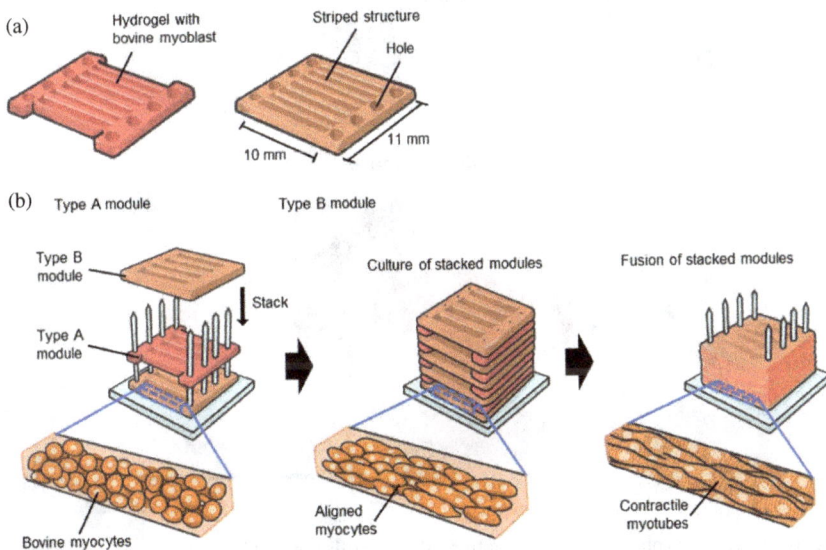

Figure 5.11. Fabrication of the millimetre-thick bovine muscle tissue. (a) Schematic drawings of the two types of myocyte-laden hydrogel modules containing holes to facilitate their immobilization to the pillars and seven/eight striped structures to encourage the alignment of myocytes. (b) Conceptual drawings displaying the fabrication process of the millimetre-thick bovine muscle tissue. Reproduced from [153], which is licensed under CC BY 4.0.

By inserting the pillars into their holes, two types of modules with varying slit positions were stacked alternately. The use of pillars to immobilize the stacked modules ensured cellular tension, which aided in myocyte differentiation. After culturing 40 assembled modules for 7 days, each module having 1-mm-wide striped structures positioned parallel to each other at 0.3 mm intervals fused into single muscle tissue. This assembly produced a millimetre-thick bovine muscle tissue with the dimensions of 8 mm × 10 mm × 7 mm (width × length × height) (Figure 5.12). The increased number of stacked modules allows large tissues to be achieved. Scaling the tissue size would be time-consuming and require many myocytes, which may be addressed by the automation involving 3D bioprinting.

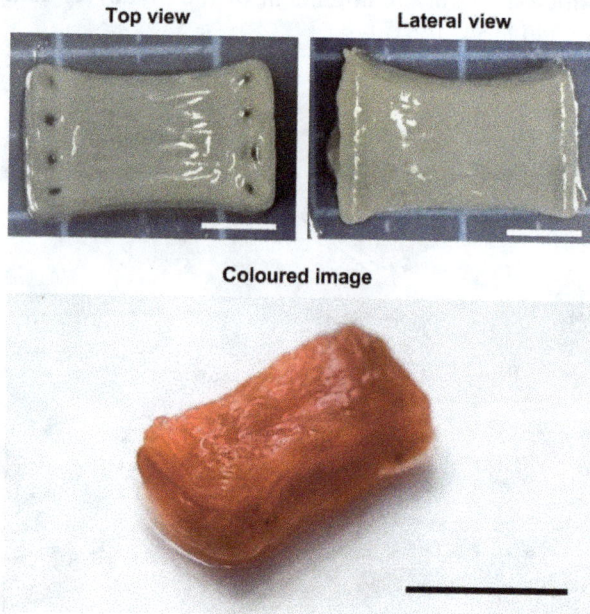

Figure 5.12.   Morphological observation of the millimetre-thick bovine muscle tissue. Top and lateral views of the millimetre-thick bovine muscle tissue after they are released from the pillars. (Scale bar = 5 mm.) The millimetre-thick bovine muscle tissue coloured with red food colouring agent. (Scale bar = 1 cm) [153]. Reproduced from [153], which is licensed under CC BY 4.0.

Subsequently, the breaking force was measured as an index of stiffness for food to determine textural differences between the cultured millimetre-thick bovine muscle tissue and commercially available steak [156]. On Day 14, the breaking force of the cultured bovine muscle tissue was similar to that of the beef tenderloin. The stiffness of the cultured bovine muscle tissue was influenced by morphological changes that occurred during the culture, hardening the bovine muscle tissue with time; a higher breaking force was measured on Day 14 than on Day 4.

### 5.3.3.2 *Tendon-gel integrated bioprinting of wagyu beef steak*

Kang *et al.* developed a method termed tendon-gel integrated bioprinting (TIP) for the creation of cell fibres differentiated into skeletal muscles, adipose tissues, and blood capillary fibres; all of which were subsequently assembled to build engineered steak-like meat by imitating the histological features of an actual beef steak [157]. TIP is a type of embedded 3D printing where the ink is extruded into a bath containing a gel or a support media with thixotropy (Figure 5.13). When subjected to shear forces, the viscosity of the support media decreases, allowing for the ink to be dispensed. The support media regain high viscosity when the shear force is removed, preserving the printed structure within the bath. Embedded 3D printing allows the overcoming of not only the limitation for the applicable ink viscosity but also the drying of the printed materials in the air-interfaced environment. Due to these advantages, multiple studies have been conducted on 3D-print complex tissues [158–164].

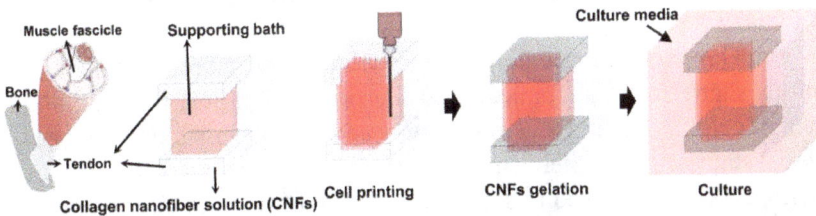

Figure 5.13.   Schematic drawing of tendon-gel integrated bioprinting (TIP) used in cell printing [157]. Reproduced from [157], which is licensed under CC BY 4.0.

Kang *et al.* demonstrated the printing of edible bovine satellite cells (bSCs) and bovine adipose-derived stem cells (bADSCs) obtained from beef meats [157]. The presence of myosin II heavy chain (MHC) confirmed the differentiation of bSCs into muscle cells while the expression of two adipogenic markers — peroxisome proliferator-activated receptor-$\gamma2$ (PPAR$\gamma2$) and fatty-acid-binding protein 4 (FABP4) — indicated adipogenesis by the bADSCs and their differentiation into adipocytes [157].

A bioink comprising bSCs, fibrinogen, and Matrigel in a culture medium was printed in a supporting bath containing granular particles of gelatin (G-Gel) or gellan gum (G-GG) and thrombin [157]. G-Gel and G-GG are edible and cell-compatible. Both materials are thixotropic and can dissolve at ~37 °C, enabling the separation from the matured muscle fibres post-printing. These properties made G-Gel and G-GG suitable candidates for the support bath in bioprinting. To avoid the collapsing of the printed filament in a globular shape, the printed filaments were anchored. Previous studies suggested that anchored structures allowed the 3D muscle tissues to retain their initially printed shape, and enhanced cellular alignment, fusion, and differentiation against the muscle fibre contractions [155, 165–169].

To enable simultaneous anchoring of the printed muscle cell fibres during printing, the printing bath was segregated into three portions: the bottom tendon-gel, the supporting bath, and the upper tendon-gel [157]. G-Gel was employed as the supporting bath, and the tendon gels were filled with 4% w/w of collagen nanofibre solution (CNFs), which possessed a reversible sol-gel transition temperature range between 4 °C and 37 °C. The bSC fibre was printed between the two tendon gels, followed by incubation.

On Day 3 of differentiation, the printed cell fibre maintained its fibrous shape and was connected to the two tendon gels with the alignment of cells, as observed in microscopy and with haematoxylin and eosin (H&E) staining. Sarcomere structures (indicating a matured state for muscle fibres) were observed in some TIP-derived bSC fibres after 14 days of differentiation. In comparison, sarcomere structures were not found in the needle-fixed culture cell fibres. These results showed that the long-term TIP culture facilitated the maturation of muscle tissues more than the needle-fixed culture.

Muscle, fat, and vascular cell fibres were fabricated separately because each of these required a specific medium for differentiation.

mRNA and protein expressions of PPARγ2 and FABP4 were measured to ensure the adipogenesis of the bADSCs-derived fat fibres by TIP. On Day 14 of the differentiation, the levels of PPARγ2 and FABP4 were higher by >6-fold and >40-fold, respectively, than the naive bADSCs. The level of corresponding mRNAs was also increased >2-fold for both PPARγ2 and FABP4.

TIP-derived cell fibres (i.e., muscle, fat, and vascular cell fibres) were assembled to show the production of the cultured steak. The cross-section image of *wagyu* was stained for sarcomeric-actinin and laminin, which indicated muscle fibres in double-positive staining, and adipose fibres in laminin-only positive staining, respectively. Based on the *wagyu* image, a cultured steak was reproduced with the estimated dimensions of 5 mm × 10 mm × 5 mm (width × length × height). To facilitate the assembly, the cell fibres were physically stacked based on the model image and then treated with transglutaminase for 2 days at 4 °C (Figure 5.14). A cross-section imaging of the assembled beef was observed to confirm the

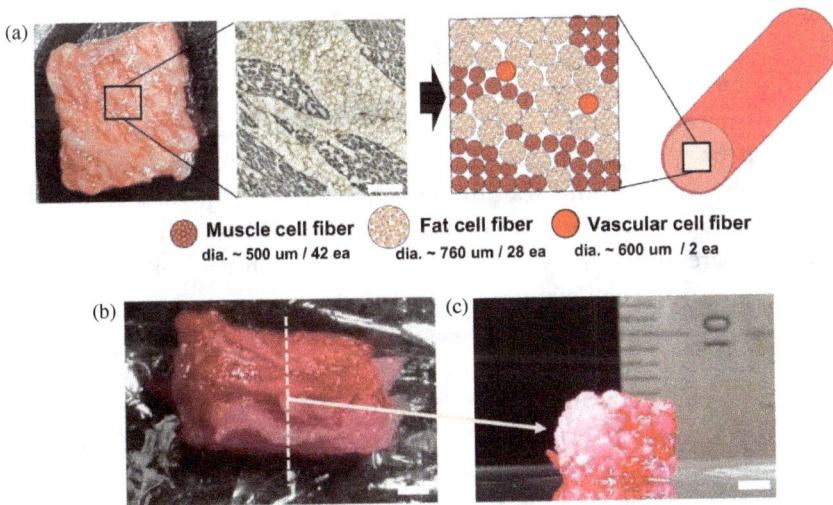

Figure 5.14. (a) Right – schematic of an assembled fibre. Left – sarcomeric α-actinin (blue) and laminin (brown) stained image of commercial meat. The diameters of the fibrous muscle, fat, and vascular tissues are estimated to be 500, 760, and 600 μm, respectively. (Scale bar = 1 mm.) Photos showing the (b) top and (c) cross-section view of the dotted-line area from (b). Carmine (red) was used to stain muscle and vascular tissues while fat tissues were not stained. (Scale bar = 2 mm) [157]. Reproduced from [157], which is licensed under CC BY 4.0.

feasibility of the TIP for *wagyu* steak fabrication. Future works consider the scalability and edibility of the printed meat with TIP. The same method would be applicable for muscle tissue engineering as well.

### 5.3.4 *Bioreactors*

Bioreactors enable the scalable production of cultured meats [170]. Technologies in the biotechnology and pharmaceutical sectors, including cell therapy, antibody production, and biologics synthesis, have demonstrated that bioreactors are effective in keeping cells alive and functional under controlled conditions. The same concepts could be applied to scale up the production of cultured meat products [171].

The production of cultured meats must be scalable and cost-effective before it can be considered a reliable alternative to conventional meats. For minced meats and full-thickness meats, the requirement for scalability is different. However, in both cases, a seed train can be used to increase the production of cells under optimal conditions. Seed train refers to the expansion of cultured cells from a small volume to a larger volume. The cultured cells are subsequently used to inoculate the main production bioreactor at the necessary cell density. The seed train can produce one ton of cultured muscle cells ($10^{13}$ cells) from the initial batch of $10^4$ cells [138]. The seed train helps keep the cells within the condition of exponential growth and avoid premature differentiation [172, 173]. As a result, the original culture is carried out in standard culture dishes or flasks; the culture is progressively transferred to bioreactors with controlled temperature and pH as the cell number increases.

Among different types of bioreactors, stirred tanks and rocking bioreactors (also called wave bioreactors) are the most used. Other types of bioreactors include airlift, hollow fibre, fluidized-bed, fixed/packed-bed, and vertical-wheel bioreactors [171, 174] (Figure 5.15). Stirred tanks are the industry standard for mammalian cell bioreactors, whereby cells are either suspended or attached to microcarriers suspended in the agitated medium [175]. Continuous stirred-tank bioreactors are the most employed bioreactors for animal cells because they provide long-term sterility through mechanical stirring, maintain a sufficiently high level of oxygen transfer, and avoid bubbling over air-lift reactors [11]. Since most mammalian cells need to be anchored to grow, microcarriers serve as suspension surfaces for cell growth. High cell densities are also possible with cell

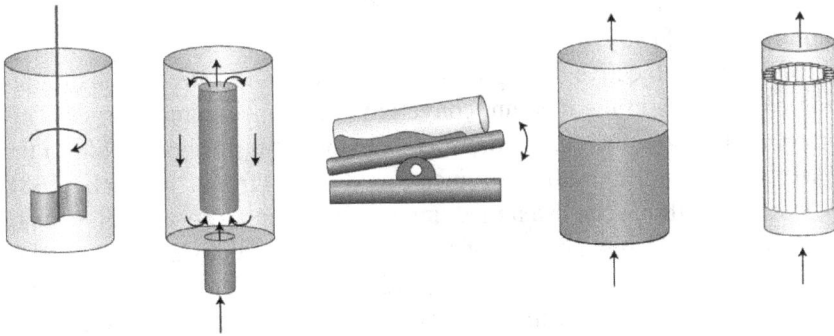

Figure 5.15.   Types of bioreactors: (from left to right) stirred tank, airlift, rocking/wave, fluidized bed or fixed/packed bed, and hollow fibre. Reproduced with permission from [138]. Copyright (2020) from Springer Nature.

suspensions, and harvesting them is rather straightforward. Bovine myoblasts, like mesenchymal stem cell (MSC), can be expanded in suspension on microcarriers [176]. It has been discovered that it is possible to alter induced pluripotent stem cells (iPSCs) so that they can thrive in aggregates [177, 178], akin to previous developments in the embryonic stem cell (ESC) derived from humans [179, 180] and mice [181]. Aggregates can form and grow with committed stem cells such as MSCs, but it is difficult to regulate the size of the cell aggregates [182]. Currently, there is an absence of any large-scale cell culture data involving aggregates [138]. Experience with huge scale culture of anchorage-dependent mammalian cells is primarily developed for the field of MSC cell therapy [183].

Each type of bioreactor has unique features. The stirred-tank reactor, for instance, is commonly employed for mammalian cell culture. It is advantageous in terms of scalability but it causes high shear stresses on the cells arising from the mechanical agitation for mixing. Hollow fibre reactors promote cell growth on the external surface of the microfibres and/or in the space between them; nutrients diffuse to the cells from the fibre lumen, reducing the shear stresses [184]. Hollow fibre reactors, however, can be used only once and may result in high operational costs. Packed/fixed-bed bioreactors face restrictions to mass transport, producing cells of varying quality and viability within the bioreactor [185]. Cell carriers can be cultured at high densities in fluidized-bed reactors as mixing is realized by fluidization with circulating media rather than mechanical mixing. Unfortunately, fluidized bed-reactors have only been ramped up to 100 L; it is unclear whether this productivity applies to larger vessels [138].

The main objective of improving bioreactors is to increase the proportion of nutrients in the medium converted into edible animal tissues. This conversion, known as the medium conversion ratio, is comparable to the feed conversion in traditional meat production from livestock [138]. Recycling methods can use a medium efficiently and improve cell density (i.e., number of cells per mL of medium). Cost-effectiveness is another essential objective of scaling up cell production. Besides increasing the cell population, cells must form functional tissues. Without a completely unified reactor system where cells can proliferate and mature into tissues after self-assembly, fabrication of tissues needs to take place in a separate bioreactor with optimized conditions for tissue formation. This requirement may result in a large variety of reactor designs based on individual tissue types and conditions. Automation is also important to replace the labour-intensive processes to reduce cost and microbial contamination risk.

Development and optimization of bioprocesses are also important for lowering production costs. For instance, to achieve consistent production in scaling up, *in silico* modelling of cell behaviour plays a critical role in the coming years. The semi-scaled up processes and trial-and-error scaling up methods are presently employed in the fields of gene and cell therapy [186, 187]. Overall, the development of bioreactors is crucial for expanding cultured cell populations. The expansion of cells is important for the 3D printing of cultured meats, as high-density cell inks require a relatively large number of cells to produce structured meats that meet the scalability at the industrial level.

## 5.4  3D Printing of Plant-Based Protein

### 5.4.1  *Introduction*

As mentioned earlier, plant-based meat is also gaining popularity as one of the approaches to generate food from non-animal sources. Plant-based meats are not entirely novel as they have existed for centuries, particularly in Asia, as foods derived from soybeans, such as tofu, tempeh, or wheat (seitan). Due to the recent advancement in food science and technology, it gives plant-based meat a "new life" by enhancing their sensory characteristics, which makes them comparable to real animal meat based on taste, texture, and nutrition [188]. Some of the key sensory properties of meat include the appearance, flavour, texture, and aroma. Hence, it is possible for the plant-based meat to achieve similar sensory properties as real animal meat. Specifically, the physical appearance of the plant-based meat can be manipulated by adding vegetable extracts (beet juice) or recombinant heme proteins to introduce the colour of real animal raw meat and transition to brown upon cooking [189]. This recreates a visual perception that the plant-based meat is identical to real animal meat. Furthermore, the addition of plant-based fats, such as coconut oil and cocoa butter will recreate the fat marbling effect of real animal meat. To recreate the flavour and aroma of real meat, it will be necessary to use food additives to either enhance the taste of meat or mask the intrinsic taste of the plant-based protein [190]. For example, certain plant proteins such as soy proteins have a strong bean and bitter taste, which can be reduced by heating [191]. Furthermore, a flavouring additive can be added to enhance the meaty flavour of the plant-based protein [192, 193]. In regard to the texture, innovative strategies such as high-moisture extrusion, shear cell technology, and 3D printing are used to apply mechanical, thermal, and shear stresses to a plant-based protein mixture, which will allow the recreation of a semi-solid fibrous structure that will have a similar mouthfeel to that of real meat [194].

In general, the manufacturing of plant-based meat consists of three primary steps [191]. Step 1: This step involves protein isolation, where target proteins are extracted from various plant sources. Furthermore, additional processing procedures such as hydrolysis may be required to improve the solubility and cross-linking of the proteins. Overall, this will improve the functionalities of the isolated plant proteins. Step 2 involves the formulation of the plant proteins, where they will be mixed with other ingredients or additives such as plant-based fat, food adhesives or flour to enhance the texture of the plant-based meat to be similar to real animal meat. Besides the textural profile, certain nutrients can be added to complement or even fortify the nutritional needs of plant-based meat that are comparable or even exceed the nutrition intake of animal meat. The last step involves the processing of the plant proteins together with other mixtures to form a texture that mimics real meat. Examples of processes may include traditional processes such as kneading, pressing folding or using innovative technologies, including shear cell technology or 3D printing [189, 194].

## 5.4.2 *Ink Formulation, Properties, and Characterization*

Plant-based meat products can be derived from several materials, including legumes (beans), cereals (barley), and oilseeds (cottonseed) [190]. The formulation of plant-based materials that mimic real meat products is challenging. Two of the most popular plant-based protein companies, namely, Impossible Foods and Beyond Burger, have effectively created and produced plant-based meat products. However, both companies used different formulations to create their own version of plant-based meat products. In the case of Impossible Foods, they use several plant-based constituents and rely on soy leghemoglobin to develop the meat colour due to the presence of the colour of protein myoglobin. On the contrary, Beyond Burger uses beet juice extract to achieve the meat colour. Other natural pigments, including red peppers, tomato paste, paprika, and red rice, can also be applied [195, 196]. Most plant proteins use either pea protein or soy protein, as they have complementary functions with other additives due to their unique attributes [195, 197]. For example, soy protein is frequently used as plant-based materials as they have several properties that can enhance gelling, water retention capacity, and fat-absorbing properties. Comparison to another commonly used plant protein, pea protein, the reported gelling capacity is much lower than soy protein; hence, it will require further processing (addition of salt or particle size modification) to overcome this issue. Wheat is also a widely used material to create a fibrous structure due to its ability to form disulphide interactions with proteins, thus improving the rheological and viscoelastic properties of plant-based protein [190, 197]. In most cases, plant-based meat products are mainly patties, sausages, or mincemeat; hence, the creation of steaks derived from plant-based materials seems to be technically challenging and requires innovative technologies [188].

In addition to the selection and formulation of plant-based protein materials to recreate the sensory properties of real meat, innovative technologies such as 3D printing have also been explored and is a heavily researched topic. Some of the popular plant-based proteins used in the ink formulations include pea, soy, and wheat protein, which provide comparable protein content levels to real meat [193, 197]. Furthermore, with the high-precision feature of 3D printing, it is possible to create fine structures of vegetable protein to resemble the textural mouthfeel of real meat. Due to the unique production method, it also allows for customization of the aesthetic effects of the meat such as colour and shape. Besides narrowing the sensory differences between plant-based protein and real meat through 3D printing, it can also cater to special nutritional needs or requirements of target individuals. For example, in order to achieve a balanced amino acid profile of the plant-based protein, mixtures of multiple plant-based protein ingredients are often used in the formulation of the plant-based product [193, 197]. Given that the formulation of the ink using the plant-based protein is one of the steps that will ensure the printability of the final product, characterization of the printability of the food ink formulation will be an important factor. A recent study investigated the printability of soybean, which is a commonly used plant-based meat ingredient. Specifically, textured soybean protein (TSP) and drawing soy protein (DSP) were used to explore their effects on various hydrocolloids and to characterize their printability [198]. A total of six different hydrocolloids at 2% (GG, Sodium Alginate, XG, Konjac Gum, Sodium Carboxymethyl Cellulose, and Hydroxyethyl Cellulose) were used and combined with 79.5% of either TSP or DSP. Interestingly, despite using a similar formulation, TSP displays much better print outcomes compared to DSP. Further characterization of the rheological properties and moisture analysis supported the printing outcome, which suggests that DSP is not suitable for printing under those conditions. A similar formulation involving the use of soy protein isolate and XG was also observed; however, in this study, a NaCl solution was also included [199]. The formulation of soy protein isolate with XG and a NaCl solution at 1 g/100 mL was effectively printed with strong correlations between the rheological properties and printability (Figure 5.16).

While the formulation of the ink is important to achieve a good printing profile, the printed product would in most cases need post-printing processes prior to consumption. Some of these post-printing processes involve cooking methods that are required to cook the food. Hence, it will be equally important to evaluate the ability of 3D printing technology to not only be able to print but also be able to withstand the cooking process. Specifically, the formulation of the plant-based food ink should be able to maintain the original shape during the cooking methods such as frying [198]. To explore if the plant-based protein has similar sensory features to chicken nuggets, the 3D-printed samples were fried and it was found that the ink formulation using XG maintained the best shape and had almost the same shape as before frying.

The texture of food is one of the parameters that give an individual the pleasure of food consumption. One of the advantages of 3D food printing

(a)

(b)

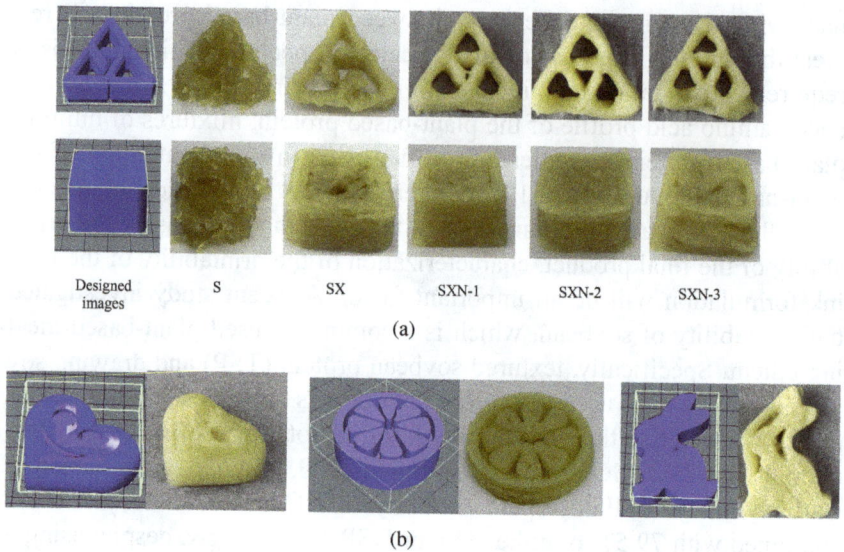

Figure 5.16.    Images showing 3D-printed soy protein isolate with different formulations of xanthan gum (XG) and NaCl solution. S = sole SPI, SX = SPI with XG (0.5 g/30 g of SPI), SXN-1 = SPI with XG (0.5 g/30 g of SPI) and 1 g/100 mL NaCl solution, SXN-2 = SPI with XG (0.5 g/30 g of SPI) and 2 g/100 mL NaCl solution, SXN-3 = SPI with XG (0.5 g/30 g of SPI) and 3 g/100 mL NaCl solution. Reproduced with permission from [199]. Copyright (2020) from Elsevier.

is the ability to modify the texture profile of food by modifying the infilling ratio and infilling pattern of the food. This will allow the creation of a plant-based meat that will have a similar textural profile (e.g., hardness, chewiness, and viscosity) to real meat [198]. Indeed, it was shown that a triangular infilling pattern combined with 60% infilling ratio of the soy protein has the closest textural profile to real chicken nuggets (Figure 5.17).

Other than soy protein, pea protein is also a commonly used plant-based protein material. Pea is a plant-based legume protein that is quickly gaining appeal as an animal protein alternative. It is a nutritious component with minimal allergenicity and plentiful essential amino acids (tryptophan and lysine) that have received a lot of interest in the food business [200]. Specifically, a study by Oyinloye *et al.* explored the rheology, thermal, and textural aspects of five different blend ratios of pea protein and hydrocolloid (alginate) solutions to determine the best formulation for a 3D printing material [201]. With the increase in pea protein content, the

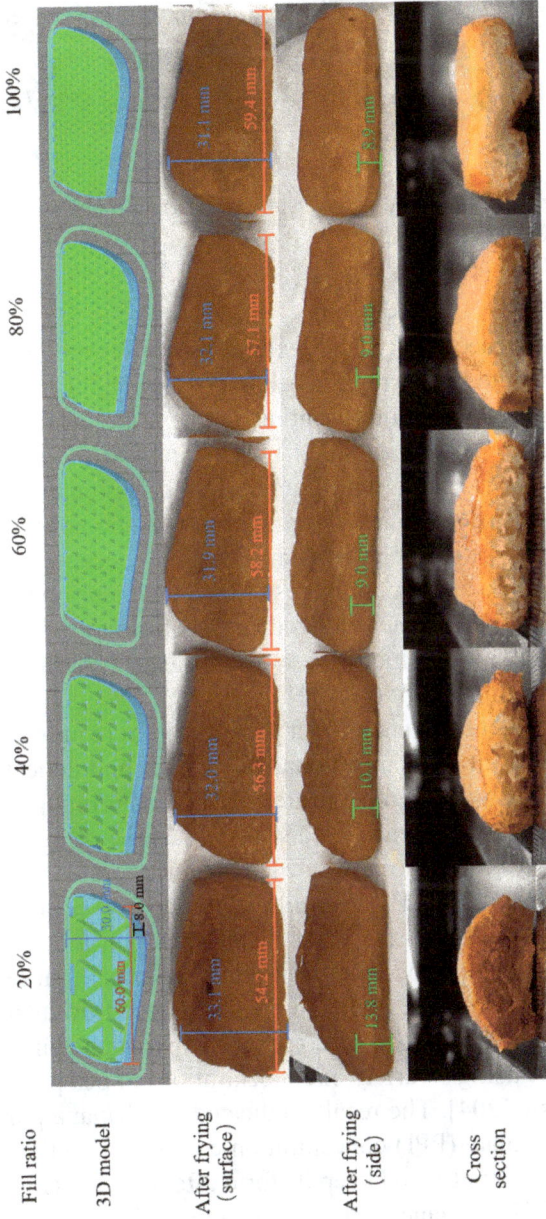

Figure 5.17. Images of fried 3D-printed soy protein with varying infill ratios (20–100%). Photos showing the 3D model: after frying surface, after frying side, and cross-section (top to bottom) [198]. Reproduced from [198], which is licensed under CC BY 4.0.

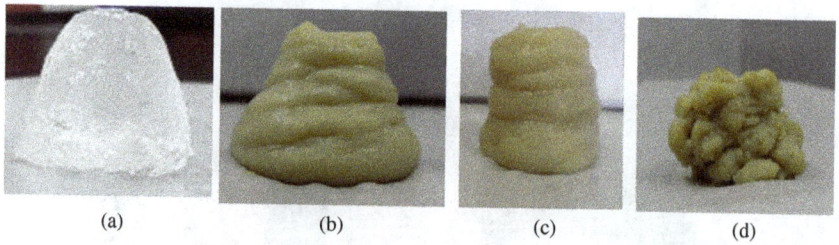

Figure 5.18.   Images of 3D-printed pea protein with different formulations of alginate. (a) Alginate gel 100%, (b) AP 90:10, (c) AP 80:20, and (d) AP 70:30. Reproduced with permission from [201]. Copyright (2021) from Elsevier.

material's thermal, rheological, and textural properties are improved. As these features impacted the extrusion behaviour during 3D printing, hence in order to examine the optimal printing settings, the ideal 3D printing material (alginate to pea protein 80:20) was employed for additive-layer manufacturing (ALM) simulation (Figure 5.18). It was demonstrated that the deposit thickness determines the residual-stress component and overall deformation across the printed sample. This is because of the load, which causes sample deformation, resulting in dimensional variations from the original model. When utilized for 3D printing of complicated shapes, ALM simulation demonstrates a significant benefit in terms of optimizing the printing conditions [201].

Fruit is a vital diet for humans, and it offers far more nutritional value than grains. As a result of its nutritional advantages and delicate texture, it has been extensively utilized in the area of 3D food printing. Extrusion-based 3D printing has been employed with a selection of fruit materials, including freeze-dried mango powder, lemon juice, orange concentrate, and strawberry juice [202, 203]. The tailing effect observed in the lengthy paste extrusion of fruits, on the other hand, results in a low resolution throughout the extrusion process, making their utilization more difficult. A recent study by Kim *et al.* explored the link between flow attributes and the tailing effect generated by discontinuous extrusion during the 3D printing procedure using a variety of concentrations of pea protein isolate in the banana paste [204]. The results indicated that banana purée with a 15% pea protein isolate (PPI) concentration could be effectively printed with a well-fitted geometry and keep its form after printing (Figure 5.19). The addition of PPI enhanced the entanglement between the banana matrix and the protein, leading to higher storage modulus, adhesion force, and loss modulus, as well as lower percentage recovery and

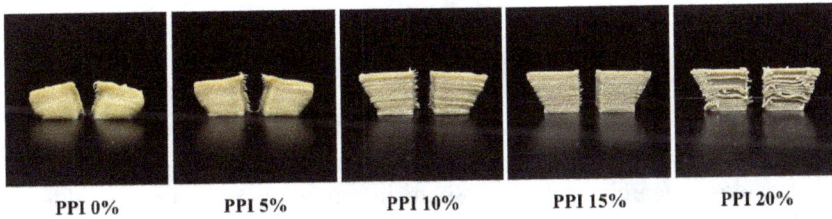

| PPI 0% | PPI 5% | PPI 10% | PPI 15% | PPI 20% |

Figure 5.19. Images of 3D-printed banana paste with different combinations of pea protein isolate (PPI). Left to right: 0–20% of PPI added to banana paste. Reproduced with permission from [204]. Copyright (2021) from Elsevier.

elongation at break. After the printing process, a collapse owing to its own weight and the production of long tails was seen at up to 10% PPI inclusion. However, a 20% increase in PPI produced protein accumulation in the matrix, leading to inadequate structural recovery and breaking of the 3D-printed line, as well as discontinuous extrusion, which was detected with poor resolution [204]. In the case where the dynamic viscoelastic characteristics $G'$ and $G''$ could not be identified, novel experimental procedures such as the tack test and the dashed line printing test can be used to investigate this phenomenon. This study revealed that when the tailing effect declined, the adhesive strength and sharp tail production between the separated lines dropped, thus compromising the printing performance [204].

Most plant-based meat formulations use important plant-based proteins (e.g., pea, soy, and wheat) to offer a total protein concentration comparable to real meat. Complementation of various plant-based proteins, on the other hand, is often required to establish a balanced amino acid profile. Proteins from legumes (rich in lysine, low in sulphur-containing amino acids) and cereals (low in lysine, high in sulphur-containing amino acids) are good complements. Structures resistant to proteolysis, antinutrients, and protein conformation have all been discovered in plant proteins as factors that may reduce nutrient bioavailability after intake (e.g., tannins, phytates, and lectins) [205]. It has been proved that some processing processes (such as soaking, heating, and sprouting) can improve digestibility [191]. Traditional and innovative plant-based meat products differ in terms of nutrition. Tofu (traditional) and Impossible™ (innovative), for example, have several advantages over real meat, like minerals and dietary fibre, while being cholesterol-free. Specifically, tofu has less calories and less fat and is sodium-free, whereas Impossible™ has a greater protein and vitamin B12 content. However, several concerns have been raised about LegH's inclusion in some of the formulations of plant-based meat, suggesting links between heme iron consumption and a higher risk of diabetes [206]. Hence, careful consideration is required in the preparation of plant-based meat and the subsequent printing and post-printing processing procedures.

Besides research groups, a Spanish company, Novameat is able to produce beef and chicken using 3D printing. Specifically, Novameat formulated their plant-based meat using pea protein, seaweed, and beetroot juice, and utilized 3D printing technology to recreate the meat in the form of a steak [207]. This 3D-printed plant-based steak is able to create microfibres (ranging from 100 to 500 μm) using micro-extrusion technology, allowing muscle fibres and fat to be entwined; hence, the complex structure of a real beef steak can be replicated with the plant-based meat [197]. Besides Novameat, another Israel-based company, Redefine Meat also uses 3D printing technology to create steaks using plant-based protein and fat to recreate the marbling effect of a real meat [208]. Besides beef, companies are also developing and formulating plant-based protein materials for other types of meats such as pork and tuna [197]. While most of the plant-based meats focus on poultry meat such as beef or pork, there is a huge gap in the area of plant-based seafood products. Several research groups have attempted to develop plant-based meat that mimics the salmon fillet. Specifically, the European company Reno-food developed a novel 3D printing technique to create plant-based fish products, which could not be achieved with the normal extrusion-based method [209]. The plant-based protein used is mainly derived from pea protein, algae extracts, plant oils, and other food additives. Furthermore, to recreate the appearance and texture of salmon fish, they utilized multi-nozzle 3D printing, where different food ingredients are used as food inks. By using this method, they are able to capture the distribution of orange meat and white connective tissue, recreating the complex appearance of the salmon fillet [210]. In addition to the visual appearance of the plant-based salmon, the nutritional aspect of the plant-based salmon is supplemented with other plant oils, which are also rich in Omega-3 [209]. More recently, a Singapore-based research group formulated plant-based salmon using red and white lentils and pea protein, and tapped into 3D printing technology to recreate the salmon fillet [211]. To date, although there are successful examples of plant-based meat and seafood alternatives, it is important to note that there is limited publicly available information on the formulation of plant-based protein materials, the cost of production, and the printing processes. According to a press release, a 50–200 g plant-based steak was made at a price of USD 1.50–4, suggesting the cost-effectiveness of the production [197, 207]. Therefore, the future and economic sustainability of plant-based products remain uncertain.

## 5.5 3D Printing of Insect-Based Protein

### 5.5.1 *Introduction*

In the earlier sections, different types of alternative proteins such as cell-cultured meat or plant-based meat were discussed. Another type of alternative protein is insect protein, which has several nutritional and environmental benefits that are often overlooked. In fact, the consumption of insects is not novel and is already present in certain cultural societies, where they are considered a delicacy and a rich source of proteins. According to the FAO, more than 2 billion people globally consume insects in 140 different countries across the Asia-Pacific, the Americas, and Africa. Furthermore, more than 2,100 species of insects are consumed and some of the commonly consumed insects include beetles, caterpillars, ants, bees, wasps, grasshoppers, locusts, and crickets [212]. While the nutritional value varies depending on the individual species, insects contain high levels of minerals and vitamins, and are low in salt content. They are also generally regarded as a high-fibre, high-protein food source; hence, they are considered to be comparable or even superior to real meats or plant-based meat in regard to protein content [212]. Hence, given the nutritional profile of insects, there are considerations to replace animal-derived proteins with insects on an equal weight basis [213]. For example, crickets contain approximately 2–3-fold more protein in comparison with the same amount of chicken or beef.

In comparison with plant-based proteins, crickets also contain most of the essential amino acids; hence, there is no additional need for supplementation of essential amino acids. From an environmental perspective, insect farming requires less feed and less land, as well as less time to reproduce, thus producing less waste and carbon footprint. Therefore, in comparison with traditional livestock, insect farming is more effective in converting from food to protein. In addition, a secondary environmental benefit would be the use of upcycling food waste or agricultural by-products (okara or spent grains) as a low-cost, high-nutrition feed for insects, hence creating a circular farming operation. Despite the nutritional and environmental benefits of insect proteins, the notion of insects as a food source for human consumption is still relatively unaccepted in most countries around the world. One of the reasons could be attributed to the common perception associated with ingestion of insects is disgust and the fact that insects are often closely associated with filth. Therefore, changing the physical form or the appearance of the insect was recommended in order to increase the acceptance of insect proteins. As such, many insects have been dehydrated and milled into powder form to reduce the "disgust" reaction experienced by the consumers [214]. Furthermore, some insects can also be potentially separated into various proteins (soluble/insoluble), lipids, and fibre fractions and can be used to make various food products such as protein bars or bakery goods. In these cases, insect proteins are often mixed with flour or other ingredients to mask the use of insects or the taste of the insect proteins. With 3D food printing technology, insect proteins can be used as the primary food ink ingredients and 3D-printed into aesthetic shapes or structures that would be acceptable for consumers. Hence, the formulation of insect-based food inks and the characterization of the textural and sensory properties of insect-based 3D-printed foods require further investigation.

## 5.5.2 *Ink Formulation, Properties, and Characterization*

A few studies have used insect proteins in combination with other ingredients in the formulation of food ink. Specifically, in one study, dried insect flour (mealworms and crickets) was combined with fondant and 3D-printed to create cake icing [215]. However, insect powders of various particle sizes combined with ingredients like hydrocolloids (e.g., gelatin or guar gum), *surimi* gel, and meat odours have yet to be studied for meat analogues. Furthermore, *Tenebrio molitor*-supplemented wheat-based foods were 3D-printed [216]. Besides ink formulations, other parameters such as the baking time and temperature, primary microstructural traits, overall quality, and nutritional qualities were also investigated. In this case, mealworm powder (20%) was added to wheat flour dough to soften raw 3D-printed snacks and prevent water loss in the baked 3D-printed final product [216]. This resulted in an overflow of dough deposition, leading to increased snack diameter, height, and weight. The overall appearance of the snacks did not change due to the baking circumstances,

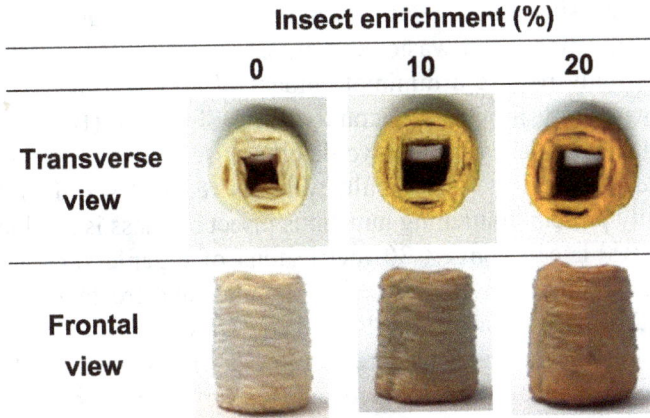

Figure 5.20.   Images of 3D-printed snacks with different enrichments of insect proteins (0–20%) baked at 200 °C for 22 min. Top to bottom: Transverse view and frontal view of the baked insect protein. Reproduced with permission from [216]. Copyright (2018) from Elsevier.

but there was a change in the primary dimensions and microstructure features owing to improved water evaporation. After optimizing the baking conditions, it was discovered that 22 min at 200 °C resulted in the maximum desirability of the printing outcome (Figure 5.20). Furthermore, printed snacks enhanced with 10% and 20% ground insects raised the total essential amino acid content from 32.5 (0% insects) to 38.2 and 41.3 g/100 g protein, respectively, when baked under these circumstances. From 0 to 20% insect enrichment, the protein digestibility-adjusted amino acid score rose from 41.6 to 65.2, with lysine and methionine+cysteine being the respective limiting amino acids [216]. When a considerable proportion of insect powders (>20%) is utilized in meat analogues, however, the impact of enhanced water dehydration from particles must be addressed. Hydrocolloids can be added to the product if dehydration occurs. Furthermore, more than one kind (composite) of a natural polymer (e.g., collagen or alginate) might possibly be used to enhance the printability of insect-based pastes. Another approach is to mix bug powders with the meat paste made from animal by-products, as well as transglutaminase [216].

Another insect protein that has gained significant interest is from Black Soldier Flies (BSFs), which offers several potential solutions to a number of pressing global issues, including environmental sustainability, resource utilization, food waste, and climate change [217, 218]. Firstly, BSF farms only require a minimal amount of area and little water and have a low-energy footprint. Secondly, the BSF larvae (BSFL) may be cultivated on organic wastes that would otherwise pollute the environment if dumped in landfills. As a result, waste is reduced significantly, and high-quality protein-containing nutritious insect biomass is produced. For example, BSFL can convert 30 metric tons of organic substrate (food waste) into around 10 metric tons of dry biomass and create 930 kg of dry biomass [218]. To top it off, the BSF is not known to cause any disease transmission; hence, it is considered a rather harmless bug. Furthermore, BSFs that have reached maturity do not cause any problems or attract human habitats, which are environmentally friendly. The larvae transform organic and food waste into fertilizer, and are extracted as insect protein to substitute unsustainable fish feed for aquaculture or crop-based animal feed. Interestingly, when the protein content was examined, it was discovered that the proportion of crude protein was much higher than that of fishmeal and soybean meal. Given that BSFL is high in protein which is used as animal feed, it is also plausible for BSFL to be considered as part of the human diet [217]. Food neophobia, as well as consumer perceptions of disgust and fear associated with insects, are the major hurdles for insect proteins; these need to be addressed before they can be considered as part of the human diet [219]. One method to overcome these obstacles is to alter the manner in which the insects are ingested, such as by having them processed or masked, or by incorporating them into cuisines that people are already acquainted with [219]. Specifically, BSFL can be mixed with a fruit matrix and formulated into food ink that can be used for 3D food printing. Besides BSFL, other insect proteins such as cricket and sericin (which is a silkworm protein discarded during cocoon processing in silk production) can be used. This can change the original form of the BSFL, which may be disgusting to people, into familiar structures and shapes such as an octopus or pyramid (Figure 5.21). In this study, food ink optimization is achieved using the response surface methodology, which is a design-of-experiment methodology used to systematically search for parameters that lead to the most desirable outcomes [220]. Using this approach, the food ink for each specific food ink is optimized to provide 3D printed structures with the best rheological properties along with the textural properties and microstructures.

Figure 5.21. Images showing original and insect proteins that are 3D-printed on Wiiboox Sweetin using a 0.84 mm pinhead. Top to bottom: BSFL, cricket, and sericin combined with carrot powder and xanthan gum that were optimized using the response surface methodology.

Furthermore, 3D-printed insect-based materials can be smoked to enhance organoleptic characteristics, offering the analogues a meat-like fragrance. As meat alternatives may require post-printing process activities such as frying or boiling, this will vary depending on the kind of meat alternative. Thus, insects were smouldered in a hot pan (without oil) rather than wood fire, which resulted in the creation of a meat-like fragrance that is currently under research [221]. While insect-supplemented snacks have been 3D-printed, reformed insect-originated products that mimic meat replacements, such as steaks, patties, or sausages, require further investigation. The alternative meat sector has a large potential market, and plant-based proteins or insect proteins are becoming more widely recognized. It is also likely to become a nutrient-dense and sustainable source of food production. Furthermore, 3D printing's customizability may be used to create steak-like foods that suit the nutritional or energy demands of different groups (gender, age), as well as customized flavour, taste, and sensations, resulting in healthful, nutritious, appetizing, and appealing 3D-printed foods.

## 5.6 Conclusion

While there are various meat replacements available, none of them have been widely accepted and consumed regularly by the general population. One of the reasons is because 3D food printing remains in its early phase, so further research into materials and printers is needed. It is crucial to concentrate on the technology concerned and the condition of meat alternatives in the food industry when imagining what the future holds for 3D-printed meat alternatives. In addition to technology development, it is necessary to investigate the food-safety aspect of bioprinting and also the materials originated from insect sources that are suitable for 3D printing of meat alternatives. From the angle of sustainability and food security, the long-term viability of meat analogues will be determined by effective food supply chain (farm-to-fork) management, which focuses on reducing the manufacturing costs and reducing the environmental impact. For example, the use of plant protein, insect proteins, and animal by-products/waste in the production of meat alternatives might increase their long-term sustainability. In an effort to enhance the consumer acceptability of 3D meat analogues, innovative processes such as 3D food printing were considered. Hence, it is necessary to build printers that are accurate, high-speed, prolific, and energy-efficient, so that there will be overall a net positive effect on sustainability, which will be discussed in Chapter 9.

## References

[1] Ritchie, H. and Roser, M. (2017). Meat and dairy production, *Our World in Data*.
[2] De Bakker, E. and Dagevos, H. (2012). Reducing meat consumption in today's consumer society: questioning the citizen-consumer gap, *Journal of Agricultural and Environmental Ethics*, 25, pp. 877–894.
[3] Gerber, P. J., Steinfeld, H., Henderson, B., Mottet, A., Opio, C., Dijkman, J., Falcucci, A. and Tempio, G. (2013). Tackling climate change through livestock: a global assessment of emissions and mitigation opportunities, *Food and Agriculture Organization of the United Nations (FAO)*.

[4] Eshel, G., Shepon, A., Makov, T. and Milo, R. (2014). Land, irrigation water, greenhouse gas, and reactive nitrogen burdens of meat, eggs, and dairy production in the United States, In *Proceedings of the National Academy of Sciences*, 111, pp. 11996–12001.

[5] Flachowsky, G. and Kamphues, J. (2012). Carbon footprints for food of animal origin: what are the most preferable criteria to measure animal yields? *Animals*, 2, pp. 108–126.

[6] McMichael, A. J., Powles, J. W., Butler, C. D. and Uauy, R. (2007). Food, livestock production, energy, climate change, and health, *The Lancet*, 370, pp. 1253–1263.

[7] Jackson, B., Lee-Woolf, C., Higginson, F., Wallace, J. and Agathou, N. (2009). Strategies for reducing the climate impacts of red meat/dairy consumption in the UK. WWF and Imperial College.

[8] Bonnedahl, K. J. and Heikkurinen, P. (2018). The case for strong sustainability. White Rose University Consortium.

[9] Bhat, Z. F., Kumar, S. and Fayaz, H. (2015). *In vitro* meat production: challenges and benefits over conventional meat production, *Journal of Integrative Agriculture*, 14, pp. 241–248.

[10] Sharma, S., Thind, S. S. and Kaur, A. (2015). *In vitro* meat production system: why and how? *Journal of Food Science and Technology*, 52, pp. 7599–7607.

[11] K. Handral, H., Hua Tay, S., Wan Chan, W. and Choudhury, D. (2020). 3D Printing of cultured meat products, *Critical Reviews in Food Science and Nutrition*, pp. 1–10.

[12] Welin, S. (2013). Introducing the new meat. Problems and prospects, *Etikk i Praksis: Nordic Journal of Applied Ethics*, pp. 24–37.

[13] Datar, I. and Betti, M. (2010). Possibilities for an *in vitro* meat production system, *Innovative Food Science & Emerging Technologies*, 11, pp. 13–22.

[14] Smetana, S., Mathys, A., Knoch, A. and Heinz, V. (2015). Meat alternatives: life cycle assessment of most known meat substitutes, *The International Journal of Life Cycle Assessment*, 20, pp. 1254–1267.

[15] Grabowska, K. J., Zhu, S., Dekkers, B. L., de Ruijter, N. C., Gieteling, J. and van der Goot, A. J. (2016). Shear-induced structuring as a tool to make anisotropic materials using soy protein concentrate, *Journal of Food Engineering*, 188, pp. 77–86.

[16] Wan, Z.-L., Wang, J.-M., Wang, L.-Y., Yuan, Y. and Yang, X.-Q. (2014). Complexation of resveratrol with soy protein and its improvement on oxidative stability of corn oil/water emulsions, *Food Chemistry*, 161, pp. 324–331.

[17] Lynch, J. and Pierrehumbert, R. (2019). Climate impacts of cultured meat and beef cattle, *Frontiers in Sustainable Food Systems*, 3, p. 5.

[18]  Conroy, S., Drennan, M., Kenny, D. and McGee, M. (2010). The relationship of various muscular and skeletal scores and ultrasound measurements in the live animal, and carcass classification scores with carcass composition and value of bulls, *Livestock Science*, 127, pp. 11–21.

[19]  Dick, A., Bhandari, B. and Prakash, S. (2019). 3D printing of meat, *Meat Science*, 153, pp. 35–44.

[20]  Ramachandraiah, K. (2021). Potential development of sustainable 3D-printed meat analogues: a review, *Sustainability*, 13, p. 938.

[21]  Jayathilakan, K., Sultana, K., Radhakrishna, K. and Bawa, A. (2012). Utilization of byproducts and waste materials from meat, poultry and fish processing industries: a review, *Journal of Food Science and Technology*, 49, pp. 278–293.

[22]  Kubberød, E., Ueland, Ø., Tronstad, Å. and Risvik, E. (2002). Attitudes towards meat and meat-eating among adolescents in Norway: a qualitative study, *Appetite*, 38, pp. 53–62.

[23]  Lupton, D. and Turner, B. (2017). *Digital Food Activism*, "'Both fascinating and disturbing': consumer responses to 3D food printing and implications for food activism" (Routledge), pp. 151–167.

[24]  Dick, A., Bhandari, B., Dong, X. and Prakash, S. (2020). Feasibility study of hydrocolloid incorporated 3D printed pork as dysphagia food, *Food Hydrocolloids*, 107, p. 105940.

[25]  Godoi, F. C., Prakash, S. and Bhandari, B. R. (2016). 3D printing technologies applied for food design: status and prospects, *Journal of Food Engineering*, 179, pp. 44–54.

[26]  Kim, H. W., Bae, H. and Park, H. J. (2017). Classification of the printability of selected food for 3D printing: development of an assessment method using hydrocolloids as reference material, *Journal of Food Engineering*, 215, pp. 23–32.

[27]  Liu, C., Ho, C. and Wang, J. (2018). The development of 3D food printer for printing fibrous meat materials. In *IOP Conference Series: Materials Science and Engineering* (IOP Publishing), p. 012019.

[28]  Feiner, G. (2006). *Meat Products Handbook: Practical Science and Technology* (Elsevier).

[29]  Lipton, J., Arnold, D., Nigl, F., Lopez, N., Cohen, D., Norén, N. and Lipson, H. (2010). Multi-material food printing with complex internal structure suitable for conventional post-processing. In *Solid Freeform Fabrication Symposium*, pp. 809–815.

[30]  Tarté, R. (2009). *Ingredients in Meat Products: Properties, Functionality and Applications* (Springer).

[31]  Boland, M., Kaur, L., Chian, F. and Astruc, T. (2018). *Encyclopedia of Food Chemistry*, eds. Melton, L., Shahidi, F. and Varelis, P. "Muscle proteins" (Elsevier), pp. 164–179.

[32] Wilson, A., Anukiruthika, T., Moses, J. and Anandharamakrishnan, C. (2020). Customized shapes for chicken meat–based products: feasibility study on 3D-printed nuggets, *Food and Bioprocess Technology*, 13, pp. 1968–1983.

[33] Li, Y., Wang, Q., Guo, L., Ho, H., Wang, B., Sun, J., Xu, X. and Huang, M. (2019). Effects of ultrafine comminution treatment on gelling properties of myofibrillar proteins from chicken breast, *Food Hydrocolloids*, 97, p. 105199.

[34] Ağar, B., Gençcelep, H., Saricaoğlu, F. T. and Turhan, S. (2016). Effect of sugar beet fiber concentrations on rheological properties of meat emulsions and their correlation with texture profile analysis, *Food and Bioproducts Processing*, 100, pp. 118–131.

[35] Yang, F., Zhang, M., Prakash, S. and Liu, Y. (2018). Physical properties of 3D printed baking dough as affected by different compositions, *Innovative Food Science & Emerging Technologies*, 49, pp. 202–210.

[36] Pulatsu, E., Su, J.-W., Lin, J. and Lin, M. (2020). Factors affecting 3D printing and post-processing capacity of cookie dough, *Innovative Food Science & Emerging Technologies*, 61, p. 102316.

[37] Zhang, L., Lou, Y. and Schutyser, M. A. (2018). 3D printing of cereal-based food structures containing probiotics, *Food Structure*, 18, pp. 14–22.

[38] Liu, Y., Yu, Y., Liu, C., Regenstein, J. M., Liu, X. and Zhou, P. (2019). Rheological and mechanical behavior of milk protein composite gel for extrusion-based 3D food printing, *LWT*, 102, pp. 338–346.

[39] Harper, W. J., Hewitt, S. A. and Huffman, L. M. (2020). *Milk Proteins*, "Model food systems and protein functionality" (Elsevier), pp. 573–598.

[40] Krishnaraj, P., Anukiruthika, T., Choudhary, P., Moses, J. and Anandharamakrishnan, C. (2019). 3D extrusion printing and post-processing of fibre-rich snack from indigenous composite flour, *Food and Bioprocess Technology*, 12, pp. 1776–1786.

[41] Tabilo-Munizaga, G. and Barbosa-Canovas, G. (2004). Ultra high pressure technology and its use in surimi manufacture: an overview, *Food Science and Technology International*, 10, pp. 207–222.

[42] Kim, S. M., Kim, H. W. and Park, H. J. (2021). Preparation and characterization of surimi-based imitation crab meat using coaxial extrusion three-dimensional food printing, *Innovative Food Science & Emerging Technologies*, 71, p. 102711.

[43] Alakhrash, F., Anyanwu, U. and Tahergorabi, R. (2016). Physicochemical properties of Alaska pollock (Theragra chalcograma) surimi gels with oat bran, *LWT*, 66, pp. 41–47.

[44] Liang, F., Lin, L., Zhu, Y., Jiang, S. and Lu, J. (2020). Comparative study between surimi gel and surimi/crabmeat mixed gel on nutritional

properties, flavor characteristics, color, and texture, *Journal of Aquatic Food Product Technology*, 29, pp. 681–692.

[45] Wang, L., Zhang, M., Bhandari, B. and Yang, C. (2018). Investigation on fish surimi gel as promising food material for 3D printing, *Journal of Food Engineering*, 220, pp. 101–108.

[46] Gudjónsdóttir, M., Napitupulu, R. J. and Petty Kristinsson, H. T. (2019). Low field NMR for quality monitoring of 3D printed surimi from cod by-products: effects of the pH-shift method compared with conventional washing, *Magnetic Resonance in Chemistry*, 57, pp. 638–648.

[47] Dong, X., Pan, Y., Zhao, W., Huang, Y., Qu, W., Pan, J., Qi, H. and Prakash, S. (2020). Impact of microbial transglutaminase on 3D printing quality of Scomberomorus niphonius surimi, *LWT*, 124, p. 109123.

[48] Cando, D., Herranz, B., Borderías, A. J. and Moreno, H. M. (2015). Effect of high pressure on reduced sodium chloride surimi gels, *Food Hydrocolloids*, 51, pp. 176–187.

[49] Tahergorabi, R. and Jaczynski, J. (2012). Physicochemical changes in surimi with salt substitute, *Food Chemistry*, 132, pp. 1281–1286.

[50] Qin, H., Xu, P., Zhou, C. and Wang, Y. (2015). Effects of L-Arginine on water holding capacity and texture of heat-induced gel of salt-soluble proteins from breast muscle, *LWT*, 63, pp. 912–918.

[51] Chaisawang, M. and Suphantharika, M. (2006). Pasting and rheological properties of native and anionic tapioca starches as modified by guar gum and xanthan gum, *Food Hydrocolloids*, 20, pp. 641–649.

[52] Dong, X., Huang, Y., Pan, Y., Wang, K., Prakash, S. and Zhu, B. (2019). Investigation of sweet potato starch as a structural enhancer for three-dimensional printing of Scomberomorus niphonius surimi, *Journal of Texture Studies*, 50, pp. 316–324.

[53] Hunt, A., Getty, K. and Park, J. W. (2009). Roles of starch in surimi seafood: a review, *Food Reviews International*, 25, pp. 299–312.

[54] Yoo, B. and Lee, C. (1993). Rheological relationships between surimi sol and gel as affected by ingredients, *Journal of Food Science*, 58, pp. 880–883.

[55] Chen, H.-H. and Huang, Y. C. (2008). Rheological properties of HPMC enhanced surimi analyzed by small-and large-strain tests — II: effect of water content and ingredients, *Food Hydrocolloids*, 22, pp. 313–322.

[56] Noda, T., Takigawa, S., Matsuura-Endo, C., Kim, S.-J., Hashimoto, N., Yamauchi, H., Hanashiro, I. and Takeda, Y. (2005). Physicochemical properties and amylopectin structures of large, small, and extremely small potato starch granules, *Carbohydrate Polymers*, 60, pp. 245–251.

[57] Tsai, M. L., Li, C. F. and Lii, C. Y. (1997). Effects of granular structures on the pasting behaviors of starches, *Cereal Chemistry*, 74, pp. 750–757.

[58] Liu, Z., Chen, H., Zheng, B., Xie, F. and Chen, L. (2020). Understanding the structure and rheological properties of potato starch induced by hot-extrusion 3D printing, *Food Hydrocolloids*, 105, p. 105812.

[59] Amiza, M. A. and Ng, S. C. (2015). Effects of surimi-to-silver catfish ratio and potato starch concentration on the properties of fish sausage, *Journal of Aquatic Food Product Technology*, 24, pp. 213–226.

[60] Kyaw, Z., Yu, S., Cheow, C., Dzulkifly, M. and Howell, N. (2001). Effect of fish to starch ratio on viscoelastic properties and microstructure of fish cracker ("keropok") dough, *International Journal of Food Science & Technology*, 36, pp. 741–747.

[61] Hamidi, A. and Tadesse, Y. (2019). Single step 3D printing of bioinspired structures via metal reinforced thermoplastic and highly stretchable elas-tomer, *Composite Structures*, 210, pp. 250–261.

[62] Yu, B., Ren, F., Zhao, H., Cui, B. and Liu, P. (2020). Effects of native starch and modified starches on the textural, rheological and microstruc-tural characteristics of soybean protein gel, *International Journal of Biological Macromolecules*, 142, pp. 237–243.

[63] Dondero, M., Figueroa, V., Morales, X. and Curotto, E. (2006). Transglutaminase effects on gelation capacity of thermally induced beef protein gels, *Food Chemistry*, 99, pp. 546–554.

[64] Seighalani, F. Z. B., Bakar, J., Saari, N. and Khoddami, A. (2016). Thermal and physicochemical properties of red tilapia (Oreochromis niloticus) surimi gel as affected by microbial transglutaminase, *Animal Production Science*, 57, pp. 993–1000.

[65] Liu, Z., Zhang, M. and Bhandari, B. (2018). Effect of gums on the rheo-logical, microstructural and extrusion printing characteristics of mashed potatoes, *International Journal of Biological Macromolecules*, 117, pp. 1179–1187.

[66] Cando, D., Borderías, A. J. and Moreno, H. (2016). Combined effect of aminoacids and microbial transglutaminase on gelation of low salt surimi content under high pressure processing, *Innovative Food Science & Emerging Technologies*, 36, pp. 10–17.

[67] Ma, X.-S., Yi, S.-M., Yu, Y.-M., Li, J.-R. and Chen, J.-R. (2015). Changes in gel properties and water properties of Nemipterus virgatus surimi gel induced by high-pressure processing, *LWT*, 61, pp. 377–384.

[68] Dick, A., Bhandari, B. and Prakash, S. (2019). Post-processing feasibility of composite-layer 3D printed beef, *Meat Science*, 153, pp. 9–18.

[69] Adewale, P., Dumont, M. J. and Ngadi, M. (2014). Rheological, thermal, and physicochemical characterization of animal fat wastes for use in bio-diesel production, *Energy Technology*, 2, pp. 634–642.

[70] Oroszvári, B. K., Rocha, C. S., Sjöholm, I. and Tornberg, E. (2006). Permeability and mass transfer as a function of the cooking temperature

during the frying of beefburgers, *Journal of Food Engineering*, 74, pp. 1–12.

[71] Braeckman, L., Ronsse, F., Hidalgo, P. C. and Pieters, J. (2009). Influence of combined IR-grilling and hot air cooking conditions on moisture and fat content, texture and colour attributes of meat patties, *Journal of Food Engineering*, 93, pp. 437–443.

[72] Dick, A., Bhandari, B. and Prakash, S. (2021). Printability and textural assessment of modified-texture cooked beef pastes for dysphagia patients, *Future Foods*, 3, p. 100006.

[73] Oliveira, S. M., Fasolin, L. H., Vicente, A. A., Fuciños, P. and Pastrana, L. M. (2020). Printability, microstructure, and flow dynamics of phase-separated edible 3D inks, *Food Hydrocolloids*, 109, p. 106120.

[74] McArdle, R. and Hamill, R. (2011). *Processed Meats*, "Utilisation of hydrocolloids in processed meat systems" (Elsevier), pp. 243–269.

[75] Varghese, C., Wolodko, J., Chen, L., Doschak, M., Srivastav, P. P. and Roopesh, M. (2020). Influence of selected product and process parameters on microstructure, rheological, and textural properties of 3D printed cookies, *Foods*, 9, p. 907.

[76] Nguyen, N. A., Bowland, C. C. and Naskar, A. K. (2018). A general method to improve 3D-printability and inter-layer adhesion in lignin-based composites, *Applied Materials Today*, 12, pp. 138–152.

[77] Trius, A., Sebranek, J. and Lanier, T. (1996). Carrageenans and their use in meat products, *Critical Reviews in Food Science & Nutrition*, 36, pp. 69–85.

[78] Shang, Y. and Xiong, Y. L. (2010). Xanthan enhances water binding and gel formation of transglutaminase-treated porcine myofibrillar proteins, *Journal of Food Science*, 75, pp. E178–E185.

[79] Foegeding, E. and Ramsey, S. (1986). Effect of gums on low-fat meat batters, *Journal of Food Science*, 51, pp. 33–36.

[80] Wu, M., Xiong, Y. L., Chen, J., Tang, X. and Zhou, G. (2009). Rheological and microstructural properties of porcine myofibrillar protein–lipid emulsion composite gels, *Journal of Food Science*, 74, pp. E207–E217.

[81] Stangierski, J. and Baranowska, H. M. (2015). The influence of heating and cooling process on the water binding in transglutaminase-modified chicken protein preparation, assessed using Low-Field NMR, *Food and Bioprocess Technology*, 8, pp. 2359–2367.

[82] Dufresne, T. and Germain, I. (2003) Method of Preparation of Adapted Foods. Patent Cooperation Treaty. World Intellectual Property Organization.

[83] Dick, A., Bhandari, B. and Prakash, S. (2021). Effect of reheating method on the post-processing characterisation of 3D printed meat products for dysphagia patients, *LWT*, p. 111915.

[84] Mizrahi, S. (2012). Mechanisms of objectionable textural changes by microwave reheating of foods: a review, *Journal of Food Science*, 77, pp. R57–R62.

[85] Gill, C., Devos, J., Badoni, M. and Yang, X. (2016). Inactivation of Escherichia coli O157: H7 in beef roasts cooked in conventional or convection ovens or in a slow cooker under selected conditions, *Journal of Food Protection*, 79, pp. 205–212.

[86] Modzelewska-Kapituła, M., Pietrzak-Fiećko, R., Tkacz, K., Draszanowska, A. and Więk, A. (2019). Influence of sous vide and steam cooking on mineral contents, fatty acid composition and tenderness of semimembranosus muscle from Holstein-Friesian bulls, *Meat Science*, 157, p. 107877.

[87] Hutton, C., Neggers, Y. and Love, T. (1981). Scanning electron microscopy, proteolytic enzyme activity, and acceptability of beef semitendinosus cooked by microwaves and conventional heat, *Journal of Food Science*, 46, pp. 1309–1314.

[88] Gropper, M., Ramon, O., Kopelman, I. and Mizrahi, S. (1997). Effects of microwave reheating on surimi gel texture, *Food Research International*, 30, pp. 761–768.

[89] Gimeno, E., Moraru, C. and Kokini, J. (2004). Effect of xanthan gum and CMC on the structure and texture of corn flour pellets expanded by microwave heating, *Cereal Chemistry*, 81, pp. 100–107.

[90] Thombare, N., Jha, U., Mishra, S. and Siddiqui, M. (2016). Guar gum as a promising starting material for diverse applications: a review, *International Journal of Biological Macromolecules*, 88, pp. 361–372.

[91] Tornberg, E. (2005). Effects of heat on meat proteins — implications on structure and quality of meat products, *Meat Science*, 70, pp. 493–508.

[92] van der Weele, C., Feindt, P., van der Goot, A. J., van Mierlo, B. and van Boekel, M. (2019). Meat alternatives: an integrative comparison, *Trends in Food Science & Technology*, 88, pp. 505–512.

[93] Ong, S., Choudhury, D. and Naing, M. W. (2020). Cell-based meat: current ambiguities with nomenclature, *Trends in Food Science & Technology*, 102, pp. 223–231.

[94] Rolland, N. C., Markus, C. R. and Post, M. J. (2020). Correction: the effect of information content on acceptance of cultured meat in a tasting context, *PLOS One*, 15, p. e0240630.

[95] Siegrist, M. and Hartmann, C. (2020). Consumer acceptance of novel food technologies, *Nature Food*, 1, pp. 343–350.

[96] Orzechowski, A. (2015). Artificial meat? Feasible approach based on the experience from cell culture studies, *Journal of Integrative Agriculture*, 14, pp. 217–221.

[97] Collins, C. A. and Partridge, T. A. (2005). Self-renewal of the adult skeletal muscle satellite cell, *Cell Cycle*, 4, pp. 1338–1341.

[98]   Sabourin, L. A. and Rudnicki, M. A. (2000). The molecular regulation of myogenesis, *Clinical Genetics*, 57, pp. 16–25.

[99]   Adams, G. R., Haddad, F. and Baldwin, K. M. (1999). Time course of changes in markers of myogenesis in overloaded rat skeletal muscles, *Journal of Applied Physiology*, 87, pp. 1705–1712.

[100]  Bonnet, M., Cassar-Malek, I., Chilliard, Y. and Picard, B. (2010). Ontogenesis of muscle and adipose tissues and their interactions in ruminants and other species, *Animal*, 4, pp. 1093–1109.

[101]  Raschke, S. and Eckel, J. (2013). Adipo-myokines: two sides of the same coin — mediators of inflammation and mediators of exercise, *Mediators of Inflammation*, 2013.

[102]  Sun, J., Zhou, W., Yan, L., Huang, D. and Lin, L.-y. (2018). Extrusion-based food printing for digitalized food design and nutrition control, *Journal of Food Engineering*, 220, pp. 1–11.

[103]  Zhang, Y. S., Oklu, R., Dokmeci, M. R. and Khademhosseini, A. (2018). Three-dimensional bioprinting strategies for tissue engineering, *Cold Spring Harbor Perspectives in Medicine*, 8, p. a025718.

[104]  Gaydhane, M. K., Mahanta, U., Sharma, C. S., Khandelwal, M. and Ramakrishna, S. (2018). Cultured meat: state of the art and future, *Biomanufacturing Reviews*, 3, pp. 1–10.

[105]  Stephens, N., Sexton, A. E. and Driessen, C. (2019). Making sense of making meat: key moments in the first 20 years of tissue engineering muscle to make food, *Frontiers in Sustainable Food Systems*, 3, p. 45.

[106]  Rorheim, A., Mannino, A., Baumann, T. and Caviola, L. (2016). Cultured meat: an ethical alternative to industrial animal farming, *Policy Paper by Sentience Politics*, 1, pp. 1–14.

[107]  Ben-Arye, T. and Levenberg, S. (2019). Tissue engineering for clean meat production, *Frontiers in Sustainable Food Systems*, 3, p. 46.

[108]  Sher, D. and Tutó, X. (2015). Review of 3D food printing, *Temes de disseny*, 31, pp. 104–117.

[109]  Portanguen, S., Tournayre, P., Sicard, J., Astruc, T. and Mirade, P.-S. (2019). Toward the design of functional foods and biobased products by 3D printing: a review, *Trends in Food Science & Technology*, 86, pp. 188–198.

[110]  Zhang, B., Gao, L., Ma, L., Luo, Y., Yang, H. and Cui, Z. (2019). 3D bioprinting: a novel avenue for manufacturing tissues and organs, *Engineering*, 5, pp. 777–794.

[111]  Munaz, A., Vadivelu, R. K., John, J. S., Barton, M., Kamble, H. and Nguyen, N.-T. (2016). Three-dimensional printing of biological matters, *Journal of Science: Advanced Materials and Devices*, 1, pp. 1–17.

[112]  Kačarević, Ž. P., Rider, P. M., Alkildani, S., Retnasingh, S., Smeets, R., Jung, O., Ivanišević, Z. and Barbeck, M. (2018). An introduction to 3D

bioprinting: possibilities, challenges and future aspects, *Materials*, 11, p. 2199.

[113] Lee, J. M. and Yeong, W. Y. (2016). Design and printing strategies in 3D bioprinting of cell-hydrogels: a review, *Advanced Healthcare Materials*, 5, pp. 2856–2865.

[114] Min, S., Ko, I. K. and Yoo, J. J. (2019). State-of-the-art strategies for the vascularization of three-dimensional engineered organs, *Vascular Specialist International*, 35, p. 77.

[115] Zhao, L., Huang, Y. and Du, M. (2019). Farm animals for studying muscle development and metabolism: dual purposes for animal production and human health, *Animal Frontiers*, 9, pp. 21–27.

[116] Kim, J. H., Seol, Y.-J., Ko, I. K., Kang, H.-W., Lee, Y. K., Yoo, J. J., Atala, A. and Lee, S. J. (2018). 3D bioprinted human skeletal muscle constructs for muscle function restoration, *Scientific Reports*, 8, pp. 1–15.

[117] Kang, H.-W., Lee, S. J., Ko, I. K., Kengla, C., Yoo, J. J. and Atala, A. (2016). A 3D bioprinting system to produce human-scale tissue constructs with structural integrity, *Nature Biotechnology*, 34, pp. 312–319.

[118] Kadim, I. T., Mahgoub, O., Baqir, S., Faye, B. and Purchas, R. (2015). Cultured meat from muscle stem cells: a review of challenges and prospects, *Journal of Integrative Agriculture*, 14, pp. 222–233.

[119] Roberts, R. M., Yuan, Y., Genovese, N. and Ezashi, T. (2015). Livestock models for exploiting the promise of pluripotent stem cells, *ILAR Journal*, 56, pp. 74–82.

[120] de Figueiredo Pessôa, L. V., Bressan, F. F. and Freude, K. K. (2019). Induced pluripotent stem cells throughout the animal kingdom: availability and applications, *World Journal of Stem Cells*, 11, p. 491.

[121] Wankhade, U. D., Shen, M., Kolhe, R. and Fulzele, S. (2016). Advances in adipose-derived stem cells isolation, characterization, and application in regenerative tissue engineering, *Stem Cells International*, 2016.

[122] Forcina, L., Miano, C., Pelosi, L. and Musarò, A. (2019). An overview about the biology of skeletal muscle satellite cells, *Current Genomics*, 20, pp. 24–37.

[123] Francesco, S., Nicolò, B., Michele, P. G. and Edoardo, R. (2019). From liposuction to adipose-derived stem cells: indications and technique, *Acta Bio Medica: Atenei Parmensis*, 90, p. 197.

[124] Dessels, C., Potgieter, M. and Pepper, M. S. (2016). Making the switch: alternatives to fetal bovine serum for adipose-derived stromal cell expansion, *Frontiers in Cell and Developmental Biology*, 4, p. 115.

[125] Shah, G. (1999). Why do we still use serum in the production of biopharmaceuticals? *Developments in Biological Standardization*, 99, pp. 17–22.

[126] Froud, S. (1999). The development, benefits and disadvantages of serum-free media, *Developments in Biological Standardization*, 99, pp. 157–166.

[127]  Andreassen, R. C., Pedersen, M. E., Kristoffersen, K. A. and Rønning, S. B. (2020). Screening of by-products from the food industry as growth promoting agents in serum-free media for skeletal muscle cell culture, *Food & Function*, 11, pp. 2477–2488.

[128]  Kolkmann, A., Post, M., Rutjens, M., Van Essen, A. and Moutsatsou, P. (2020). Serum-free media for the growth of primary bovine myoblasts, *Cytotechnology*, 72, pp. 111–120.

[129]  McFarland, D. C., Pesall, J. E., Norberg, J. M. and Dvoracek, M. A. (1991). Proliferation of the turkey myogenic satellite cell in a serum-free medium, *Comparative Biochemistry and Physiology Part A: Physiology*, 99, pp. 163–167.

[130]  Benjaminson, M. A., Gilchriest, J. A. and Lorenz, M. (2002). *In vitro* edible muscle protein production system (MPPS): stage 1, fish, *Acta Astronautica*, 51, pp. 879–889.

[131]  Dodson, M., McFarland, D., Grant, A., Doumit, M. and Velleman, S. (1996). Extrinsic regulation of domestic animal-derived satellite cells, *Domestic Animal Endocrinology*, 13, pp. 107–126.

[132]  Baquero-Perez, B., Kuchipudi, S. V., Nelli, R. K. and Chang, K.-C. (2012). A simplified but robust method for the isolation of avian and mammalian muscle satellite cells, *BMC Cell Biology*, 13, pp. 1–12.

[133]  Dodson, M. and Mathison, B. (1988). Comparison of ovine and rat muscle-derived satellite cells: response to insulin, *Tissue and Cell*, 20, pp. 909–918.

[134]  Doumit, M. E., Cook, D. R. and Merkel, R. A. (1993). Fibroblast growth factor, epidermal growth factor, insulin-like growth factors, and platelet-derived growth factor-BB stimulate proliferation of clonally derived porcine myogenic satellite cells, *Journal of Cellular Physiology*, 157, pp. 326–332.

[135]  Duque, P., Gómez, E., Díaz, E., Facal, N., Hidalgo, C. and Díez, C. (2003). Use of two replacements of serum during bovine embryo culture in vitro, *Theriogenology*, 59, pp. 889–899.

[136]  Benders, A. A., van Kuppevelt, T. H., Oosterhof, A. and Veerkamp, J. H. (1991). The biochemical and structural maturation of human skeletal muscle cells in culture: the effect of the serum substitute Ultroser G, *Experimental Cell Research*, 195, pp. 284–294.

[137]  Jairath, G., Mal, G., Gopinath, D. and Singh, B. (2021). An holistic approach to access the viability of cultured meat: a review, *Trends in Food Science & Technology*.

[138]  Post, M. J., Levenberg, S., Kaplan, D. L., Genovese, N., Fu, J., Bryant, C. J., Negowetti, N., Verzijden, K. and Moutsatsou, P. (2020). Scientific, sustainability and regulatory challenges of cultured meat, *Nature Food*, 1, pp. 403–415.

[139] Narayanan, N., Jiang, C., Wang, C., Uzunalli, G., Whittern, N., Chen, D., Jones, O. G., Kuang, S. and Deng, M. (2020). Harnessing fiber diameter-dependent effects of myoblasts toward biomimetic scaffold-based skeletal muscle regeneration, *Frontiers in Bioengineering and Biotechnology*, 8, p. 203.

[140] Jun, I., Jeong, S. and Shin, H. (2009). The stimulation of myoblast differentiation by electrically conductive sub-micron fibers, *Biomaterials*, 30, pp. 2038–2047.

[141] MacQueen, L. A., Alver, C. G., Chantre, C. O., Ahn, S., Cera, L., Gonzalez, G. M., O'Connor, B. B., Drennan, D. J., Peters, M. M. and Motta, S. E. (2019). Muscle tissue engineering in fibrous gelatin: implications for meat analogs, *npj Science of Food*, 3, pp. 1–12.

[142] Ben-Arye, T., Shandalov, Y., Ben-Shaul, S., Landau, S., Zagury, Y., Ianovici, I., Lavon, N. and Levenberg, S. (2020). Textured soy protein scaffolds enable the generation of three-dimensional bovine skeletal muscle tissue for cell-based meat, *Nature Food*, 1, pp. 210–220.

[143] Abbasi, J. (2020). Soy scaffoldings poised to make cultured meat more affordable, *JAMA*, 323, pp. 1764–1764.

[144] Canavan, H. E., Cheng, X., Graham, D. J., Ratner, B. D. and Castner, D. G. (2005). Cell sheet detachment affects the extracellular matrix: a surface science study comparing thermal liftoff, enzymatic, and mechanical methods, *Journal of Biomedical Materials Research Part A: An Official Journal of The Society for Biomaterials, The Japanese Society for Biomaterials, and The Australian Society for Biomaterials and the Korean Society for Biomaterials*, 75, pp. 1–13.

[145] Da Silva, R. M., Mano, J. F. and Reis, R. L. (2007). Smart thermoresponsive coatings and surfaces for tissue engineering: switching cell–material boundaries, *Trends in Biotechnology*, 25, pp. 577–583.

[146] Lam, M. T., Huang, Y.-C., Birla, R. K. and Takayama, S. (2009). Microfeature guided skeletal muscle tissue engineering for highly organized 3-dimensional free-standing constructs, *Biomaterials*, 30, pp. 1150–1155.

[147] Duan, B., Hockaday, L. A., Kang, K. H. and Butcher, J. T. (2013). 3D bioprinting of heterogeneous aortic valve conduits with alginate/gelatin hydrogels, *Journal of Biomedical Materials Research Part A*, 101, pp. 1255–1264.

[148] Capel, A. J., Rimington, R. P., Fleming, J. W., Player, D. J., Baker, L. A., Turner, M. C., Jones, J. M., Martin, N. R., Ferguson, R. A. and Mudera, V. C. (2019). Scalable 3D printed molds for human tissue engineered skeletal muscle, *Frontiers in Bioengineering and Biotechnology*, 7, p. 20.

[149] Post, M. J. (2014). Cultured beef: medical technology to produce food, *Journal of the Science of Food and Agriculture*, 94, pp. 1039–1041.

[150] Simsa, R., Yuen, J., Stout, A., Rubio, N., Fogelstrand, P. and Kaplan, D. L. (2019). Extracellular heme proteins influence bovine myosatellite cell proliferation and the color of cell-based meat, *Foods*, 8, p. 521.

[151] Choudhury, D., Tseng, T. W. and Swartz, E. (2020). The business of cultured meat, *Trends in Biotechnology*, 38, pp. 573–577.

[152] Koohmaraie, M., Kent, M. P., Shackelford, S. D., Veiseth, E. and Wheeler, T. L. (2002). Meat tenderness and muscle growth: is there any relationship? *Meat Science*, 62, pp. 345–352.

[153] Furuhashi, M., Morimoto, Y., Shima, A., Nakamura, F., Ishikawa, H. and Takeuchi, S. (2021). Formation of contractile 3D bovine muscle tissue for construction of millimetre-thick cultured steak, *npj Science of Food*, 5, pp. 1–8.

[154] Morimoto, Y., Mori, S., Sakai, F. and Takeuchi, S. (2016). Human induced pluripotent stem cell-derived fiber-shaped cardiac tissue on a chip, *Lab on a Chip*, 16, pp. 2295–2301.

[155] Morimoto, Y., Kato-Negishi, M., Onoe, H. and Takeuchi, S. (2013). Three-dimensional neuron–muscle constructs with neuromuscular junctions, *Biomaterials*, 34, pp. 9413–9419.

[156] Mittal, G., Nadulski, R., Barbut, S. and Negi, S. (1992). Textural profile analysis test conditions for meat products, *Food Research International*, 25, pp. 411–417.

[157] Kang, D.-H., Louis, F., Liu, H., Shimoda, H., Nishiyama, Y., Nozawa, H., Kakitani, M., Takagi, D., Kasa, D. and Nagamori, E. (2021). Engineered whole cut meat-like tissue by the assembly of cell fibers using tendon-gel integrated bioprinting, *Nature Communications*, 12, pp. 1–12.

[158] Bhattacharjee, T., Zehnder, S. M., Rowe, K. G., Jain, S., Nixon, R. M., Sawyer, W. G. and Angelini, T. E. (2015). Writing in the granular gel medium, *Science Advances*, 1, p. e1500655.

[159] Hinton, T. J., Jallerat, Q., Palchesko, R. N., Park, J. H., Grodzicki, M. S., Shue, H.-J., Ramadan, M. H., Hudson, A. R. and Feinberg, A. W. (2015). Three-dimensional printing of complex biological structures by freeform reversible embedding of suspended hydrogels, *Science Advances*, 1, p. e1500758.

[160] Choi, Y.-J., Jun, Y.-J., Kim, D. Y., Yi, H.-G., Chae, S.-H., Kang, J., Lee, J., Gao, G., Kong, J.-S. and Jang, J. (2019). A 3D cell printed muscle construct with tissue-derived bioink for the treatment of volumetric muscle loss, *Biomaterials*, 206, pp. 160–169.

[161] Jeon, O., Lee, Y. B., Jeong, H., Lee, S. J., Wells, D. and Alsberg, E. (2019). Individual cell-only bioink and photocurable supporting medium for 3D printing and generation of engineered tissues with complex geometries, *Materials Horizons*, 6, pp. 1625–1631.

[162] Lee, A., Hudson, A., Shiwarski, D., Tashman, J., Hinton, T., Yerneni, S., Bliley, J., Campbell, P. and Feinberg, A. (2019). 3D bioprinting of collagen to rebuild components of the human heart, *Science*, 365, pp. 482–487.

[163] Noor, N., Shapira, A., Edri, R., Gal, I., Wertheim, L. and Dvir, T. (2019). 3D printing of personalized thick and perfusable cardiac patches and hearts, *Advanced Science*, 6, p. 1900344.

[164] Skylar-Scott, M. A., Uzel, S. G., Nam, L. L., Ahrens, J. H., Truby, R. L., Damaraju, S. and Lewis, J. A. (2019). Biomanufacturing of organ-specific tissues with high cellular density and embedded vascular channels, *Science Advances*, 5, p. eaaw2459.

[165] Choi, Y. J., Kim, T. G., Jeong, J., Yi, H. G., Park, J. W., Hwang, W. and Cho, D. W. (2016). 3D cell printing of functional skeletal muscle constructs using skeletal muscle-derived bioink, *Advanced Healthcare Materials*, 5, pp. 2636–2645.

[166] Heher, P., Maleiner, B., Prüller, J., Teuschl, A. H., Kollmitzer, J., Monforte, X., Wolbank, S., Redl, H., Rünzler, D. and Fuchs, C. (2015). A novel bioreactor for the generation of highly aligned 3D skeletal muscle-like constructs through orientation of fibrin via application of static strain, *Acta Biomaterialia*, 24, pp. 251–265.

[167] Jones, J. M., Player, D. J., Martin, N. R., Capel, A. J., Lewis, M. P. and Mudera, V. (2018). An assessment of myotube morphology, matrix deformation, and myogenic mRNA expression in custom-built and commercially available engineered muscle chamber configurations, *Frontiers in Physiology*, 9, p. 483.

[168] Prüller, J., Mannhardt, I., Eschenhagen, T., Zammit, P. S. and Figeac, N. (2018). Satellite cells delivered in their niche efficiently generate functional myotubes in three-dimensional cell culture, *PLOS One*, 13, p. e0202574.

[169] Rao, L., Qian, Y., Khodabukus, A., Ribar, T. and Bursac, N. (2018). Engineering human pluripotent stem cells into a functional skeletal muscle tissue, *Nature Communications*, 9, pp. 1–12.

[170] Li, X., Zhang, G., Zhao, X., Zhou, J., Du, G. and Chen, J. (2020). A conceptual air-lift reactor design for large scale animal cell cultivation in the context of *in vitro* meat production, *Chemical Engineering Science*, 211, p. 115269.

[171] Specht, E. A., Welch, D. R., Clayton, E. M. R. and Lagally, C. D. (2018). Opportunities for applying biomedical production and manufacturing methods to the development of the clean meat industry, *Biochemical Engineering Journal*, 132, pp. 161–168.

[172] Heng, B. C., Bezerra, P. P., Preiser, P. R., Alex Law, S., Xia, Y., Boey, F. and Venkatraman, S. S. (2011). Effect of cell-seeding density on the

proliferation and gene expression profile of human umbilical vein endothelial cells within ex vivo culture, *Cytotherapy*, 13, pp. 606–617.

[173]　Larson, B. L., Ylöstalo, J. and Prockop, D. J. (2008). Human multipotent stromal cells undergo sharp transition from division to development in culture, *Stem Cells*, 26, pp. 193–201.

[174]　Stephenson, M. and Grayson, W. (2018). Recent advances in bioreactors for cell-based therapies, *F1000Research*, 7.

[175]　Moritz, M. S., Verbruggen, S. E. and Post, M. J. (2015). Alternatives for large-scale production of cultured beef: a review, *Journal of Integrative Agriculture*, 14, pp. 208–216.

[176]　Verbruggen, S., Luining, D., van Essen, A. and Post, M. J. (2018). Bovine myoblast cell production in a microcarriers-based system, *Cytotechnology*, 70, pp. 503–512.

[177]　Burrell, K., Dardari, R., Goldsmith, T., Toms, D., Villagomez, D. A., King, W. A., Ungrin, M., West, F. D. and Dobrinski, I. (2019). Stirred suspension bioreactor culture of porcine induced pluripotent stem cells, *Stem Cells and Development*, 28, pp. 1264–1275.

[178]　Lipsitz, Y. Y., Woodford, C., Yin, T., Hanna, J. H. and Zandstra, P. W. (2018). Modulating cell state to enhance suspension expansion of human pluripotent stem cells, *Proceedings of the National Academy of Sciences*, 115, pp. 6369–6374.

[179]　Abbasalizadeh, S., Larijani, M. R., Samadian, A. and Baharvand, H. (2012). Bioprocess development for mass production of size-controlled human pluripotent stem cell aggregates in stirred suspension bioreactor, *Tissue Engineering Part C: Methods*, 18, pp. 831–851.

[180]　Chen, V. C., Couture, S. M., Ye, J., Lin, Z., Hua, G., Huang, H.-I. P., Wu, J., Hsu, D., Carpenter, M. K. and Couture, L. A. (2012). Scalable GMP compliant suspension culture system for human ES cells, *Stem Cell Research*, 8, pp. 388–402.

[181]　Fok, E. Y. and Zandstra, P. W. (2005). Shear-controlled single-step mouse embryonic stem cell expansion and embryoid body-based differentiation, *Stem Cells*, 23, pp. 1333–1342.

[182]　Tsai, A. C., Liu, Y., Yuan, X., Chella, R. and Ma, T. (2017). Aggregation kinetics of human mesenchymal stem cells under wave motion, *Biotechnology Journal*, 12, p. 1600448.

[183]　Schnitzler, A. C., Verma, A., Kehoe, D. E., Jing, D., Murrell, J. R., Der, K. A., Aysola, M., Rapiejko, P. J., Punreddy, S. and Rook, M. S. (2016). Bioprocessing of human mesenchymal stem/stromal cells for therapeutic use: current technologies and challenges, *Biochemical Engineering Journal*, 108, pp. 3–13.

[184] Wung, N., Acott, S. M., Tosh, D. and Ellis, M. J. (2014). Hollow fibre membrane bioreactors for tissue engineering applications, *Biotechnology Letters*, 36, pp. 2357–2366.

[185] Chaudhuri, J. and Al-Rubeai, M. (2005). *Bioreactors for Tissue Engineering* (Springer).

[186] Morrow, D., Ussi, A. and Migliaccio, G. (2017). Addressing pressing needs in the development of advanced therapies, *Frontiers in Bioengineering and Biotechnology*, 5, p. 55.

[187] Moutsatsou, P., Ochs, J., Schmitt, R., Hewitt, C. and Hanga, M. (2019). Automation in cell and gene therapy manufacturing: from past to future, *Biotechnology Letters*, 41, pp. 1245–1253.

[188] Ismail, I., Hwang, Y.-H. and Joo, S.-T. (2020). Meat analog as future food: a review, *Journal of Animal Science Technology*, 62, pp. 111–120.

[189] Fraser, R. Z., Shitut, M., Agrawal, P., Mendes, O. and Klapholz, S. (2018). Safety evaluation of soy leghemoglobin protein preparation derived from pichia pastoris, intended for use as a flavor catalyst in plant-based meat, *International Journal of Toxicology*, 37, pp. 241–262.

[190] Asgar, M. A., Fazilah, A., Huda, N., Bhat, R. and Karim, A. A. (2010). Nonmeat protein alternatives as meat extenders and meat analogs, *Comprehensive Reviews in Food Science and Food Safety*, 9, pp. 513–529.

[191] Joshi, V. K. and Kumar, S. (2015). Meat analogues: plant based alternatives to meat products — a review, *International Journal of Food and Fermentation Technology*, 5, pp. 107–119.

[192] Hsieh, Y. P. C., Pearson, A. M. and Magee, W. T. (1980). Development of a synthetic meat flavor mixture by using surface response methodology, *Journal Food Science*, 45, p. 1125–1130.

[193] Rubio, N. R., Xiang, N. and Kaplan, D. L. (2020). Plant-based and cell-based approaches to meat production, *Nature Communications*, 11, p. 6276. https://doi.org/6210.1038/s41467-41020-20061-y.

[194] Krintiras, G. A., Gadea Diaz, J., Van Der Goot, A. J., Stankiewicz, A. I. and Stefanidis, G. D. (2016). On the use of the Couette Cell technology for large scale production of textured soy-based meat replacers, *Journal of Food Engineering*, 169, pp. 205–213.

[195] Bohrer, B. M. (2019). An investigation of the formulation and nutritional composition of modern meat analogue products, *Food Science and Human Wellness*, 8, pp. 320–329.

[196] Ramachandraiah, K., Choi, M.-J. and Hong, G.-P. (2018). Micro- and nano-scaled materials for strategy-based applications in innovative livestock products: a review, *Trends in Food Science and Technology*, 71, pp. 25–35.

[197]   Ramachandraiah, K. (2021). Potential development of sustainable 3D-printed meat analogues: a review, *Sustainability*, 13, p. 938. https://doi.org/910.3390/su13020938.

[198]   Chen, Y., Zhang, M. and Bhandari, B. (2021). 3D printing of steak-like foods based on textured soybean protein, *Foods*, 10, p. 2011. https://doi.org/2010.3390/foods10092011.

[199]   Phuhongsung, P., Zhang, M. and Devahastin, S. (2020). Investigation on 3D printing ability of soybean protein isolate gels and correlations with their rheological and textural properties via LF-NMR spectroscopic characteristics, *LWT*, p. 109019. https://doi.org/10.1016/j.lwt.2020.109019.

[200]   Roy, F., Boye, J. I. and Simpson, B. K. (2010). Bioactive proteins and peptides in pulse crops: pea, chickpea and lentil, *Food Research International*, 43, pp. 432–442.

[201]   Oyinloye, T. M. and Yoon, W. B. (2021). Stability of 3D printing using a mixture of pea protein and alginate: precision and application of additive layer manufacturing simulation approach for stress distribution, *Journal of Food Engineering*, p. 110127. https://doi.org/10.1016/j.jfoodeng.2020.110127.

[202]   Azam, R. S., Zhang, M., Bhandari, B. and Yang, C. (2018). Effect of different gums on features of 3D printed object based on vitamin-D enriched orange concentrate, *Food Biophysics*, 13, pp. 250–262.

[203]   Liu, Y., Liang, X., Saeed, A., Lan, W. and Qin, W. (2019). Properties of 3D printed dough and optimization of printing parameters, *Innovative Food Science & Emerging Technologies*, 54, pp. 9–18.

[204]   Kim, Y., Kim, H. W. and Park, H. J. (2021). Effect of pea protein isolate incorporation on 3D printing performance and tailing effect of banana paste, *LWT*, 150, p. 111916.

[205]   Neacsu, M., Mcbey, D. and Johnstone, A. M. (2017). Meat reduction and plant-based food: replacement of meat: nutritional, health, and social aspects. The Rowett Institute of Nutrition and Health Aberdeen Centre For Environmental Sustainability School of Medicine, Medical Sciences & Nutrition.

[206]   Hu, F. B., Otis, B. O. and McCarthy, G. (2019). Can plant-based meat alternatives be part of a healthy and sustainable diet? *JAMA*, pp. 1547–1548. https://doi.org/10.1001/jama.2019.13187.

[207]   Novameat. (2021). Available online: https://www.novameat.com/. Accessed on: 2 Dec. 2021.

[208]   Redefinemeat. (2021). Available online: https://www.redefinemeat.com/. Accessed on: 2 Dec. 2021.

[209]   Renofoods. (2021). Availble online: https://revo-foods.com/. Accessed on: 2 Dec. 2021.

[210] Foodnavigator. (2020). Available online: https://www.foodnavigator.com/ Article/2020/07/06/3D-printed-fish-Plant-based-salmon-with-complex-structure-under-development-for-EU-market. Accessed on: 2 Dec. 2021.

[211] The Straits Times. (2021). Available online: https://www.straitstimes.com/singapore/environment/graduate-who-made-3d-printed-salmon-from-lentils-among-69-astar-scholars. Accessed on: 2 Dec 2021.

[212] Sun-Waterhouse, D., Waterhouse, G. I. N., You, L., Zhang, J., Liu, Y., Ma, L., and Dong, Y. (2016). Transforming insect biomass into consumer wellness foods: a review, *Food Research International*, 89, pp. 129–151.

[213] Alexander, P., Brown, C., Arneth, A., Dias, C., Finnigan, J., Moran, D. and Rounsevell, M. D. A. (2017). Could consumption of insects, cultured meat or imitation meat reduce global agricultural land use? *Global Food Security*, 15, pp. 22–32.

[214] Azzollini, D., Wibisaphira, T., Lakemond, C. M. M. and Fogliano, V. (2019). Toward the design of insect-based meat analogue: the role of calcium and temperature in coagulation behavior of Alphitobius diaperinus proteins, *LWT*, 100, pp. 75–82.

[215] Soares, S. and Forkes, A. (2014). Insects au gratin — an investigation into the experiences of developing a 3D printer that uses insect protein based flour as a building medium for the production of sustainable food, In *Proceedings of the 16th International Conference on Engineering and Product Design Education*, pp. 426–431.

[216] Severini, C., Azzollini, D., Albenzio, M. and Derossi, A. (2018). On printability, quality and nutritional properties of 3D printed cereal based snacks enriched with edible insects, *Food Research International*, 106, pp. 666–676.

[217] Bessa, L. W., Pieterse, E., Marais, J. and Hoffman, L. C. (2020). Why for feed and not for human consumption? The black soldier fly larvae, *Comprehensive Reviews in Food Science and Food Safe*, 19, pp. 2747–2763.

[218] Kim, C. H., Ryu, J. H., Lee, J. K., Ko, K. Y., Lee, J. Y., Park, K. Y. and Chung, H. G. (2021). Use of black soldier fly larvae for food waste treatment and energy production in asian countries: a review, *Processes*, 9, p. 161. https://doi.org/110.3390/pr9010161.

[219] Hartmann, C. and Siegrist, M. (2017). Consumer perception and behaviour regarding sustainable protein consumption: a systematic review, *Trends in Food Science & Technology*, 61, pp. 11–25.

[220] Al-Ansi, W., Ali Mahdi, A., Li, Y., Qian, H. and Wang, L. (2018). Optimization and acceptability evaluation of shapporah biscuits formulated by different ingredients: using Response Surface Methodology (RSM), *Journal of Food and Nutrition Research*, 6, pp. 192–199.

[221]  Fombong, F. T., Van Der Borght, M. and Vanden Broeck, J. (2017). Influence of freeze-drying and oven-drying post blanching on the nutrient composition of the edible insect Ruspolia differens, *Insects*, 8, p. 102.

# Problems

1. Discuss the reasons for the 3D printing of meat, plant proteins, and alternative proteins. Do you agree with them? Why or why not?
2. What are some of the real meats that have been 3D-printed?
3. What are the additives used in meat inks? Why are they added?
4. Define cultured meat. How does 3D printing value-add to the culturing of meat *in vitro*?
5. Discuss the technologies used in the culturing of beef and pork.
6. Discuss the considerations necessary to develop 3D-printed meat for dysphagic patients.
7. Discuss examples of bioreactors. How can they be developed further to advance 3D printing of meats?
8. Define plant-based protein. How are they different from animal meat?
9. What are the 3D printing methods that make plant-based protein to feel like real meat?
10. What are the advantages of insect protein over animal protein?
11. What are the challenges for insect proteins to be considered as part of the human diet?
12. Compare and contrast the pros and cons of using 3D printing for alternative proteins (cultured meat, plant-based meat, and insect proteins).

# Chapter 6

# 3D Printing of Pharmaceuticals and Nutraceuticals

## 6.1 Introduction

3D printing technologies are enabling new advances in the field of pharmaceuticals and nutraceuticals. In this chapter, we discuss the new advances in the manufacture of pharmaceuticals and nutraceuticals made possible by 3D printing processes. The motivations of utilizing 3D printing technologies are presented, and their applications are analyzed in terms of materials and designs in a tabular format. In addition, four companies are highlighted in this chapter, where we discuss the principles and processes employed by these companies to print pharmaceuticals and nutraceuticals. The regulatory concerns and outlook of pharmaceutical and nutraceuticals are also presented.

A pharmaceutical is a compound manufactured for use as a medicinal drug, whereas a nutraceutical is a substance that may be considered a food or part of a food and that provides medical or health benefits. Nutraceuticals and pharmaceuticals exhibit high similarities and overlapping among their properties and functionalities [1].

Traditional manufacturing processes, such as tablet compression, are well established and developed with regulatory pathways. However, these conventional processes are lacking in terms of design capability and manufacturing flexibility.

Besides pharmaceuticals, nutraceuticals have also attracted considerable attention due to their potential nutritional, safety, and therapeutic effects [2].

Nutraceuticals are substances which are mostly believed to prevent diseases and can be found in many fruits and vegetables. Personalized supplementation has recently gained momentum, with an estimated global market size of USD 1.6 billion in 2019 [3]. With the increasing interest in nutraceutical consumption, it is an inevitable trend that the production process of nutraceuticals will undergo dynamic changes.

As a new innovative technology for manufacturing pharmaceuticals and nutraceuticals, 3D printing provides the following advantages:

- Capable of developing medicines in a variety of dosages, shapes, sizes, and forms of medicines.
- Enables unique formulations or spatial distributions of the active and inactive ingredients within a product.
- Can customize the drug release profile to achieve the ideal treatment or health effects.
- Allows for the development of personalized medicines such as polypills containing multiple pharmaceutical and nutraceutical substances.
- Allows for the rapid modification of a product to optimize it for the needs of specific age populations such as children or the ageing population or to treat rare diseases.
- Enables the flexible and rapid manufacture of small batches for clinical trials, patient acceptability, taste evaluation, and consumer trials with potentially lower cost.
- Contributes towards on-demand production of pharmaceuticals and nutraceuticals and remote production.
- Allows flexible combinations of active substances in the printed product, thus avoiding the issue of being allergic to specific substances in the conventional product.
- Rapid production will improve the speed of consumer or patient feedback, thereby improving the quality of care.

Several additive manufacturing or 3D printing technologies have been shown to be effective in the production of pharmaceuticals and nutraceuticals for research, pre-clinical, and clinical studies (Figure 6.1).

These processes, namely material extrusion, vat photopolymerization, binder jetting, and powder-bed fusion, are listed according to the definition in the ISO/ASTM Standards on Additive Manufacturing (ISO/ASTM 52900:2015).

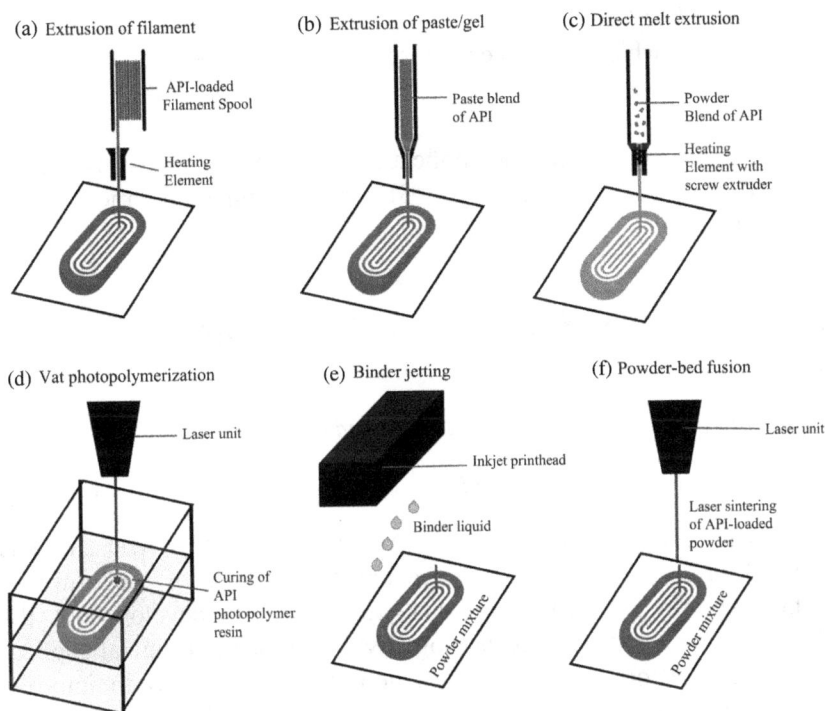

Figure 6.1.   3D printing processes of pharmaceuticals and nutraceuticals. API: active pharmaceutical ingredient.

(a) *Material extrusion: Extrusion of filament*
   Process in which melted or softened filament material is selectively dispensed through a nozzle or orifice.

(b) *Material extrusion: Extrusion of paste*
   Process in which a paste or gel-like material is selectively dispensed through a nozzle or orifice.

(c) *Material extrusion: Direct melt extrusion*
   Process in which a material is selectively dispensed through a nozzle or orifice without the use of a filament.

(d) *Vat photopolymerization*
   Process in which a liquid photopolymer in a vat is selectively cured using light-activated polymerization.

(e) *Binder jetting*
   Process in which a liquid bonding agent is selectively deposited to join powder materials.

(f)  *Powder-bed fusion*
Process in which thermal energy selectively fuses regions of a powder bed.

Table 6.1 summarizes the applications, design, and materials used in several articles published on 3D-printed pharmaceuticals and nutraceuticals.

## 6.2  Companies' Spotlight on 3D Printing of Pharmaceuticals

### 6.2.1  *Aprecia Pharmaceuticals Company (US)*

#### 6.2.1.1  *Company background*

Aprecia Pharmaceuticals Company (Aprecia) is privately held and was founded in 2003. Aprecia is the exclusive licensee of the powder–liquid 3D printing technology that was originally developed at the Massachusetts Institute of Technology (MIT) for pharmaceutical applications. Initial efforts emphasized on new machine designs to improve throughput. The company focused on the "fast-melt" style of dosage forms as its first multi-product platform application, which is branded as ZipDose technology. In 2015, the company received Food and Drug Administration (FDA) approval for its 3D-printed SPRITAM (levetiracetam) medication, an anti-epileptic drug. This also marks the first 3D-printed pharmaceutical to be approved by the FDA. The porous tablet achieves full disintegration within seconds after coming in contact with a liquid. Additionally, variable doses, from ultra-low to 1,000 mg or higher, are achievable (Figure 6.2) (https://www.aprecia.com/).

#### 6.2.1.2  *Process and materials*

ZipDose technology is the brand name of Aprecia's formulation technology related to orodispersible dosage forms (Figure 6.3). ZipDose is a platform approach that can be used in the formulation of both pharmaceutical and nutraceutical compounds [14]. ZipDose technology requires knowledge of several key factors. These factors include the attributes of the target drug compound, selection of excipients materials, and determination of the optimal 3D printing process parameters.

Table 6.1. Summary of 3D printing technologies, applications, designs, and materials used in the fabrication of pharmaceuticals and nutraceuticals.

| | Technology | Application | Design | Materials | Ref. |
|---|---|---|---|---|---|
| a) | **Extrusion of filament** | To develop mini-tablets as flexible personalized dosage forms for proper management of hypertension | Mini-tablets with channel design | Mixtures of nifedipine, Aqualon ethylcellulose EC N10, PEG 4000, magnesium stearate (MgS), Kollidon VA 64 (KVA64), and Klucel HPC LF extruded into filaments | [4] |
| | | To achieve extended-release prednisolone tablets | Ellipse-shaped tablet | Prednisolone-loaded poly(vinyl alcohol) (PVA)-based (1.75 mm) filaments | [5] |
| b) | **Extrusion of paste or gel** | To develop oral polypills with multiple release profiles | Core-shell design with 2 compartments, a pill diameter of 16 mm, and a height of 6 mm | Different dietary ingredients (vitamins B1, B3, and B6 and caffeine) and a mixture of pharmaceutical excipients | [3] |
| | | To create new nutraceutical oral dosage forms | Oval shape printed with 7 layers of materials | Mixtures of monoglycerides, oleogels, and phytosterols (PS) as printing materials | [6] |
| c) | **Direct melt extrusion** | To achieve sustained release of itraconazole To enhance the dissolution and absorption of the drug along the gastrointestinal (GI) tract | Cylindrical printlet (10 mm diameter × 3.6 mm height) | Itraconazole as a model drug of low solubility and hydroxypropyl cellulose (HPC) blend | [7] |

*(Continued)*

Table 6.1.   *(Continued)*

| Technology | Application | Design | Materials | Ref. |
|---|---|---|---|---|
| | To create a novel product development approach by design | Core-shell structure with a multi-layered drug compartment with different surface areas and pH-responsive layers | Model API used are metoprolol, tofacitinib, levodopa, topiramate, and clonidine; different materials for specific functions such as matrix, plasticizer, model API, and disintegrant | [8] |
| d) *Vat photopolymerization* Stereolithography (SLA) method | To create multi-layer constructs (polypills) with variable drug content | Cylinder, ring shape, and ring shape with a soluble filler | Six different model drugs (paracetamol, naproxen, caffeine, aspirin, prednisolone, and chloramphenicol); polyethylene glycol diacrylate (PEGda) as the photopolymerizable monomer and diphenyl (2,4,6-trimethylbenzoyl) phosphine oxide (TPO) as the photoinitiator | [9] |
| | To develop novel resin chemistry using PEGDMA as the encapsulating polymer network and riboflavin as the PI for SLA 3D printing that is non-toxic | Different geometries (coaxial annulus, 4-circle pattern and honeycomb pattern) with specific surface area to volume ratios | Ascorbic acid (AA) encapsulated in a poly(ethylene glycol) dimethacrylate (PEGDMA)-based polymer network and polymerized using riboflavin as a photoinitiator | |

| | | | | |
|---|---|---|---|---|
| e) **Binder jetting** on powder bed | To develop near zero-order release dosage forms | Multi-chamber pharmaceutical dosage forms with a drug core and a release rate-regulating drug-free shell | Mixtures of Kollidon SR and hydroxypropylmethyl cellulose (HPMC) as powder drug carriers. Aqueous pseudoephedrine hydrochloride PEH binder as core, and ethanolic triethyl citrate (TEC) binder as shell | [10] |
| | To support proof-of-concept formulation experiments | Tablet | Powder mixture of lactose monohydrate and Kollidon® VA64 (KL) and an aqueous binder containing 5% KL with polyvinylpyrrolidone (PVP) as a binder | [11] |
| f) **Powder-bed fusion:** Selective laser sintering | To achieve customized drug release properties in a cost-effective manner by changing the geometrical design without the need to alter the formulation composition | Cylindrical, gyroid lattice, and bi-layer structures having customizable release characteristics | Paracetamol loaded in four different pharmaceutical-grade polymers including polyethylene oxide, Eudragit (L100-55 and RL), and ethyl cellulose | [12] |
| | To create orodispersible printlets with characteristics similar to a commercial orally disintegrating tablet (ODT) | Cylindrical printlet (12.4 mm diameter × 3.6 mm height) | Ondansetron was first incorporated into drug-cyclodextrin complexes and then combined with the filler mannitol | [13] |

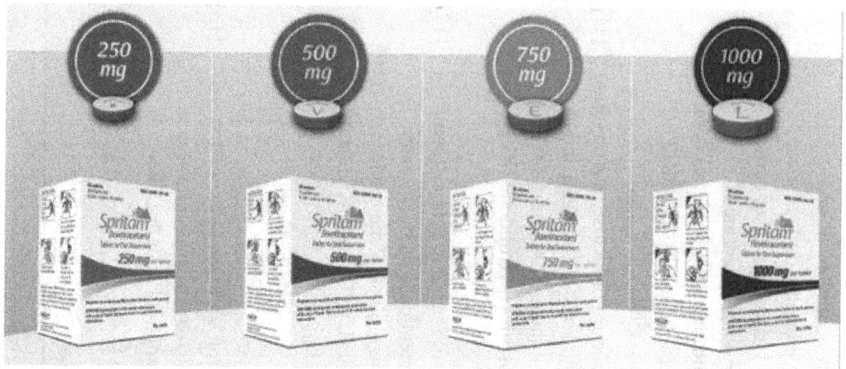

Figure 6.2.   Variable doses, from ultra-low to 1,000 mg or higher, are achievable. (Image courtesy of Aprecia.)

Figure 6.3.   ZipDose technology allows disintegration of particles in a liquid within seconds. (Image courtesy of Aprecia.)

As ZipDose technology is a fast-melt dosage form, the excipients are chosen to achieve multiple functions, namely, to support initial wetting and binding during printing, to aid rewetting and disintegration of the dosage unit upon administration, and to provide the necessary taste-masking effect while in the oral cavity [14]. The fast disintegration of ZipDose formulations is a function of both the ingredient selection and the

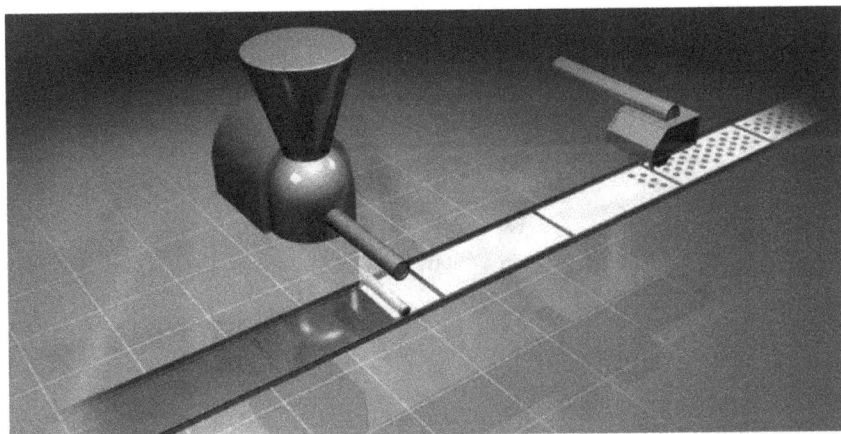

Figure 6.4. Powder–liquid (binder jetting) fabrication process of ZipDose technology. (Image courtesy of Aprecia.)

interconnected porous network within the ZipDose dosage forms as inherent from the 3D printing process [14].

The powder–liquid (binder jetting) fabrication process of ZipDose technology includes three general steps, namely 3D printing, drying, and harvesting [14]. In the 3D printing step, the binder jetting technique is utilized, where the droplets of a liquid binding agent are deposited by a print nozzle onto a pharmaceutical powder bed (Figure 6.4). The powder bed is then lowered to allow for another layer of powder to be added. Further droplets are introduced, and the printing process continues to form a 3D pill. The surrounding powder particles act as a support to prevent the collapse of the highly porous pill structure.

A drying step is conducted to solidify the wet printed unit doses in a controlled environment. The key parameters of this step include temperature, relative humidity, and drying time optimized for each product.

In the harvesting step, the printed units are placed on perforated plates on a rotary shaker. The vibration at the rotary shaker will separate the printed units from the unbounded particles. The unprinted powder from the build cycles is continuously collected and recirculated back to the 3D printing step together with a fresh feed of powder.

The 3D printing process also allows for freedom in designing the unit doses because different configurations of unit doses can be achieved as demonstrated by Aprecia (Figure 6.5).

Figure 6.5.  Different configurations of unit doses can be achieved. (Image courtesy of Aprecia.)

### 6.2.1.3 *Advantages*

(1) The powder-based 3D printing process used by Aprecia enables versatility in terms of materials, using well-established materials that are compendial and generally recognized as safe (GRAS).

(2) Due to flexibility in material selection, a wide range of taste-masking and modified release capabilities can be achieved.

(3) The 3D printing process in ZipDose technology allows flexibility in controlling dosage. There are two methods for incorporating the active ingredient, either by blending the active ingredient as part of the powder bed for higher dosage or by printing the active ingredient in the binding fluid for lower dosage [14].

(4) ZipDose formulation technology can accommodate large amounts of the target compound while still retaining the ability to disintegrate within seconds when taken with a sip of liquid. 3D printing enables significantly higher drug loading than traditional fast-melt dosage forms. A dose loading of 1,000 mg has been shown commercially.

(5) Different configurations or designs of unit doses can be achieved including fixed-dose combinations, complex release profiles, buccal or sublingual absorption, anticounterfeit features, ultra-low-precision dosing, and targeted delivery for enhanced absorption.

## 6.2.2  FabRx (UK)

### 6.2.2.1  Company background

The 3D-printed pharmaceuticals firm FabRx is a specialist biotech company based in the United Kingdom (UK) that focuses on developing 3D printing technologies for fabricating pharmaceuticals and medical devices. The company was created by leading academics from University College London (UCL). Using their patented technologies, they provide a novel and flexible platform to produce personalized medicines (https://www.fabrx.co.uk/).

### 6.2.2.2  Process and materials

FabRx launched its M3DIMAKER 3D printer in April 2020. The M3DIMAKER system is an extrusion-based multi-nozzle printing system (Figures 6.6 and 6.7). The system is controlled by specialized software to allow the pharmacist to select the required dose according to the prescription given by the clinician. The system is fitted with advanced in-line

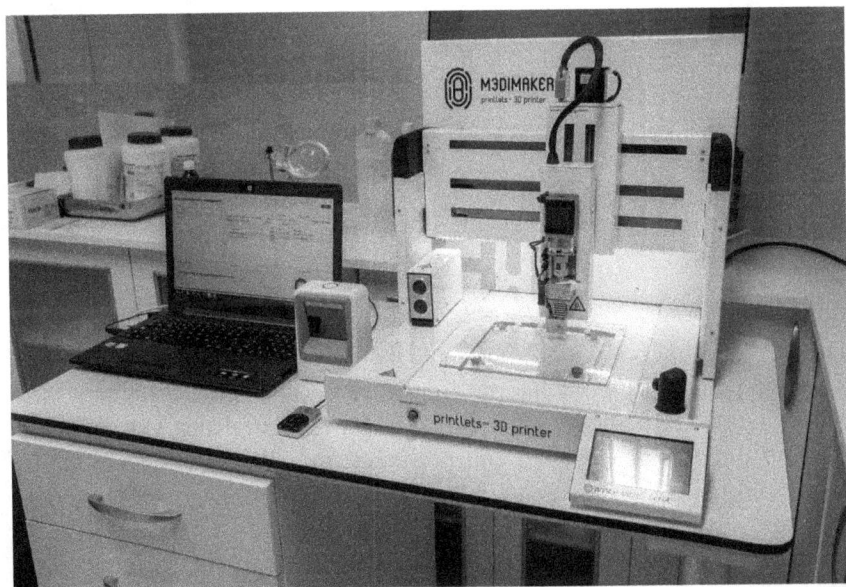

Figure 6.6.   The M3DIMAKER system is a multi-nozzle printing system based on three different extrusion-based technologies. (Image courtesy of FabRx.)

Figure 6.7.    3D printing of tablets using the M3DIMAKER system. (Image courtesy of FabRx.)

quality control procedures alongside camera monitoring of the printing process to track the progress and detect any faults during manufacture. The system also incorporates fingerprint access control alongside a data matrix reader to ensure manufacturing reliability, which limits the access to only qualified personnel. The total construction volume (X/Y/Z) of the system is $200 \times 200 \times 100$ mm$^3$.

The M3DIMAKER system is based on three different extrusion-based technologies, with the ability to carry out (i) extrusion of drug-loaded filaments, (ii) direct powder extrusion (DPE), and (iii) semi-solid extrusion (SSE).

## (1) *Extrusion of drug-loaded filament*

This involves the extrusion of a melted/softened thermoplastic filament from the printer nozzle, which is controlled by dedicated computer software. The filament material is a mix of API with excipients. During printing, the filament is extruded and deposited in a layer-by-layer manner, solidifying in less than a second. The speed of this printing process depends on the resolution required and the size of the tablet. In general, the printing time of one printlet (3D-printed tablet) is approximately 3 min. This technology has been widely investigated for pharmaceutical use, with the ability to manufacture multiple-drug combinations (polyp-ills) as well as sustained- or delayed-release tablets.

## (2)  *Direct powder extrusion*

In this process, a mixture of excipients and drugs in powder form is added directly into the printing nozzle. This process eliminates the need to use any filament. Therefore, this technique is advantageous as it allows printing of materials when filaments cannot be obtained. The printing time for this technology is similar to that of the filament-based extrusion technique. Direct powder extrusion has the capability of manufacturing polyppills and sustained- or delayed-release tablets.

## (3)  *Semi-solid extrusion 3D printing*

Also known as gel extrusion, this technique uses gelling agents as the starting material. In the pharmaceutical field, this technology can be used to prepare chewable formulations such as gummies, which are well accepted by children. FabRx has completed a clinical study in children using this technology. With this technology, a month's supply of medicine, up to 28 printlets, can be prepared in approximately 8 min.

Figure 6.8 shows the flexibility of 3D printing in creating printlets of different shapes and designs.

Figure 6.8.   Printlets of different shapes and designs are demonstrated by FabRx. (Image courtesy of FabRx.)

### 6.2.2.3 *Advantages*

(1) Multiple extrusion technologies are included, allowing the processing of a wide range of materials and active ingredients.
(2) M3DIMAKER™ has a large array of applications within drug development, including the ability to manufacture small batches of medicine for pre-clinical and clinical studies, as well as the potential to create personalized medicines for clinical practice.
(3) It has the ability to manufacture multiple-drug combinations (polypills) as well as sustained- or delayed-release tablets.
(4) Depending on the medicine being made, the preparation of one month's medication (28 printlets) can be carried out in ~8 min, thus revolutionizing the drug manufacture timeline.

## 6.3  Industry Spotlight on 3D Printing of Nutraceuticals

### 6.3.1  *Craft Health (Singapore)*

#### 6.3.1.1  *Company background*

Craft Health Pte Ltd is a pharmaceutical drug delivery company with the vision of simplifying drug delivery with 3D-printed solutions. Craft Health won the title of "Most Promising Startup" at the National Additive Manufacturing Innovation Cluster (NAMIC) Start-up Innovation Forum, 2019, a partner event of Slingshot@Switch. Craft Health successfully closed approximately SGD 1 million in its seed funding round in November 2019, and runs a formulation-equipped laboratory together with a class 100K clean room manufacturing facility (https://www. crafthealth.me/).

#### 6.3.1.2  *Process and materials*

The company has developed the following platforms:

- Craft*Make*: a 3D printer developed in-house that specializes in the 3D printing of nutraceuticals and pharmaceuticals.

- Craft*Control*: 3D printing software optimized for Craft*Make*.
- Craft*Blends*: a series of proprietary 3D-printable formulations for different controlled release profiles (immediate/sustained/delayed release) that can be combined into a single 3D-printed tablet.

Using paste extrusion, CraftMake™ (Figure 6.9) involves no heating or ultraviolet light curing, thus allowing a wide variety of different active ingredients to be incorporated. CraftMake™ allows for accurate deposition of paste materials through a combination of pneumatic and mechanical techniques. With a dual-controlled $Z$ axis, CraftMake™ achieves

Figure 6.9. The CraftMake™ system for the 3D printing of nutraceuticals and pharmaceuticals. (Image courtesy of Craft Health Pte Ltd.)

concurrent 3D printing of two materials, reducing the time required for each print job.

Craft Health has developed a proprietary formulation base using excipients sourced from the US FDA GRAS (Generally Regarded As Safe) list to make the base

- Safe
- 3D-printable
- Controlled release profiles enabled (immediate, sustained, or delayed release profile)

The required excipients are mixed with the active ingredient of the supplement or medicine and rendered into a paste before loading into CraftMake™. The paste material is then 3D-printed into the required shape or geometry. Different configurations can be achieved, including single-material pill, multi-material pill, and pill-in-pill configuration (Figure 6.10). Each material indicates a single release profile, and more than one material can be 3D-printed into a single polypill, achieving multiple controlled release profiles in a single polypill. The 3D-printed tablets are then left to dry in a dry cabinet.

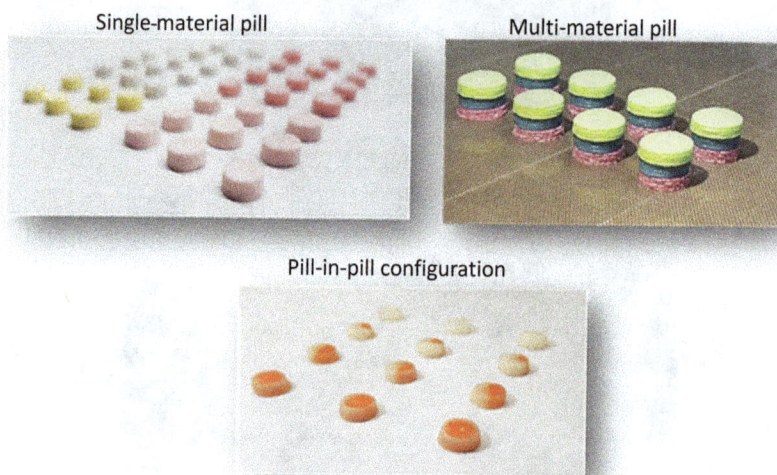

Figure 6.10. Different pill configurations can be achieved. (Image courtesy of Craft Health Pte Ltd.)

### 6.3.1.3 *Advantages*

(1) The process of paste extrusion does not involve heat or UV curing, thus enabling a wide range of active ingredients to be incorporated.
(2) Ability to 3D-print two different materials concurrently, and 3D-print multi-layer polypills.
(3) Ability to 3D-print other geometries and shapes, such as pill-in-pill configurations, without the need for a mould.
(4) Designed with Good Manufacturing Practice (GMP) in mind, leading to faster adoption in the manufacturing space.
(5) High-throughput printer, taking approximately 10 sec for a two-material 100 mg tablet (approximately 3,000 to 5,000 tablets in a 12-hour period).

## 6.3.2 *Nourished (UK)*

### 6.3.2.1 *Company background*

Nourished was launched in January 2020 in the UK, and in November 2020 in the United States (US), under the parent health-tech company founded in 2019, Remedy Health. Nourished is a customized vitamin company whose aim is to fit high-impact nutrition into an enjoyable medium for daily consumption (https://get-nourished.com/).

Customers start by taking a quiz on the company website to share their profile. The quiz is algorithm-powered, and the answers will generate a recommendation for an optimum combination of 7 vitamins, minerals, or supplements out of 34 possible ingredients. The recommended combination is then fabricated by the company's patented 3D printing technology into a vegan, sugar-free gummy vitamin stack before shipping to the customers.

### 6.3.2.2 *Process and materials*

The 3D printing process is carried out by the machines in Nourished's factory in Birmingham, England. Seven chosen nutrients are placed inside 3D printer cartridges (Figure 6.11) and extruded into hexagonal layers [15]. The printed product is a multi-layered gummy, with seven evenly stacked layers atop one another (Figure 6.12).

Figure 6.11.   Seven chosen nutrients are placed inside 3D printer cartridges and extruded into hexagonal layers. (Image courtesy of Nourished.)

Figure 6.12.   The printed product is a multi-layered gummy, with seven evenly stacked layers atop one another. (Image courtesy of Nourished.)

### 6.3.2.3 *Advantages*

(1)  The recommended vitamins are personalized to the consumer's individual requirements.
(2)  Customized multi-layered stack eliminates the need to take multiple pills for different vitamins every day.
(3)  The Nourished product is vegan in nature.
(4)  3D printing allows on-demand ordering to ensure the efficacy of active ingredients and direct delivery to the customer, resulting in a higher impact when consumed.

## 6.4  Regulatory Considerations of Pharmaceuticals and 3D Printing

Drug regulation is the control of drug use by international agreement or by regulatory authorities. This includes regulations on the development, approval, manufacturing, and marketing of drugs. Regulatory standards require each product in each batch to be within specifications with respect to purity, impurities, and quality [16].

3D printing technology has been used to produce medical devices since 2007, with approximately 200 devices fabricated by 3D printing approved by the US Food and Drug Administration (FDA) [17]. These include hearing aid devices, dental crowns, bone implants, skull plates, spinal plating systems, facial implants, surgical guides and instruments, and dental products. However, only as recently as 2015 did Aprecia Pharmaceuticals receive FDA approval for its 3D-printed medication under the 505b(2) pathway. This shows that 3D printing is an emerging technology in the manufacture of pharmaceutical products.

In the US, drugs can be approved by the FDA through three possible regulatory pathways. For new drugs, there are two primary pathways, namely 505b(1) and 505b(2). For generic drugs, the regulatory pathway is 505(j) [18].

Regulation of pharmaceutical products in the European Union (EU) is governed by the European Medicines Agency (EMA). The EMA provides scientific guidelines on the quality of human medicines, which represents a harmonized approach of the EU member states on a list of topics in the development of medicine. The topics include active substance, manufacturing, impurities, specifications (analytical process and validations), excipients, packaging, stability, pharmaceutical development, quality by design, specific types of products, and lifecycle management [19].

3D printing could play the role of an advanced manufacturing tool in the production of pharmaceutical products. The application of 3D printing in pharmaceutical or medical technologies is being evaluated by regulatory agencies following the same framework of scientific principles and risk-based approaches, which are used for the development and approval of products manufactured by other conventional manufacturing methods.

### 6.4.1 *Key Considerations of 3D Printing in Pharmaceutical Manufacturing*

There are several key considerations in the application of 3D printing for manufacturing pharmaceuticals [17]. Some of the considerations are as follows:

(1) *Equipment and process considerations*:

- As 3D printers were previously not standardized or fit-for-purpose (pertaining to Good Manufacturing Practice (GMP)) for the production of pharmaceutical products, unique machine features should be tailored for the GMP requirement in the pharmaceutical industry.
- Critical process parameters for each 3D printing technology must be investigated clearly.
- A comprehensive validation plan for the equipment and process is critical.
- Issues on data security and accessibility need to be well controlled and maintained.

(2) *Design, formulation, and printing materials*:

- Considerations include understanding and optimizing the 3D geometric design.
- Consistency of the drug loading step when processing filament or powder.
- Key characteristics of the raw materials, including filament, resin, and powder, must be well documented

(3) *Performance of the 3D-printed product*:

- The interlinked effects of material attributes, 3D geometric designs, and 3D printing process parameters on the performance of 3D-printed solid dosage forms need to be investigated.
- The suitability of existing *in vitro* testing to characterize the performance of 3D-printed products should be verified.

## 6.5 Outlook and Conclusions

*Digital manufacturing*: The digital nature of 3D printing allows easy integration with other future technologies including cyber technologies, such as artificial intelligence (AI), machine learning, and advanced simulation. There is great potential for creating personalized digital pharmaceutical dosage forms enabled by data-driven approaches. In addition, new 3D-enabled intelligent manufacturing platforms could be developed with the integration of real-time process analytical technology (PAT), where PAT is a key characteristic in pharmaceutical production.

A research team has recently reported the development of a 3D printer that operates using a mobile smartphone [20]. In this work, the SLA-based printer uses the light from the smartphone's screen to photopolymerize liquid resins and create solid printed dosage forms. The shape of the printed structure is determined using a customized application on the smartphone. This work demonstrates the potential of 3D printing towards point-of-care manufacturing of personalized medications.

*Social impact*: In addition, 3D printing can be applied to create social impact in the production of pharmaceutical products. In a recent report, researchers have demonstrated a new application to produce pills readable to the visually impaired population [21]. In this report, printlets were designed with Braille and Moon patterns on their surface, thus enabling the patients to identify medications when taken out of their original packaging. This new approach aims to reduce medication errors and improve medication adherence in patients with visual impairment.

*Patient acceptability*: Though the possibilities of technological advances using 3D printing are promising, it is critical to address the issue of patient acceptability of the 3D-printed product. Considerations such as the perceived taste, texture, and familiarity of the 3D-printed pharmaceutical and nutraceutical products should be studied alongside technical issues.

*Regulatory framework*: Regulatory concerns about 3D-printed pharmaceuticals and nutraceuticals are still a challenge and will require further collaboration between the research community, regulatory specialists, and industry players.

In summary, 3D printing will remain an important manufacturing technology in the future pharmaceutical industry, adding value from research to pre-clinical and clinical applications. The technologies will be able to complement traditional manufacturing of pharmaceuticals and nutraceuticals.

# References

[1] Kata Trifković, M. B. (n.d.) *Nutraceuticals and Natural Product Pharmaceuticals*, ed. Galanakis, C. M., "Introduction to nutraceuticals and pharmaceuticals" (Academic Press).

[2] Nasri, H. R., Baradaran, A., Shirzad, H. and Rafieian-kopaei, M. (2014). New concepts in nutraceuticals as alternative for pharmaceuticals, *International Journal of Preventive Medicine*, 5, pp. 1487–1499.

[3] Goh, W. J., Tan, S. X., Pastorin, G., Ho, P. C. L., Hu, J. and Lim, S. H. (2021). 3D printing of four-in-one oral polypill with multiple release profiles for personalized delivery of caffeine and vitamin B analogues, *International Journal of Pharmaceutics*, 598, p. 120360.

[4] Ayyoubi, S., Cerda, J. R., Fernández-García, R., Knief, P., Lalatsa, A., Healy, A. M. and Serrano, D. R. (2021). 3D printed spherical mini-tablets: geometry versus composition effects in controlling dissolution from personalised solid dosage forms, *International Journal of Pharmaceutics*, 597, p. 120336.

[5] Skowyra, J., Pietrzak, K. and Alhnan, M. A. (2015). Fabrication of extended-release patient-tailored prednisolone tablets via fused deposition modelling (FDM) 3D printing, *European Journal of Pharmaceutical Sciences*, 68, pp. 11–17.

[6] Cotabarren, I. M., Cruces, S. and Palla, C. A. (2019). Extrusion 3D printing of nutraceutical oral dosage forms formulated with monoglycerides oleogels and phytosterols mixtures, *Food Research International*, 126, p. 108676.

[7] Goyanes, A., Allahham, N., Trenfield, S. J., Stoyanov, E., Gaisford, S. and Basit, A. W. (2019). Direct powder extrusion 3D printing: fabrication of

drug products using a novel single-step process, *International Journal of Pharmaceutics*, 567, p. 118471.

[8] Zheng, Y., Deng, F., Wang, B., Wu, Y., Luo, Q., Zuo, X., Liu, X., Cao, L., Li, M., Lu, H., Cheng, S. and Li, X. (2021). Melt extrusion deposition (MED™) 3D printing technology — a paradigm shift in design and development of modified release drug products, *International Journal of Pharmaceutics*, 602, p. 120639.

[9] Robles-Martinez, P., Xu, X., Trenfield, S. J., Awad, A., Goyanes, A., Telford, R., Basit, A. W. and Gaisford, S. (2019). 3D printing of a multi-layered polypill containing six drugs using a novel stereolithographic method, *Pharmaceutics*, 11.

[10] Wang, C.-C., Tejwani, M. R., Roach, W. J., Kay, J. L., Yoo, J., Surprenant, H. L., Monkhouse, D. C. and Pryor, T. J. (2006). Development of near zero-order release dosage forms using three-dimensional printing (3-DP™) technology, *Drug Development and Industrial Pharmacy*, 32, pp. 367–376.

[11] Chang, S.-Y., Jun, J., Jun, Y., Xin, D., Chaudhuri, B, Nagapudi, K. and Ma, A. W. K. (2021). Development of a pilot-scale HuskyJet binder jet 3D printer for additive manufacturing of pharmaceutical tablets, *International Journal of Pharmaceutics*, 605.

[12] Fina, F., Goyanes, A., Madla, C. M., Awad, A., Trenfield, S. J., Kuek, J. M., Patel, P., Gaisford, S. and Basit, A. W. (2018). 3D printing of drug-loaded gyroid lattices using selective laser sintering, *International Journal of Pharmaceutics*, 547, pp. 44–52.

[13] Allahham, N., Fina, F., Marcuta, C., Kraschew, L., Mohr, W., Gaisford, S., Basit, A. W. and Goyanes, A. (2020). Selective laser sintering 3D printing of orally disintegrating printlets containing ondansetron, *Pharmaceutics*, 12.

[14] West, T. G. and Bradbury, T. J. (2019). *3D and 4D Printing in Biomedical Applications*, "3D printing: a case of ZipDose® technology — world's first 3D printing platform to obtain FDA approval for a pharmaceutical product" (Wiley), pp. 53–79.

[15] Visram, T. (2021). This company 3D-prints your favorite vitamins into a single gummy. *World Changing Ideas*. Retrieved from https://www.fastcompany.com/90634244/this-company-3d-prints-your-favorite-vitamins-into-a-single-gummy

[16] Khairuzzaman, A. (2018). *3D Printing of Pharmaceuticals*, eds. Gaisford, S. and Basit, A. W., "Regulatory perspectives on 3D printing in pharmaceuticals" (Springer), pp. 215–236.

[17] Zidan, A. (2017). CDER researchers explore the promise and potential of 3D printed pharmaceuticals. Retrieved from https://www.fda.gov/drugs/news-events-human-drugs/cder-researchers-explore-promise-and-potential-3d-printed-pharmaceuticals

[18] FDA. (2019). Development & approval process: drugs. Retrieved from https://www.fda.gov/drugs/development-approval-process-drugs

[19] EMA. (n.d.) Quality guidelines. *Human Regulatory.* Retrieved from https://www.ema.europa.eu/en/human-regulatory/research-development/scientific-guidelines/quality-guidelines

[20] Xu, X., Seijo-Rabina, A., Awad, A., Rial, C., Gaisford, S., Basit, A. W. and Goyanes, A. (2021). Smartphone-enabled 3D printing of medicines, *International Journal of Pharmaceutics*, 609, p. 121199.

[21] Awad, A., Yao, A., Trenfield, S. J., Goyanes, A., Gaisford, S. and Basit, A. W. (2020). 3D printed tablets (printlets) with Braille and Moon patterns for visually impaired patients, *Pharmaceutics*, 12, p. 172.

## Problems

1. Discuss the advantages of 3D printing in the manufacture of pharmaceutical and nutraceutical products.
2. Name the 3D printing processes that have been utilized in the printing of medication or pills.
3. The M3DIMAKER system is based on three different extrusion-based technologies. What are the three extrusion-based methods?
4. What are the advantages of using paste material extrusion in the 3D printing of pharmaceuticals and nutraceuticals?
5. The 3D printing process of ZipDose technology includes three general steps, namely 3D printing, drying, and harvesting. Describe each step briefly.
6. What are the key functions of excipient materials in a fast-melt dosage form?
7. What are the motivations or advantages of creating a customized nutraceutical product?
8. What are the key regulatory considerations of 3D printing in pharmaceutical manufacturing?
9. How does 3D printing create social impact in the pharmaceutical and nutraceutical industry?
10. Explain the potential of 3D-printed pharmaceutical and nutraceutical in the future of Industry 4.0.
11. Discuss the challenge of security issues, in terms of data, equipment and cybersecurity, in the application of 3D printing for pharmaceuticals and nutraceuticals.

# Chapter 7

# Food Safety and Regulation of 3D-Printed Food

## 7.1 Introduction

Food is any edible ingredient consumed to provide the necessary energy requirements and nutritional support, which are essential for the survival of any living organisms. Food, which can originate from plants, animals, or fungi, contains essential macromolecules such as carbohydrates, fats, proteins, and other nutrients including vitamins or minerals. Although food is considered one of the important factors for living organisms to survive, there are both positive and detrimental effects associated with food consumption. Specifically, there are a variety of naturally occurring phytochemicals present in the food that may diminish nutritional value or even be detrimental to health (e.g., toxicants). In addition to the intrinsic properties of the food, external factors such as the preparation and processing of foods are equally important, as they can influence the quality of the food, resulting in food hazards.

There are several different ways in which food hazards can be introduced. For example, addition of food additives during the food processing stage or contamination with pathogenic bacteria or their toxins from the environment. The consequences of food hazards are disastrous as they play a critical role in the transmission and onset of a variety of foodborne diseases such as typhoid and diarrhoea, causing fatalities all over the world. Specifically, an estimated 600 million people fall ill after consuming contaminated food and 420,000 die annually [1]. Hence, the microbiological standard is regarded as the most crucial food safety factor [2].

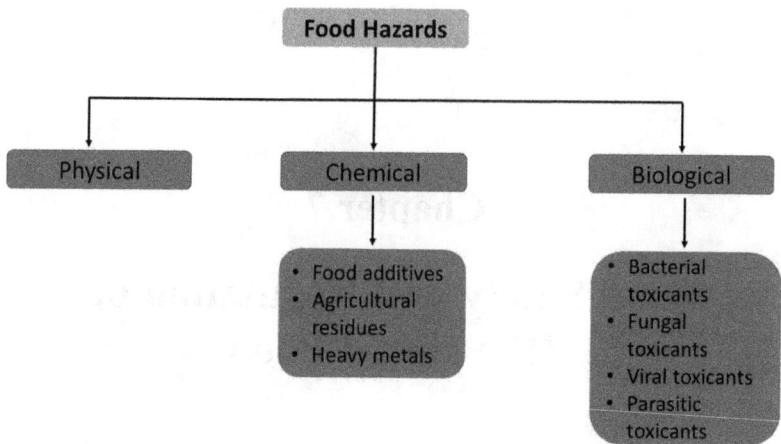

Figure 7.1.   Overview of various categories of food safety hazards.

In addition to foodborne diseases, the presence of food contaminants, including bacterial contaminants, pesticides, drug deposits, and industrial chemicals, is a major concern that can have a significant influence on public health. This chapter aims to give an outline of the different types of food hazards, such as physical, chemical, and biological (Figure 7.1).

## 7.2  Various Categories of Food Safety Hazards

A food safety hazard can be broadly defined as the presence of any means in the food that can cause harmful health outcomes for users. Food safety hazards arise when food has interacted with unsafe agents, which can exist in various forms such as microbes, allergens, and pesticides. According to the National Advisory Committee on Microbiological Criteria for Foods, "any changes to the physical, chemical, or biological properties of food that results in an unwanted health risk for consumers are regarded as hazards" [3]. Therefore, regardless of the type of hazardous agents, food safety hazards are classified into three broad categories: physical, chemical, and biological. Physical hazards include sharp or solid objects such as bones or plastics that are found in the food due to food processing, which can result in injuries like cuts. Although not all external materials found in food products, such as hair or sand, can cause injuries, they are also classified as physical hazards. Chemical hazards occur when too high concentrations of certain substances used during food preparation or processing are consumed and cause illness or injury. Lastly, contamination of food with microbes is classified as a biological hazard. Certain microorganisms are potent pathogens, capable of causing diseases of varying severity.

## 7.2.1 *Physical Hazards*

In general, physical hazards are dangerous exterior substances, which include all physical materials that are usually not regarded as part of food. When these substances are consumed, they are considered hazardous if they cause injuries to the body. However, not all substances are hazardous, as some substances are non-hazardous and are often associated with unhygienic practices that occur during food processing, food production, and food distribution. In most cases, physical hazards are relatively easy to detect as they cause injury instantaneously. For example, hazardous substances are usually hard, sharp, or large and may cause cuts and wounds to our body or become a choking hazard. This is particularly risky for infant food or food prepared for the elderly, especially dysphagia patients, who require puréed food for their diet requirements; hence, the presence of any hard particles in the food would cause serious injuries to this group of people.

Exterior substances can be divided into two groups: unavoidable and avoidable. Unavoidable substances are materials that arise in the food as a by-product during food processing. Such unavoidable exterior substances could include, for example, plant parts (stem or leaf) from fruits or dirt from carrots found in the final food product during food processing. On the other hand, avoidable exterior substances are materials found in food that can be prevented using proper methods. Examples of avoidable exterior substances include small parts of plastics or glass fragments that are present in food, which can be avoided if the food is prepared properly (Figure 7.2).

## 7.2.2 *Chemical Hazards*

### 7.2.2.1 *Food additives*

According to the US Food and Drug Administration (FDA), "food additive is defined as any ingredient that affects the properties of any food whether it is added directly or indirectly". There are three broad categories

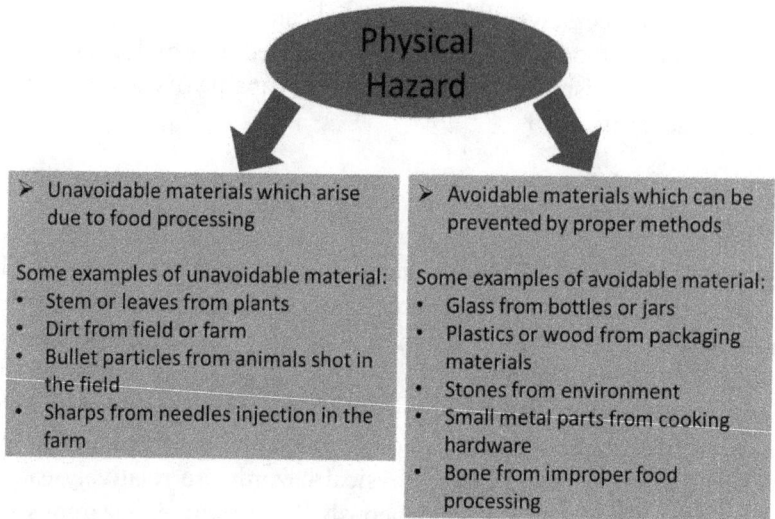

Figure 7.2.   Summary of the two types of physical hazards and examples of where they can be found.

of food additives: (1) ingredients that are added intentionally, (2) ingredients that form the natural components of food, and (3) ingredients that contaminate food [4]. During various stages of food preparation or processing, these ingredients can be included in the food to induce chemical changes in the raw ingredient or food product. Generally, the primary motivation of using food additives is to obtain definite technical or aesthetic effects for the food, such as enhancing flavour, changing colour, or increasing food preservation. For example, food additives are utilized to preserve or enhance taste, texture or appearance, or nutritional content and extend shelf life (Figure 7.3).

### 7.2.2.1.1  Food colouring

Food colour is one of the important visible sensory characteristics that enhance flavour perception [5]. Hence, food colour is recognized to be one of the most critical features of food products, which could in part directly control the selection and eating desires of consumers [5]. As a result, many food manufacturers use a variety of colours in the production of food products, ranging from beverages to bakery products, dairy, candies, etc. The FDA regulates and defines a colour additive as any material

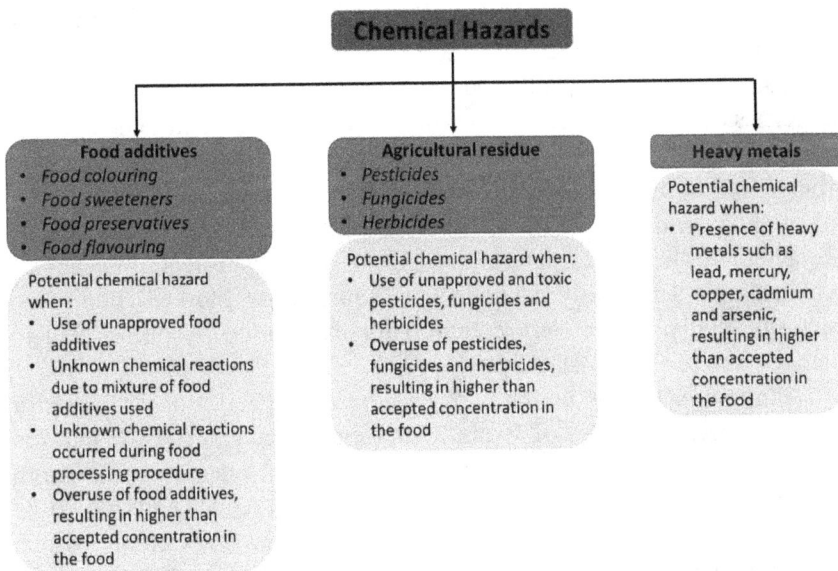

Figure 7.3.   Summary of the three categories of chemical hazards and examples of how they may cause food hazards.

that gives colour to a food or drink product. Food colouring materials, whether chemically synthesized or naturally occurring, have the chemical property of absorbing colour due to the presence of double bonds [6]. According to the Pure Food and Drug Act of 1906, seven synthetic colouring agents are permitted as food additives. These colouring agents are Ponceau 3R, Amaranth, Erythrosine, Indigotine, Naphthol Yellow 1, Light Green SF, and Orange 1 [7]. These manmade food colourants are predominantly used to provide various colours to food, thereby improving the visual effect of numerous foodstuffs. These approved artificially synthesized colouring agents have the advantage of being resistant to chemical modifications on exposure to external stimuli such as heat or pH. Furthermore, they are more cost-efficient and can be easily combined to generate a variety of colours. Although naturally occurring colouring agents such as pigments derived from plants or animals do not require FDA approval for use, their chemical structures are unstable under conditions of pH, heat, and light, which may cause alteration in their colours. Caramel, which is yellow and turns to tan, and dehydrated beets used in soft drinks, which are bluish red and turn to brown, are examples of natural food colourants altering their colours. Furthermore, the natural sources of colour additives are more expensive than synthetic colours, but they also come with added benefits such as improving the flavour of food, being safer, providing health benefits, and contributing to the functional properties of food products [8, 9]. For example, several natural sources of food colourants, including anthocyanin, beet, or carotenoids, are commonly used in numerous food products. A common natural food colourant, anthocyanin, can be obtained from plant parts, and its favourable characteristics have been determined. As the colourant effect of the anthocyanin pigment is highly dependent on temperature, storage conditions, humidity, and pH, it is very common for the colour of anthocyanin to change from red to bluish purple [10, 11].

Although synthetic colours are less expensive and more reliable for creating different shades of foods, some reports have suggested that the ingestion of certain artificial food colourants could cause harmful health impacts [12]. Specifically, some foods such as coloured ice balls, coloured candies, and cold drinks are prepared using unapproved colours, which are much cheaper than approved colours. Hence, these foods tend to contain higher quantities of toxic chemicals, which are likely to affect the population with lower income. One of the manmade colourants, titanium dioxide, causes allergic reactions after consumption and affects children's behaviour [13]. Furthermore, certain manmade food colourants (e.g., Tartrazine or Allura Red AC) have been identified by Council Regulation (EC 1333/2008) to cause attention deficit hyperactivity disorder in consumers.

During food processing, it is common for food colourants to be mixed to form different colours [14]. However, in certain cases, a mixture of two or more dyes has been demonstrated to cause toxicity in animal models [15, 16]. For example, Tartrazine (yellow colour) and Carmoisine (red colour) are used in several food products such as chips, ice cream, soft drinks, and cakes. After consumption of these azo dyes, the gut and liver enzymes in the body can break down the azo dyes to form aromatic amines [16, 17]. The accumulation of the colour chemicals and their metabolites are reported to be carcinogenic to humans. Prolonged exposure to azo dyes also causes several chronic diseases, including lesions in key organs (kidney, brain, liver, and spleen), tumour formation, and mental disorders [18]. In addition to causing health problems, some synthetic food dyes transfer their intense colouring to effluents, leading to environmental pollution [19].

### 7.2.2.1.2 Food sweeteners

In the 20th century, metabolic diseases such as diabetes and obesity have become major health problems [20–22], as they increase the risk for many other life-threatening complications, including hypertension as well as renal and vascular diseases [23, 24]. One of the common factors contributing to metabolic diseases is the increasing consumption of sugars in the diet; hence, there is a demand to extract sweeteners from natural sources to study their chemical structure and biology. This will lead to the development of synthetic sweeteners that could potentially replace sugar partially or totally. Sweeteners are very distinct chemical compounds, which are either natural or synthetic in origin. Due to their structural similarities with sucrose, sweeteners can target the sweet-tasting receptors, resulting in a sweet-tasting experience without inducing any effect on caloric intake [25]. Thus, these are employed to produce food products with lower calories, which could help curb the growing obesity pandemic worldwide.

In the market today, sweeteners are commonly used in many foods or drinks as sugar substitutes, which are much sweeter than sucrose but do not have any calories [26]. Some of the commonly used sweeteners which have zero calories are aspartame, saccharin, and sucralose [27]. Unlike sugar, it is reported that physiological satiety mechanisms are not initiated by these sweeteners [28, 29]. Despite the reported benefits of sweeteners, the use of several artificial sweeteners is limited as they have certain harmful side effects [25]. Some side effects include the stability, cost, quality of taste, and more importantly, safety of the sweeteners. For example, when sweeteners are used in high concentrations, it can lead to a bad taste which can change from sweet to bitter or metallic [30]. Furthermore, the safety of long-term consumption of sweeteners remains controversial as there have been reports of harmful effects on health, such as heart problems, mental disorders, and cancer [31].

### 7.2.2.1.3 Food flavouring

The flavours present in food products are often due to the combination of volatile and non-volatile compounds that exhibit distinct physical and chemical profiles. Non-volatile elements primarily provide flavour, while volatile compounds provide both flavour and fragrance. Certain compounds (e.g., dicarbonyls, aldehydes, or alcohols) may contribute to the aromatic effects of food [32]. In the case of fruits and vegetables, the ripening process generates most of the flavouring compounds as secondary by-products of the breakdown of macromolecules such as carbohydrates, lipids, and proteins. Currently, food flavouring covers almost a quarter of the food additives market and can be derived from natural sources or chemically synthesized. In earlier days, flavouring compounds were usually extracted from natural sources; however, when modern chemistry advanced, chemical synthesis was used to generate synthetic flavouring compounds. Furthermore, certain chemical syntheses may use toxic solvents, which increases the purification costs.

## 7.2.2.1.4 Food preservatives

A food preservative is a common additive that is included in food products to prevent degradation, maintain nutritional content, and extend shelf life [33]. Specifically, preservatives are applied to kill the germs present in the food to prevent their growth, which could produce numerous toxins that cause food poisoning [34]. This can lead to the extension of food product shelf life and prolonged food safety. Therefore, the three key goals of food preservation are to maintain the appearance of the food, safeguard the nutritional content of the food, and prolong the shelf life of food. For food preservation, several types of preservatives are used, such as antimicrobial, chemical, or antioxidant [34].

Antimicrobial additives are mainly used to impede the expansion of microbes in food. Furthermore, other traditional methods alter the environment, making it unsuitable for the survival of microorganisms [34]. These include high salt content or reduced moisture in food, which are known to inhibit spoilage due to microbial growth. Furthermore, air can be removed from meals by sealing them within containers, or by freezing or pickling them. Allergies, asthma, and skin rashes have been linked to antimicrobial additives, including benzoates or sorbic acid [35]. Nuclear radiation was used to irradiate food products to disrupt bacterial deoxyribonucleic acid (DNA), inhibiting the growth of microbes. Although ionizing radiation does not make foods radioactive, one drawback of using this method for preservation is that it may cause alteration in food colour or texture [36].

Besides controlling the environment of food, chemical or traditional preservatives are also commonly used to preserve food. Some of the commonly used chemical preservatives are nitrites, sulphites, calcium propionate, and disodium ethylenediaminetetraacetic acid [37]. On the other hand, natural substances such as salt or vinegar are frequently used as traditional preservatives. However, some chemical preservatives also cause detrimental effects on health. Specifically, a common wine preservative, sulphur dioxide, was shown to induce allergies in the lungs of asthmatic patients. Furthermore, nitrates, which are used as meat preservatives, are also regarded as potent carcinogens [38, 39]. Sulphites, which are used as a chemical additive in certain fruits, can cause migraines, palpitations, allergies, and even cancer [35, 40]. More recently, extracts from plant sources have also begun to be widely considered as natural food preservatives due to high levels of nitrates [41].

In the last few decades, antioxidants have gained lots of attention in the food industry as they are reported to scavenge free radicals and prevent lipid oxidation, which are known to cause oxidative damage to living systems during various metabolic processes [42]. Hence, antioxidant additives are utilized to inhibit food products from oxidation [42]. Specifically, there is a large body of evidence to suggest that oxidative stress is an important contributor to many processes such as ageing [43] and pathological conditions, including diabetes [44], hypertension [45], atherosclerosis [46], and neurological [47] and cardiovascular diseases [48–50]. Hence, numerous studies have demonstrated that antioxidants can inhibit oxidative stress by removing free radicals and improve health outcomes for many diseases [51–53]. Antioxidants can originate from natural or synthetic sources [54, 55]. Some examples of the natural sources of antioxidant additives include phenolic, flavonoids, minerals (zinc, selenium), enzymes (catalase, glutathione peroxidase, superoxide dismutase), and vitamin E [56, 57]. While chemically synthesized antioxidants are used in food to prevent lipid peroxidation, they are generally considered less safe than naturally occurring antioxidants obtained from fruits and vegetables. For instance, tert-butyl hydroquinone or esters of gallic acid are chemically synthesized antioxidants that are currently used worldwide [58]. Although the positive health effects of antioxidants are widely reported, there are also harmful side effects associated with them. Specifically, vitamin C, a well-known antioxidant, is used for the common cold but its consumption in high concentrations has been associated with cardiovascular disease or certain types of cancers [59]. Similarly, polyphenolic antioxidants, which can reduce oxidative damage and limit atherosclerosis [56, 60], have also been reported to promote cancer [61]. Furthermore, while synthetic antioxidants have a typical suggested consumption of roughly 0.1 mg/kg, larger amounts cause changes in various enzyme or lipid functions, as well as carcinogenic consequences [62]. As a result, antioxidant additives are thoroughly examined for safety, and only approved antioxidants are allowed to be used in food. These antioxidants should be safe even at high dosages, avoiding any potentially hazardous side effects.

### 7.2.2.2   *Agricultural residues*

### 7.2.2.2.1   Pesticides

With the projected rise in global population to 9.2 billion by 2050, there will be a parallel increase in the demand for food production to cater for the growing population [63]. Concurrently, it is estimated that approximately 35% of the harvest yield is destroyed due to pests during preharvest worldwide [64]. This has resulted in the application of agricultural pesticides to increase agricultural food production worldwide, contributing to food security [65]. Pesticides are defined as chemicals that can kill insects and pests from agricultural crops [66] and are regarded as critical for the continued production of food [67]. To decrease agricultural losses due to insects and pests and dramatically improve the supply of food to growing populations at a fair price in growing economies, modern agricultural methods rely on the widespread use of pesticides [68]. Therefore, the use of pesticides is regarded as a critical component of agriculture that stops unwanted species in crops from growing [69]. Some examples of popular pesticides used are organochlorines and organophosphates. While the use of pesticides brings about economic benefits and alleviates food security issues, it can cause damage to non-crop vegetation and non-target organisms within the ecosystem and result in environmental pollution [70]. For instance, excessive levels of pesticides present in food are toxic and cause a dangerous impact on human health. Specifically, extensive toxicological studies in animals have demonstrated that exposure to several commonly used pesticides causes toxicity and, possibly, cancer. Pesticide poisoning have a severe impact on protein metabolism [71], affecting the endocrine [72] and reproductive systems [73]. In some cases, it can also cause neurodegenerative diseases in humans [74]. Given the harmful side effects of pesticide use, it is important to develop pesticides that are effective but at the same time have fewer toxic effects on humans — for example, the development of pyrethroids and *Bacillus thuringiensis*-based pesticides, which are reported to have a milder impact on the health of humans and the environment [75].

## 7.2.2.2.2  Fungicides

Fungicides, like pesticides, are chemicals that stop fungus or fungal spores from growing. Fungicides have a wide range of uses in agriculture, including the avoidance of fungal infection in livestock. Fungicides are categorized according to how they are used, where they come from, and how they are made. Fungicides are divided into two categories based on their origin: biological and chemical. The active components of bio-fungicides are live microorganisms such as bacteria and fungi, which are efficient against pathogens that cause turf disease. The bio-fungicides Ecoguard and Bio-Trek 22G, for example, include *Bacillus licheniformis* and *Trichoderma harzianum*, which are commonly used in agriculture. Chemical fungicides, on the other hand, are made up of both organic and inorganic compounds. Fungicide residues can linger in the soil, move through the food chain, and finally end up in people since chemical fungicides are not biodegradable [76]. Fungicides have a low to moderate toxicity for humans, but are thought to be strongly carcinogenic agents compared to other pesticides [77]. In more severe cases, the use of some fungicides is extremely dangerous to humans — for example, vinclozolin, which has been totally banned [78]. Overall, different fungicides have unique modes of action that cause specific reproductive, mutagenic, and carcinogenic effects on the affected population due to the ingestion of fungicides [79]. Globally, there is an increasing awareness of the potential environmental and health threats [80] linked to the accumulation of toxic by-products in food products [81]. This has led to the development of newer classes of fungicides that generally have lower toxicity [79, 82].

## 7.2.2.2.3  Herbicides

Consistent with pests and fungi, weeds cause biotic stress and have a substantial effect on global crop quality and output. Thus, much effort is focused on removing weeds by mechanical operations, which will give a conducive ecosystem for crops to grow. While crude chemicals such as sulphuric acid, oil wastes, and copper salts are effective in killing all weeds [83], the exposed area can remain toxic to plants for a prolonged period; hence, such chemicals are not suitable for agricultural purposes. Ideally, chemicals that do not damage crop plants and can selectively kill weeds will be suitable for agricultural purposes. As a result, a range of efficient weed killers (herbicides) have been created in the contemporary agricultural age to combat the impact of weeds on crop productivity. The manner of application, chemical structure, and mechanism of action of these herbicides are all classified [83]. For example, the herbicide cyclohexanedione inhibits the enzyme acetyl-CoA carboxylase, which is present in both plants and mammals, and it selectively kills plants but has no effect in mammals by suppressing *de novo* synthesis of fatty acids [84]. Furthermore, several other herbicides such as glyphosate, atrazine, bialaphos, and bromoxynil fall into different categories and have different modes of action. Humans may be exposed to herbicides applied to plants and crops, which can cause mild to severe harmful effects. Herbicides may also infiltrate the human food chain as residues in fruits, vegetables, and dairy products, resulting in unintentional fatalities of those who are exposed.

### 7.2.2.3 *Heavy metals*

Heavy metals refer to elements with high atomic numbers that are found in nature [85]. Heavy metals can be present in a range of food products such as raw or cooked food products [86] and beverages [87]. Heavy metals are toxic in very low amounts and regarded as an important health hazard. Some examples of heavy metals that are known to cause human toxicity include arsenic or mercury. The ingestion or deposition of these metals in the liver or kidney of humans can cause various diseases [88]. Specifically, ingestion of nitrates from food sources causes methemoglobinemia (blue baby syndrome) and is associated with various types of cancers.

### 7.2.3 *Biological Hazards*

Biological toxicants are microorganism-produced compounds that pose a risk to human health and well-being. They are crucial during food processing since they are to blame for the majority of foodborne disease outbreaks. Unhygienic practices or cross-contamination during the shipping, handling, processing, and storage of foods from the environment can

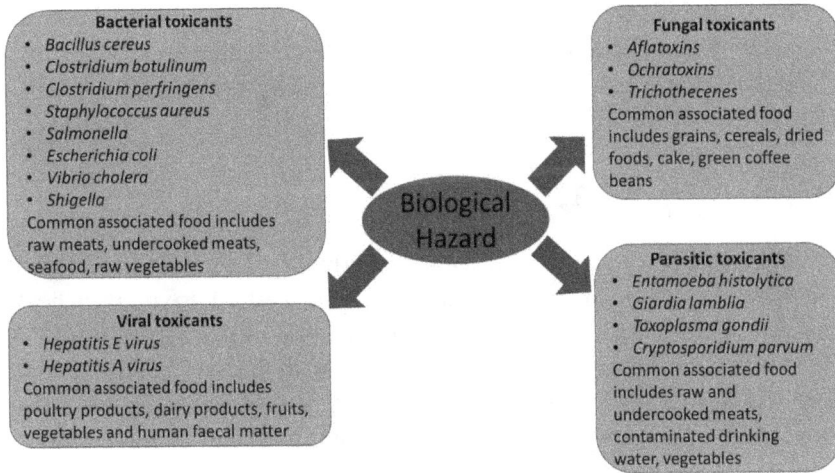

**Bacterial toxicants**
- *Bacillus cereus*
- *Clostridium botulinum*
- *Clostridium perfringens*
- *Staphylococcus aureus*
- *Salmonella*
- *Escherichia coli*
- *Vibrio cholera*
- *Shigella*

Common associated food includes raw meats, undercooked meats, seafood, raw vegetables

**Fungal toxicants**
- *Aflatoxins*
- *Ochratoxins*
- *Trichothecenes*

Common associated food includes grains, cereals, dried foods, cake, green coffee beans

**Biological Hazard**

**Viral toxicants**
- *Hepatitis E virus*
- *Hepatitis A virus*

Common associated food includes poultry products, dairy products, fruits, vegetables and human faecal matter

**Parasitic toxicants**
- *Entamoeba histolytica*
- *Giardia lamblia*
- *Toxoplasma gondii*
- *Cryptosporidium parvum*

Common associated food includes raw and undercooked meats, contaminated drinking water, vegetables

Figure 7.4. Summary of the four categories of biological hazards and examples of where they are detected in food sources.

present biological dangers to food. Any sickness induced by exposure to contaminated food owing to the presence of pathogenic bacteria, fungi, parasites, and viruses in food is referred to as a microbial foodborne disease. Foodborne illness is caused by the ingestion of contaminated food, which causes pathogenic microorganisms to develop in the host's body. These pathogenic microorganisms can grow and colonize the digestive tract, producing infections in the mucosa and surrounding tissues. Toxins produced by these bacteria have a negative impact on the activities of many organs and tissues. Viruses, bacteria, and parasitic worms are the most common pathogens found in food (Figure 7.4).

### 7.2.3.1 *Toxicants derived from bacteria*

Bacteria are unicellular prokaryotes that are assumed to be the source of foodborne illnesses. Dairy products (milk and eggs) and chicken are two popular foods that encourage the development of bacteria (fish, meat, and shellfish). Bacterial toxicants are poisonous substances generated by bacteria that cause harm to the host. Toxins produced by bacteria can take the form of proteins, peptides, or lipopolysaccharides. Bacterial toxins with distinct structures, sources, immunological identities, or modes of action can be distinguished from one another based on the physical or chemical characteristics, the organism of origin, and the clinical results of the host. The descriptions of some of the most frequent bacteria implicated in food-borne illnesses are included here.

### 7.2.3.1.1 *Bacillus cereus*

*Bacillus cereus* is a rod-shaped, Gram-positive, spore-forming bacterium that generates two kinds of toxins. The first sort of toxin is a heat-sensitive, high-molecular-weight peptide that causes toxic infection [89]. This toxin causes diarrhoea and is associated with the indications of vomiting and stomach cramps [89]. The second sort of toxin, on the other hand, is a heat-resistant, low-molecular-weight protein that induces vomiting and stomach discomfort. *B. cereus* poisoning can be caused by a variety of foods, including chicken, vegetables, and dairy products, and the symptoms can appear anywhere from one to six hours after consuming infected food [89].

### 7.2.3.1.2 *Clostridium botulinum*

*Clostridium botulinum* is a rod-shaped, motile, Gram-positive anaerobic bacterium which forms spores that are commonly found in fruit products, vegetables, and poultry products [90]. The spores of *C. botulinum* can thrive in harsh environmental conditions, which makes it a crucial contaminant of food sources. Consumption of food contaminated with botulinum toxin is extremely deadly. Specifically, a few nanograms of the toxin can cause severe paralysis. *C. botulinum* is known to produce seven immunogenic types (A–G) of toxins, and it has enterotoxic, hemotoxic, and neurotoxic properties that have different effects depending on the species. For instance, the botulinum neurotoxin is known to inhibit the release of the neurotransmitter acetylcholine at the synapse. *C. botulinum* toxin types A, B, E, and F account for botulism in humans, whereas types B, C, and D are attributed to botulism in cattle and types C and E have an effect on birds [91]. Upon consumption of contaminated food, symptoms can occur between 12 and 72 hours, which include tiredness, nausea, vomiting, cerebral pain, and muscular paralysis. In most cases, inadequate heating during food preparation is responsible for *C. botulinum* poisoning;

hence, at least 30 min of heating to 80 °C will destroy the toxins present in the food [90].

### 7.2.3.1.3 *Clostridium perfringens*

*Clostridium perfringens* is an anaerobic, Gram-positive, non-motile, rod-shaped bacterium that can generate spores. The spores can withstand high temperatures and may be found in soil, human, and animal digestive systems and a variety of foods such as poultry and vegetables [92]. The cooling and warming cycle of food allows the spores of the bacteria to germinate and grow, which produces toxins that cause food poisoning and diseases. Meat products and gravy are the most prevalent foods linked to *C. perfringens* food illness. An enterotoxin is released in the gut after *C. perfringens* is consumed, producing fluid buildup in the intestinal lumen [92]. Hence, *C. perfringens* poisoning often occurs 8–22 hours after consuming *C. perfringens*-contaminated foods, and the associated symptoms include pain in the abdominal, nausea, and diarrhoea. In a normal situation, the symptoms should subside within 1 day; however, they may also persist for a prolonged period (1–2 weeks) for some individuals if their symptoms are mild.

### 7.2.3.1.4 *Staphylococcus aureus*

*Staphylococcus aureus* is a Gram-positive, spherical-shaped, non-motile anaerobic bacterium, which produces enterotoxins that cause food poisoning. Although staphylococcal food poisoning is usually not lethal, it is extremely aggressive. Dairy products, chicken products, and salads are some of the foods linked to staphylococcal food poisoning [93]. There are five known types (A–E) of *S. aureus*-producing enterotoxins [94]. Although the majority of these toxins are cytolysins, which tear down the membrane phospholipids in the gut, their mechanisms of action differ. Enterotoxin types A and D, in particular, block water absorption from the intestinal lumen and target the emetic receptor, causing diarrhoea and vomiting, respectively. Enterotoxin type B is an enzyme that induces membrane phospholipid hydrolysis, causing damage to the intestinal epithelium and colitis. Symptoms such as fatigue, nausea, vomiting, stomach aches, diarrhoea, and fever might appear within 1–6 hours of consuming infected food [93, 94].

## 7.2.3.1.5 *Salmonella*

*Salmonella* species are anaerobic bacteria that are Gram-negative and rod-shaped and do not produce spores. *Salmonella* species, in general, are pathogenic to humans and are spread by other animals or insects. The severity of *Salmonella* infection in people is determined on the basis of the strain and the individual's susceptibility to the bacterium. Salmonellosis is a *Salmonella*-related foodborne sickness that can result in septicaemia, typhoid fever, and enteric disease [95]. Improper cooking resulting in undercooked foods such as poultry products or dairy products is recognized as a key contributor to human infections. Upon ingestion of contaminated food, symptoms of salmonellosis can develop between 1 and 6 hours, and include nausea, vomiting, abdominal pains, diarrhoea, headache, and fever.

## 7.2.3.1.6 *Escherichia coli*

Human gastroenteritis is caused by *Escherichia coli*, a Gram-negative, rod-shaped, motile, non-spore-forming bacterium. Pathogenic features, clinical symptoms, epidemiology, and serogroups are used to divide *E. coli* into four primary classes [96]. Gastroenteritis or infantile diarrhoea is caused by enteropathogenic *E. coli*, whereas bacillary dysentery is caused by enteroinvasive *E. coli*. Furthermore, enterotoxigenic *E. coli* is responsible for diarrhoea linked with travel, whereas enterohemorrhagic *E. coli* is responsible for hemorrhagic colitis. Enterohemorrhagic strains are the most dangerous and can cause serious illness. Infection with these strains, for example, is linked to haemolytic uremic syndrome in young infants [97]. Raw chicken, milk, vegetables, and fruits, as well as faecal contamination of food or water, are some of the foods linked to *E. coli*. The symptoms of *E. coli* infection include diarrhoea, fever, dehydration, and vomiting, and vary depending on the kind of sickness [96].

## 7.2.3.1.7 *Vibrio cholerae*

*Vibrio* is a genus of Gram-negative, non-encapsulated, motile anaerobic bacteria that generates cholera toxin [98]. Cholera is generally spread by the intake of raw or undercooked seafood, such as shellfish, or through the contamination of food and water with human faeces. Individuals who have been infected with this disease may experience moderate or watery diarrhoea, vomiting, and stomach discomfort [98].

## 7.2.3.1.8 *Shigella*

The shigella is a Gram-negative, rod-shaped, non-sporulating, non-motile anaerobic bacterium which causes infection of the intestines (shigellosis). There are four species within the *Shigella* genus, known as *Shigella sonnei, Shigella flexneri, Shigella dysenteriae*, and *Shigella boydii*, which are capable of producing various toxins [92]. The shiga toxin, the most widely studied form, is known to be enterotoxic, neurotoxic, and cytotoxic. The spread of shigellosis may be via close contact with an infected individual, houseflies, faecal-oral route, or ingestion of contaminated food and water. Some of the food sources associated with shigella include raw meat, fruits, vegetables, and dairy products [99]. The severity of shigellosis depends on the individual and may take 1–7 days to develop. In some cases, infected individuals may not develop symptoms and may progress from slight diarrhoea to diarrhoea with blood or mucus, with accompanying abdominal pains. In extreme cases, individuals may experience dehydration, fever, vomiting, and bloody stools [99].

### 7.2.3.2 Fungal toxicants

Moulds and fungi can create a wide variety of physiologically active fungal metabolites. While some of these fungal metabolites are useful in the manufacture of foods like cheese and medications (antibiotics), certain filamentous fungi can create harmful metabolites. Mycotoxins are poisonous metabolites that have a variety of forms and toxicological features [100]. Mycotoxicosis refers to the adverse impact that humans or animals experience when foods exposed to mycotoxins are ingested. These mycotoxins are organic molecules with a low molecular weight that are powerful acute or chronic toxins, carcinogens, mutagens, and teratogens [101]. The generation of mycotoxins and their contamination of foods will vary depending on environmental factors such as the weather and moisture.

### 7.2.3.2.1 Aflatoxins

*Aspergillus flavus, Aspergillus nominus*, and *Aspergillus parasiticus* synthesize aflatoxins, a family of extremely hazardous bisfuran polycyclic fungal metabolites, carcinogens, and hepatotoxic compounds [102]. Under the right conditions, these fungi may grow on a wide range of agricultural items. *A. flavus*, for example, is found in a variety of foods and feeds, including stored grains, beans, nuts, and dried fruits. In general, maize is one of the most common foods infected with aflatoxin, which causes aflatoxicosis. Because aflatoxins are lipid soluble and heat stable, they can withstand most food processing methods. Furthermore, aflatoxins are divided into four categories (B1, B2, G1, and G2). The health consequences of aflatoxins on living beings vary greatly, depending on a variety of parameters such as species, dose level, and age [103]. Although the liver is the organ most impacted by aflatoxins' toxicity and carcinogenicity, other organs may be harmed as well. Liver edema, biliary proliferation, jaundice, vomiting, and anorexia are some of the clinical signs of ingesting aflatoxin-contaminated foods, and they might appear within 3 weeks of ingestion.

### 7.2.3.2.2 Ochratoxins A

*Aspergillus ochraceus* produces ochratoxins, which are harmful secondary metabolites that thrive in wet environments with moderate temperatures [100, 101]. Ochratoxins have been discovered in cereals, beans, and a variety of other animal-based foods. In mammals, ochratoxin A causes chronic kidney damage, and in rats, it disrupts foetal development [100, 101].

### 7.2.3.2.3 Trichothecenes

Mycotoxins generated mostly by members of the *Fusarium* genus are known as trichothecenes. Toxins may infect a variety of foods, including cereals, bread, snacks, and cake [100]. Acute trichothecene poisoning is characterized by neurological symptoms; however, long-term toxicity can harm other organs such as the bone marrow, thymus, spleen, and gastrointestinal system. Inflammation, nausea, vomiting, and diarrhoea are some of the other symptoms [104].

### 7.2.3.3 *Viral toxicants*

Viruses are microscopic particles that are present everywhere, including in food sources. Without multiplying, these viruses do not require any resources for their survival. However, as they are regarded as obligatory intracellular parasites, viruses require an appropriate host such as human beings for replication and multiplication [105]. In most cases, foodborne viruses consist of single-stranded RNA and are surrounded by structural proteins. Although viruses do not cause food spoilage, they infect and use the host cell's resources to multiply, causing diseases in living organisms [105]. Viruses are spread enterically and released through human excrement, and contaminate upon ingestion. Food contamination by viruses is usually a result of poor hygienic practices during food handling and preparation [106]. For example, hepatitis A and E viruses are accountable for hepatitis and are transmitted via the faecal-oral route. Furthermore, some insects such as flies are possible carriers of the hepatitis A virus. Some likely origins of the hepatitis A virus can be found in poultry products, dairy products, fruits, vegetables, and human faecal matter. Similarly, the transmission of the hepatitis E virus could be attributed to the consumption of contaminated water and food or physical interaction. Hepatitis A virus infection can take anywhere from 15 to 20 days to develop, but hepatitis E virus infection might take anywhere from 2 to 9 weeks to develop [106]. Fever, vomiting, stomach pain, liver inflammation and enlargement, and jaundice are all frequent hepatitis virus symptoms. Rotavirus, like hepatitis A and E viruses, is transferred by food contaminated by an infected individual via the faecal-oral pathway. Fever, vomiting, and watery diarrhoea are some of the symptoms of rotavirus infection [107].

### 7.2.3.4 *Parasitic toxicants*

Parasites, like viruses, are entities that require a host to survive and are spread by contaminated food or water or through the faeces of infected hosts [108]. In most cases, parasite infection is linked to raw or half-cooked meals, such as meat. Parasitic protozoa and worms are two forms of parasites that may infect people through food or water. (1) Protozoa are single-celled organisms that cause disease. *Entamoeba histolytica, Giardia lamblia, Toxoplasma gondii*, and *Cryptosporidium parvum* are some of the most prevalent disease-causing protozoan species [108]. They develop cysts during the infectious stage, which may be found in plants, polluted water, and badly cooked foods. Diarrhoea, stomach pain, nausea, weakness, and fever are some of the signs of sickness caused by these protozoa. (2) Parasitic worms include roundworms (nematodes), tapeworms (cestodes), and flukes, which are multicellular creatures (trematodes). In most cases, food contamination occurs as a result of inadequate sanitation, and disease is transferred through the larval cysts of these worms. Half-cooked meals, raw fish, shellfish, vegetables, and fruits are all potential sources of worm infection [109]. Abdominal discomfort, bloody stool, nausea, and muscle and joint aches are examples of illness symptoms caused by worm infestation.

### 7.2.4 *Toxicity of Nutrients*

In a balanced diet, it is necessary to have a satisfactory quantity of nutrient ingestion for maintaining health and well-being. Nutrients play an important contributing role in the maintenance of good health, which is underpinned by a group of macromolecules and nutrients. Food fortification describes the supplementation of vitamins, minerals, or amino acids to a food product to enhance the nutritional value of the food for health purposes. While these nutrients are known for their health benefits, some nutrients can be toxic and potentially detrimental to health when taken in high concentrations. For example, an excessive amount of vitamin A causes birth defects and inhibits bone formation [110], and excessive ingestion of vitamin D results in high-level calcium in the blood, leading to tissue deposits of calcium. Furthermore, increasing dietary calcium can result in kidney stones and prevents the uptake of other essential ions and nutrients. Moreover, high sodium ingestion is associated with hypertension and subsequent adverse impact on the cardiovascular system [111]. Carbohydrates, proteins, and lipids are all essential sources of energy in the diet, with carbohydrates being the most important. Excess carbs, on the other hand, are retained in the adipose tissue as glycogen or fatty acids once energy demands are fulfilled, resulting in undesirable weight gain and obesity [112]. Furthermore, because the human liver has a limited capacity to digest proteins, a high protein diet (35% of total calories) might result in elevated levels of amino acids, ammonium, and insulin in the blood, which can lead to mortality in severe cases [113]. Systemic infections, renal illnesses, heart disease, high blood pressure, diabetes, and cancer are all common health concerns linked to increased protein excretion through urine.

## 7.3 Food Safety Management Systems

Food safety encompasses issues associated with basic hygiene practices and standard operating procedures during food preparation. Hence, the set of procedures and measures used to properly control the risks and hazards of food from farm to table is defined as a food safety management system. The Food Safety Management System is related to the Codex Alimentarius, also known as the "Food Code", which is set up by the Food and Agriculture Organization of the United Nations (FAO) and the World Health Organization (WHO) [1]. The "food code" consists of a set of internationally recognized standards and guidelines for most government authorities to reference, aiming to ensure the production of safe food for consumers [114]. For example, the "food code" and food hygiene committees explain the principles and rationale of food hygiene and provide examples on how to apply these principles [1, 114]. Specifically, food safety management systems explain active managerial control, which actively manages food safety risk by identifying several risk factors that are commonly associated with foodborne pathogens. The risk factors include procuring food from unsafe sources, failure to cook food adequately, storing food at improper temperatures, using contaminated equipment, and poor personal hygiene during food preparation. However, different countries would refer to the main guidelines and adopt them in the context that would suit their individual needs; hence, it is important to refer to a country's own regulatory system [115].

As food safety is an important concern for health, several systems are put in place in different countries to ensure food safety. Such systems include Hazard Analysis Critical Control Points (HACCP), which ensures quality of service and product delivery in the whole food production process [116], and the International Organization for Standardization (ISO), which is the world's largest creator of voluntary international standards. Within the variety of ISO standards, ISO 22000 is related to the Food Safety Management Standard. ISO 22000 was developed based on the Codex principles for food hygiene; this allows various authorities to reference ISO 22000 to ensure that all the criteria for food safety are met and in accordance with their individual national requirements and government inspections [117]. Furthermore, ISO provides certification to institutions and has developed a set of standards for the management of food safety-related activities in an organization. Therefore, an ISO 22000-recognized accredited auditor can formally register an organization if it is compliant with the requirements of the system. The system has a comprehensive description of food safety requirements that is applicable for different stakeholders within the food production process. This includes every step from food production to processing, distribution, storage, and handling of all food ingredients. Some of these stakeholders involved in the food production process include (1) primary producers (farmers), (2) processors (poultry, vegetables), (3) manufacturers (snacks, canned, or frozen food), (4) food service providers (restaurants, hospitals), and (5) other service providers (food storage provider, caterer, or delivery service provider).

The HACCP has been regarded as one of the best food safety measures and practices, and the ISO 22000 also uses their policies and structures. The HACCP system includes the methods and procedures to identify, prevent, and control potential food safety hazards. Specifically, this system will allow each stakeholder to conduct a food safety hazard analysis, establish critical control points, identify critical control limits during the process, and develop protocols to monitor the critical control points. Specifically, critical control limits refer to a set of values that separate what is acceptable from what is not acceptable based on the recommended threshold of tolerance for that particular hazard to confirm the safety of food products. In other words, critical control points provide a tolerance level in the food flow. For example, critical control point analyses check on the optimal temperature and the duration of food during the food production step and their impact on a particular food hazard. Hence, the analyses would include 3Ws and 1H — how, where, when, and who. Generally, "how" refers to the method that was used to monitor the critical limit, "where" defines the location of the activity, "when" denotes the duration of the intended process, and "who" refers to the responsibility of the individual that will be performing the monitoring procedure [115, 116]. Finally, action plans can be designed to handle and correct the critical control points, which will enable food safety record keeping and validation that the entire workflow has met the requirements of food safety.

While numerous food hygiene standards, including the General Food Safety Hygiene regulations and FAO/WHO Codex Alimentarius Commission Standards, have been in operation in many organizations, there are currently limited to no reported ISO standards related to the 3D food printing process. This could be due to the nascent stage of the 3D food printing technology in the areas of pre-printing (e.g., food ink formulation), printing (e.g., development of 3D food printers and 3D printing processes), and post-printing (e.g., baking of the printed food product).

## 7.4  Considerations for Food Safety during 3D Food Printing

It is evident from the earlier chapters that 3D food technology has the potential to process and produce different unique food designs using a variety of ingredients, including meat, chocolate, candy, vegetables and pizza dough [118]. Furthermore, 3D food printing allows the customization of food production by controlling the type and quantity of ingredients that have specific nutrient content and flavour characteristics that are suited to the needs of particular individuals [119]. Customization of foods is a very delicate and creative process, which requires food design and implementation of the procedure algorithm before food printing. Once the procedure is programmed, the 3D food printer carries out continuous printing for layer accumulation [118] using the predetermined ingredients to print out a specific structure and shape of food based on individual preferences [120]. This level of food customization will be suitable for individuals who require special diets such as the elderly with dysphagia, children, or athletes [121].

In the current market, there are a few food printers which depend on a few main 3D printing technologies such as fused deposition modelling (FDM) [122], colour jet printing (CJP), and selective laser sintering (SLS) [123]. Briefly, FDM is recognized as one of the extrusion methods by pushing primarily liquid materials into holes at high temperature and high pressures and stacking them one layer at a time. On the other hand, the CJP method uses a print head to selectively distribute the binder into a powder layer. The SLS method is applicable for powder/dry ingredients, where a laser is used to harden the specific part to cause it to form a shape. Specifically, foods with more diverse colours and flavours can be easily printed using sugar-like powders and adding food additives via the SLS method [124]. The SLS method involves melting powder particles using laser irradiation, which forms and binds a new layer with a hardened layer until the required structure is created. In some cases, once the food is 3D-printed, further steps may be required to cook the food before it is suitable for consumption [124]. Hence, ongoing research is exploring ways to integrate cooking elements during the printing process. Specifically, the introduction of infrared lamps or laser processes has been shown to cook food uniformly during the printing process. For example, an infrared lamp is a single lamp-and-reflector heating system that is integrated with the 3D printer for simultaneous infrared cooking during printing. The infrared lamp can generate contactless heat up to 1,800 F onto small targets for localized cooking within the specific areas (0.25 inch) [125].

Therefore, although 3D food printing technology enables the creation of new processes for food manufacturing, food sustainability, and food customization for specific individual preferences and needs, food safety testing and regulation for 3D food printing processes are imperative. As discussed earlier, certain 3D food printing processes may pose certain food safety risks and involve dangerous equipment and/or processes that could affect the quality of the food. For example, the SLS method is potentially dangerous as it involves lasers during the printing process. This also applies to 3D printers that offer simultaneous cooking using laser methods [126]. According to the FDA, there are four major hazard classes (I–IV) of lasers, which have a varying degree of class and output power, and when using the printer, the user should pay attention to the warning symbol [127]. While there are safety regulations pertaining to the use of 3D printers with lasers, there is unfortunately limited information on the FDA regulation regarding the safety of food products using a 3D food printer using a laser beam, as the use of lasers during food printing could cause chemical reactions in food ingredients. Similarly, this may also be true for other 3D printing processes or infrared lamps that use high temperatures, which results in unwanted chemical reactions within the food ingredients. As the nature of food hazards is highly dependent on the 3D printer, the 3D printing process, and food ingredients, the following section proposes some food safety considerations for 3D food printing.

### 7.4.1 *Pre-printing Process*

The formulation of food ink depends on the food material itself and the addition of food additives to ensure the printability of a food product. Hence, the usual edible food materials used for printing would be no different from any standard edible raw materials whose food safety would have been otherwise tested by organizations. However, due to the addition of other food additives such as hydrocolloids or flavourings, it is possible that this potentially introduces certain chemical or biological hazards during the food production process. Hence, in the development of novel food inks, it would be important to identify and establish the critical control limits for the potential hazards in accordance with the guidelines. Furthermore, the physical state of the food ink — whether it is dry or wet food ink and cooked or uncooked food ink — may also expose the food ink to certain risks for bacterial contamination; thus, the temperature and duration of the storage of food ink will be another consideration factor.

The degree to which chemical reactions occur in food components is influenced by a number of factors, including laser dose, dose duration, temperature, and the kind of food ingredient. This is the result of a chemical interaction involving major dietary components such as carbohydrates, fat, proteins, and vitamins. Several chemicals are produced during laser irradiation and high temperature heating of the food, including 2-alkylcyclobutanones (2-ACBs), acrylamide, polycyclic aromatic hydrocarbons (PAHs), and furan. When carbohydrates and proteins are subjected to irradiation, chemical reactions can occur. The types of responses are mostly determined by a number of elements, including protein amino acid content, protein structure type, and the presence of other chemicals in the diet. Amino acids with aromatic rings or sulphur groups, for example, are more susceptible to irradiation alteration. Oxidation, denaturation, and aggregation are just a few of the chemical changes that proteins go through [128]. Irradiation of protein may also cause the formation of other small radiolytic products such as keto acids or amide-like intermediates, which may be potentially harmful to health [129]. Furthermore, as enzymes are classified as protein, irradiation also denatures the activity of several key enzymes such as α-amylase and pectinase, which are important for breaking down carbohydrates and pectin, respectively. Similarly, for carbohydrates, irradiation of various foods results in the production of radiolytic products [130].

Irradiation-induced chemical responses of lipids depend on the concentration, physiochemical status (saturated or unsaturated), and environmental conditions [129]. Lipid peroxidation is the chain of chemical reactions of oxidative degradation of lipids, which is enhanced by the irradiation process through the generation of free radicals [130]. This happens a lot in foods with a lot of unsaturated fatty acids and a lot of fat.

Furthermore, naturally occurring plant sterols, which have structural similarity with cholesterol, may be found in cereals, fruits, nuts, vegetables, and seeds. Given their structural similarities with cholesterol, they are reported to have health benefits by helping with a cholesterol-lowering effect. However, plant sterols are also prone to oxidation by irradiation and heating, causing them to lose their beneficial health outcomes. Studies have demonstrated that lipid peroxidation was reduced through the use of low temperature or addition of antioxidant food additives [131]. For example, oregano or rosemary antioxidant extracts are able to reduce lipid oxidation in beef burgers [132], suggesting that antioxidant food additives may be useful for the 3D printing of meat products with simultaneous laser cooking to limit lipid peroxidation.

Besides large macromolecules, irradiation also has an effect on nutrients and vitamins. Generally, irradiation decreases the activities of most vitamins. The extent of bioactivity loss is subjected to the chemical properties of vitamins (fat-soluble vs. water-soluble vitamins). Vitamin E, for example, is the most irradiation-sensitive of the fat-soluble vitamins, but vitamin D loss is significantly lower. Freezing temperatures and a lack of oxygen are two conditions that have been observed to reduce vitamin loss due to radiation [133]. Certain food additives can produce toxic radiolytic products upon irradiation; hence, it is important to consider the addition of food additives to food ink for 3D food printing. For example, a common antimicrobial agent, potassium benzoate, can be converted into the volatile compound benzene through the decarboxylation of potassium benzoate [134].

Furan is a well-regarded carcinogen that can be derived from proteins, lipids, and carbohydrates [135]. Furan and the related compounds 2- and 3-methylfurans are chemical contaminants that naturally form during thermal processing. Several studies have shown that the concentration of furan increases proportionally with an increasing irradiation dose [136]. Long-term exposure to furans and methylfurans in food has the potential to harm the liver. Similarly, irradiation-induced triglyceride breaking produces 2-alkycyclobutanones from fat-containing foods. 2-ACBs can only be found in irradiated fat-containing foods, but not in non-irradiated foods. Furthermore, the amount of 2-ACBs produced in irradiated meals is linked to fat content and absorbed dosage [135]. The concentration of 2-ACBs in irradiated food is estimated to range from 0.2 to 2 g/g of fat, depending on the dosage received [137].

When meat is cooked at a high temperature, harmful compounds called heterocyclic amines (HCAs) and polycyclic aromatic hydrocarbons (PAHs) are formed [138]. HCAs are produced when sugars and proteins present in the meat react with each other at high temperatures. PAHs, on the other hand, are formed when meat is cooked at high temperatures, causing PAHs in the smoke to adhere to the flesh [139]. It is evident that HCAs and PAHs are toxic and mutagenic in laboratory experiments because studies have shown that they are able to alter the DNA sequences, resulting in an increased risk of cancer [140]. Specifically, a number of studies have demonstrated that PAHs are associated with an increased risk of breast and prostate tumours [141], kidney cancer [142], and colon cancer [143]. Therefore, this is an important food safety consideration for the development of 3D food printing of meat products with simultaneous cooking capability.

Certain 3D food printing processes, individually or combined with infrared cooking elements, may result in the use of high temperatures. Acrylamide is an unscented solid with a melting point of 84.5 °C that is produced in food rich in carbohydrates cooked at extreme temperatures and under low-moisture conditions [144]. Acrylamide is a well-known toxic chemical with mutagenic and carcinogenic properties, as shown in some experimental studies. Acrylamide has been identified in a variety of high-temperature-processed carbohydrate diets and demonstrated to be a carcinogen and genotoxic in animal experiments [145], and neurotoxic in humans [146]. Hence, it is important to consider determining the level of acrylamide for 3D food printing in order to establish food safety.

### 7.4.2 *3D Printer and Printing Process*

The development of 3D printers and the 3D printing process will be an important part of food manufacturing and food service providers if a 3D printer is used to produce food in a restaurant or hospital setting. Like those in any other industry or as kitchen appliances, the materials and parts used to build the 3D food printer would be food safe, preventing migration from its components to the food products. There are already existing standards in place for the selection of approved material. Furthermore, the materials and parts should be resistant to cleaning products and preferably suitable for dishwashers, making them attractive for industry adoption. From the food safety viewpoint, 3D printers should have a hygienic design that allows certain critical parts of the printer to be disassembled for cleaning purposes in order to reduce the risk of potential bacterial contamination [147]. Furthermore, brittle materials that could be broken from the printer should be avoided in the printer design in order to minimize the risk of physical hazards. Therefore, for the development of 3D food printers, it is important that any components that might come into contact with the 3D printing material or the parts are food-grade materials that do not contain or leach harmful chemicals. Furthermore, some 3D printing materials are not food safe and might contain toxic chemicals, which may contaminate food. Therefore, 3D printer parts intended for food contact should be made by using materials that are certified for food safety. Besides selecting food-grade materials for the development of food printers, the design of the printer should cater to easy cleaning protocols. Certain 3D printing materials have a low heat deflection temperature, which means that the 3D printer parts might become brittle and crack, or deform and warp at elevated temperatures. This may expose certain parts of the materials that may potentially leach harmful chemicals. If proper cleaning protocols are not followed, the 3D printer or parts can become a breeding ground for bacterial growth if such parts of the printers are not easily accessible for cleaning [148]. This may result in food particles becoming trapped in certain parts of the printers such as tubes or edges, allowing dangerous bacteria such as *E. coli* and *Salmonella* to grow. Some toxic moulds find favourable growth conditions on several types of plastic and are hard to remove. Furthermore, the risk of migration of chemicals or biological contaminants is higher if the food is exposed to the 3D printer part for an extended time period.

Besides using appropriate printer materials and cleaning protocols, the speed of food printing may be another factor to consider. Specifically, the 3D food printing process will determine whether the 3D-printed food would be cooked concurrently using heating elements or laser, or separately cooked after the printing process. Hence, based on the formulation of the food ink, different types of chemical hazards could be introduced when combined with the type of 3D printing process used for the particular food ingredients. Hence, it is likely that identifying and establishing critical control limits for potential hazards would be important for a particular food ink and the required printing process. This will subsequently result in the development of food safety standards for the particular food and printing process. Foodborne bacteria (such as *S. aureus* and *E. coli*) can grow to harmful levels and cause sickness if kept at room temperature for too long. Bacteria grow quickly in the temperature range of 5–60 °C, which the US Food Safety and Inspection Service has designated as the risk zone. Potentially dangerous foods should not be stored at this temperature, and food that has been stored at this temperature for more than 2 hours should not be consumed [149]. Furthermore, given that foodborne microorganisms grow rapidly between 21 °C and 47 °C, it is recommended that food not be stored for more than 60 min if the temperature is above 32 °C. To limit the bacterial growth and contamination, the duration for which food is kept in the hazard zone must be reduced. Therefore, controlling the time and temperature during the 3D food printing process, especially if raw food products are used, will be a very crucial factor for food safety. Lastly, shelf-life studies should be conducted to assess if 3D printing methods of food production affect product shelf-life time and the risk to foodborne pathogen contamination, which may have a potential impact on food security and storage.

### 7.4.3 *Post-printing Process*

In addition to the possible food hazards mentioned above, any changes to the quality of the printed food over time through post-processing and storage should also be monitored if it is not intended to be consumed immediately. Specifically, the safety of the printed products needs to be quantified by monitoring the microbial content in the food during preparation, printing, and post-printing. Furthermore, certain foods may require a post-printing process such as cooking or baking so that they can be consumed. The potential introduction of potential chemical hazards during the post-printing process should also be monitored as part of the food safety management system. Importantly, the post-printing process should not affect the physical appearance and the texture of food.

As mentioned in the earlier chapter, one of the well-regarded applications of 3D food printing is for patients suffering from dysphagia. Therefore, the final food product after the post-printing process must be regulated and suitable for dysphagia patients. Specifically, the 3D-printed food must comply with and follow the International Dysphagia Diet Standardization Initiative (IDDSI) framework. The IDDSI framework consists of an eight-level continuum (0–7), where drinks are assessed from levels 0 to 4, and foods are measured from levels 3 to 7 [150]. The IDDSI framework establishes a standard for describing food textures and liquid thickness. Therefore, the final 3D-printed foods should be subjected to the IDDSI testing methods before they can be deemed suitable for dysphagia patients. The IDDSI testing methods are designed to confirm a product's flow or textural properties at the time of testing. Foods and beverages should be tested in the same conditions that they will be served in (especially temperature). Based on their detailed clinical assessment, the clinician is responsible for making food or drink recommendations for a specific patient [150]. The IDDSI framework consists of testing methods that make use of everyday dining utensils, reducing the requirement for subjectivity that typically comes with description-based procedures. Forks and spoons were chosen because they are affordable and easy to obtain, and can be used in most meal preparation and dining situations. To identify which level a meal belongs in, a combination of tests may be necessary. The fork drip test, spoon tilt test, and fork or spoon pressure test are all testing procedures for puréed, soft, firm, and solid meals [150].

For example, the fork drip test is used to ensure that meals in levels 3–5 have the right thickness and cohesiveness by analyzing whether they flow through or stay together on the slots/prongs of a fork and then comparing the results to the thorough descriptions of each level. The spoon tilt test is used to measure the stickiness (adhesiveness) of meals as well as their capacity to stay together (cohesiveness). The fork pressure test is best used to assess foods in levels 4–7 and transitional foods to determine how firm or hard they are by observing how the food changes when pressure is applied to it with the tines/prongs of a fork or the back of a spoon. A regular metal fork's slots/gaps between the tines/prongs generally measure 4 mm, making the fork a handy instrument to measure the particle size of meals in level 5 — minced and moist [150]. For example, in a study that evaluated the 3D food printing of various vegetables [151], a fork and spoon test were evaluated, which complied with the IDDSI standard that these foods are suitable for dysphagia (Figure 7.5).

While 3D food printing has been widely considered a potential technology to improve the quality of life and provide nutritious food to elderly

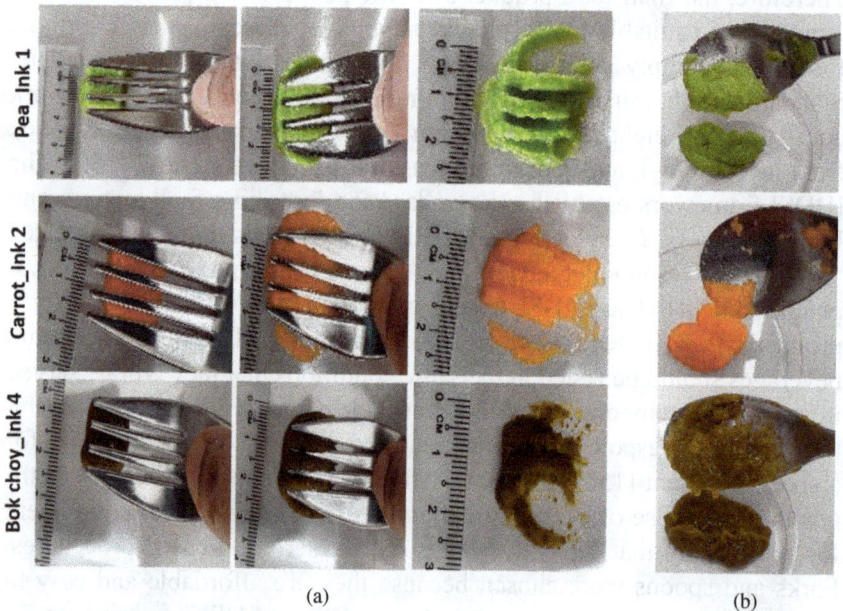

Figure 7.5.    Representation of the IDDSI fork pressure test (a) and the spoon tilt test (b) on various 3D-printed vegetable inks. Top to bottom: Pea ink, carrot ink, and bok choy ink. Reproduced with permission from [151]. Copyright (2021) from Elsevier.

people with dysphagia, there is currently no standard or regulation under the various frameworks. Therefore, it would be important to monitor the food safety regulations and include the IDDSI framework testing for 3D-printed foods that are suitable for patients with dysphagia.

## 7.5 Conclusion

Food is an essential component of life as it satisfies our nutritional needs by supplying energy, as well as providing growth and maintenance of human health. In the era of digitalization, 3D printing technology will enable the production and personalization of food in the years to come. The disruption of traditional food production with 3D printing technology poses many challenges, particularly in terms of food safety. Various food hazards (physical, chemical, and biological) may be introduced into food either intentionally or unintentionally during the stages of food ink formulation, printer development, or during the food printing process or storage. The HACCP concept for food safety requires the identification of potential risks and the establishment of critical control points in the 3D food printing process. As there are few to no ISO standards for 3D food printing, the primary objective is to determine and eliminate the possibility of critical biological, chemical, and physical hazards in the 3D food printing process to ensure food safety. While there are identifiable food safety threats, these are outweighed by the exciting innovative opportunities to prepare food, with customized sensory qualities and appealing shapes, in the desired serving size for our nutrition needs, which we call digital gastronomy. Identified food safety risks can be prevented or overcome if food safety standards or regulations are in place for 3D food printing in the near future. Besides food safety, other food testing methods such as the IDDSI food test should also be implemented as part of the regulatory process to ensure that the 3D-printed food is suitable for the target population groups (e.g., dysphagia patients).

## References

[1] WHO. (2020). Available online: https://www.who.int/news-room/fact-sheets/detail/food-safety. Accessed on: 2 Dec. 2021.
[2] Scallan, E., Hoekstra, R. M., Angulo, F. J., Tauxe, R. V., Widdowson, M. A., Roy, S. L., Jones, J. L. and Griffin, P. M. (2011). Foodborne illness acquired in the United States — major pathogens, *Emerging Infectious Diseases*, 17, pp. 7–15.

[3]  NACMCF. (2021). Available online: https://www.fsis.usda.gov/policy/advisory-committees/national-advisory-committee-microbiological-criteria-foods-nacmcf. Accessed on: 2 Dec. 2021.

[4]  FDA. (2020). Available online: https://www.fda.gov/food/food-ingredients-packaging/food-additives-petitions. Accessed on: 2 Dec. 2021.

[5]  Shim, S. M., Seo, S. H., Lee, Y., Moon, G. I., Kim, M. S. and Park, J. H. (2011). Consumers' knowledge and safety perceptions of food additives: evaluation on the effectiveness of transmitting information on preservatives, *Food Control*, 22, pp. 1054–1060.

[6]  Schoefs, B. (2002). Chlorophyll and carotenoid analysis in food products. Properties of the pigments and methods of analysis, *Trends in Food Science and Technology*, 13, pp. 361–371.

[7]  Young, J. H. (1989). *Pure Food: Securing the Pure Food and Drug Act of 1906* (Princeton University Press).

[8]  Rodriguez-Amaya, D. B. (2016). Natural food pigments and colorants, *Current Opinion in Food Science*, 7, pp. 20–26.

[9]  Carocho, M., Barreiro, M. F., Morales, P. and Ferreira, I. C. F. R. (2014). Adding molecules to food, pros and cons: a review on synthetic and natural food additives, *Comprehensive Reviews in Food Science and Food Safety*, 13, pp. 377–399.

[10]  Nontasan, S., Moongngarm, A. and Deeseenthum, S. (2012). Application of functional colorant prepared from black rice bran in yogurt, *Procedia APCBEE*, 2, pp. 62–67.

[11]  Jimenez-Aguilar, D. M., Ortega-Regules, A. E., Lozada-Ramirez, J. D., Perez-Perez, M. C. I., Vernon-Carter, E. J. and Welti-Chanes, J. (2011). Color and chemical stability of spray-dried blueberry extract using mesquite gum as wall material, *Journal of Food Composition and Analysis*, 24, pp. 889–894.

[12]  Dilrukshi, P. G. T., Munasinghe, H., Buddhika, A., Silva, G. and De Silva, P. G. S. M. (2019). Identification of synthetic food colours in selected confectioneries and beverages in Jaffna District, Sri Lanka, *Journal of Food Quality*, 2019. https://doi.org/2010.1155/2019/7453169.

[13]  Gostner, J. M., Becker, K., Ueberall, F. and Fuchs, D. (2015). The good and bad of antioxidant foods: an immunological perspective, *Food and Chemical Toxicology*, 80, pp. 72–79.

[14]  Sharma, S., Goyal, R. P., Chakravarty, G. and Sharma, A. (2008). Toxicity of tomato red, a popular food dye blend on male albino mice, *Experimental and Toxicologic Pathology*, 60, pp. 51–57.

[15]  Singh, R. L. (1989). Metabolic disposition of (14-C) Metanill Yellow in rats, *Biochemistry International*, 19, pp. 1109–1116.

[16]  Singh, R. L., Singh, S., Khanna, S. K. and Singh, G. B. (1991). Metabolic disposition of Metanil Yellow, Orange II and their blend by caecal

microflora, *International Journal of Occupational and Environmental Health*, 1, p. 250.

[17]   Chequer, F. M. D., Lizier, T. M., de Felicio, R., Zanoni, M. V. B., Debonsi, H. M., Lopes, N. P., Marcos, R. and Palma de Oliveira, D. (2011). Analyses of the genotoxic and mutagenic potential of the products formed after the biotransformation of the azo dye Disperse Red 1, *Toxicology in Vitro*, 25, pp. 2054–2063.

[18]   Sayed, H. M., Fouad, D., Ataya, F. S., Hassan, N. H. and Fahmy, M. A. (2012). The modifying effect of selenium and vitamins A, C, and E on the genotoxicity induced by sunset yellow in male mice, *Mutation Research*, 744, pp. 145–153.

[19]   Singh, R. L., Singh, P. K. and Singh, R. P. (2015). Enzymatic decolorization and degradation of azo dyes: a review, *International Biodeterioration and Biodegradation*, 104, pp. 21–31.

[20]   Jelinic, M., Kahlberg, N., Leo, C. H., Ng, H. H., Rosli, S., Deo, M., Li, M., Finlayson, S., Walsh, J., Parry, L. J., Ritchie, R. H. and Qin, C. X. (2020). Annexin-A1 deficiency exacerbates pathological remodelling of the mesenteric vasculature in insulin-resistance but not insulin-deficiency, *British Journal of Pharmacology*, 177, pp. 1677–1691.

[21]   Kahlberg, N., Qin, C. X., Anthonisz, J., Jap, E., Ng, H. H., Jelinic, M., Parry, L. J., Kemp-Harper, B. K., Ritchie, R. H. and Leo, C. H. (2016). Adverse vascular remodelling is more sensitive than endothelial dysfunction to hyperglycaemia in diabetic rat mesenteric arteries, *Pharmacological Research*, 111, pp. 325–335.

[22]   Leo, C. H., Jelinic, M., Ng, H. H., Marshall, S. A., Novak, J., Tare, M., Conrad, K. P. and Parry, L. J. (2017). Vascular actions of relaxin: nitric oxide and beyond, *British Journal of Pharmacology*, 174, pp. 1002–1014.

[23]   Leo, C. H., Jelinic, M., Ng, H. H., Parry, L. J. and Tare, M. (2019). Recent developments in relaxin mimetics as therapeutics for cardiovascular diseases, *Current Opinion in Pharmacology*, 45, pp. 42–48.

[24]   Leo, C. H., Jelinic, M., Ng, H. H., Tare, M. and Parry, L. J. (2016). Serelaxin: a novel therapeutic for vascular diseases, *Trends in Pharmacological Sciences*, 37, pp. 498–507.

[25]   Tandel, K. R. (2011). Sugar substitutes I: health controversy over perceived benefits, *Journal of Pharmacology and Pharmacotherapeutics*, 2, pp. 236–243.

[26]   Gardner, C., Wylie-Rosett, J., Gidding, S. S., Steffen, L. M., Johnson, R. K., Reader, D. and Lichtenstein, A. H. (2012). Nonnutritive Sweeteners I: current use and health perspectives, *Diabetes Care*, 35, pp. 1798–1808.

[27]   FDA. (2018). Available online: https://www.fda.gov/food/food-additives-petitions/additional-information-about-high-intensity-sweeteners-permitted-use-food-united-states. Accessed on: 7 Dec. 2021.

[28]  Swithers, S. E., Martin, A. A. and Davidson, T. L. (2010). High intensity sweeteners and energy balance, *Physiology and Behavior*, 100, pp. 55–62.

[29]  Raben, A., Vasilaras, T. H., Moller, A. C. and Astrup, A. (2002). Sucrose compared with artificial sweeteners I: different effects on ad libitum food intake and body weight after 10 week of supplementation in overweight subjects, *American Journal of Clinical Nutrition*, 76, pp. 721–729.

[30]  Riera, C. E., Vogel, H., Simon, S. A. and le Coutre, J. (2007). Artificial sweeteners and salts producing a metallic taste sensation activate TRPV1 receptors, *American Journal of Physiology. Regulatory, Integrative and Comparative Physiology*, 293, pp. 626–634.

[31]  Sun, H., Cui, M. L., Ma, B. and Ezura, H. (2006). Functional expression of the taste modifying protein miraculin, in transgenic lettuce, *FEBS Journal*, 580, pp. 620–626.

[32]  Urbach, G. (1997). The flavor of milk and dairy products: II. Cheese: contribution of volatile compounds, *International Journal of Dairy Technology*, 50, pp. 79–89.

[33]  Abdulmumeen, H. A., Ahmed, N. R. and Agboola, R. S. (2012). Food: its preservatives, additives and applications, *International Journal of Chemical and Biochemical Sciences*, 1, pp. 36–47.

[34]  Gokoglu, N. (2019). Novel natural food preservatives and applications in seafood preservation: a review, *Journal of the Science of Food and Agriculture*, 99, pp. 2068–2077.

[35]  Parke, D. V. and Lewis, D. F. (1992). Safety aspects of food preservatives, *Food Additives and Contaminants*, 9, pp. 561–577.

[36]  John, E. M. (2003). Ionizing radiations sources, *Biol. Effects Emission Exposures*, 83, p. 1766.

[37]  Dalton, L. (2002). Food preservatives, *Critical Reviews in Food Science and Nutrition*, 50, p. 40.

[38]  Gassara, F., Kouassi, A. P., Brar, S. K. and Belkacemi, K. (2016). Green alternatives to nitrates and nitrites in meat-based products — a review, *Critical Reviews in Food Science and Nutrition*, 56, pp. 33–48.

[39]  Poulsen, E. (1980). Use of nitrates and nitrites as food additives in Nordic countries, *Oncology*, 37, pp. 299–301.

[40]  Mischek, D. and Krapfenbauer-Cermak, C. (2012). Exposure assessment of food preservatives (sulphites, benzoic and sorbic acid) in Austria, *Food Additives and Contaminants: Part A: Chemistry, Analysis, Control, Exposure and Risk Assessment*, 29, pp. 371–382.

[41]  Efenberger-Szmechtyk, M., Nowak, A. and Czyzowska, A. (2021). Plant extracts rich in polyphenols: antibacterial agents and natural preservatives for meat and meat products, *Critical Reviews in Food Science and Nutrition*, 61, pp. 149–178.

[42] Franco, R., Navarro, G. and Martínez-Pinilla, E. (2019). Antioxidants versus food antioxidant additives and food preservatives, *Antioxidants (Basel)*, 8, p. 542. doi: 510.3390/antiox8110542.

[43] Ng, H. H., Jelinic, M., Parry, L. J. and Leo, C. H. (2015). Increased superoxide production and altered nitric oxide-mediated relaxation in the aorta of young but not old male relaxin-deficient mice, *American Journal of Physiology-Heart and Circulatory Physiology*, 309, pp. H285–H296.

[44] Marshall, S. A., Qin, C. X., Jelinic, M., O'Sullivan, K., Deo, M., Walsh, J., Li, M., Parry, L. J., Ritchie, R. H. and Leo, C. H. (2020). The novel small-molecule annexin-a1 mimetic, compound 17b, elicits vasoprotective actions in streptozotocin-induced diabetic mice, *International Journal of Molecular Sciences*, 21, p. E1384. doi: 1310.3390/ijms21041384.

[45] Leo, C. H., Ng, H. H., Marshall, S. A., Jelinic, M., Rupasinghe, T., Qin, C., Roessner, U., Ritchie, R. H., Tare, M. and Parry, L. J. (2020). Relaxing reduces endothelium-derived vasoconstriction in hypertension: revealing new therapeutic insights, *British Journal of Pharmacology*, 177, pp. 217–233.

[46] Qin, C. X., Anthonisz, J., Leo, C. H., Kahlberg, N., Velagic, A., Li, M., Jap, E., Woodman, O. L., Parry, L. J., Horowitz, J. D., Kemp-Harper, B. K. and Ritchie, R. H. (2020). NO resistance, induced in the myocardium by diabetes is circumvented by the NO redox sibling, nitroxyl, *Antioxidants and Redox Signaling*, 32, pp. 60–77.

[47] Rana, I., Badoer, E., Alahmadi, E., Leo, C. H., Woodman, O. L. and Stebbing, M. J. (2014). Microglia are selectively activated in endocrine and cardiovascular control centres in atreptozotocin-induced diabetic rats, *Journal of Neuroendocrinology*, 26, pp. 413–425.

[48] Leo, C. H., Hart, J. L. and Woodman, O. L. (2011). Impairment of both nitric oxide-mediated and EDHF-type relaxation in small mesenteric arteries from rats with streptozotocin-induced diabetes, *British Journal of Pharmacology*, 162, pp. 365–377.

[49] Marshall, S. A., Leo, C. H., Girling, J. E., Tare, M., Beard, S., Hannan, N. J. and Parry, L. J. (2017). Relaxin treatment reduces angiotensin II-induced vasoconstriction in pregnancy and protects against endothelial dysfunction, *Biology of Reproduction*, 96, pp. 895–906.

[50] Ng, H. H., Leo, C. H., Prakoso, D., Qin, C., Ritchie, R. H. and Parry, L. J. (2017). Serelaxin treatment reverses vascular dysfunction and left ventricular hypertrophy in a mouse model of Type 1 diabetes, *Scientific Reports*, 7, p. 39604. doi: 39610.31038/srep39604.

[51] Leo, C. H., Fernando, D. T., Tran, L., Ng, H. H., Marshall, S. A. and Parry, L. J. (2017). Serelaxin treatment reduces oxidative stress and increases aldehyde dehydrogenase-2 to attenuate nitrate tolerance, *Frontiers in Pharmacology*, 8, p. 141. doi: 110.3389/fphar.2017.00141.

[52] Ng, H. H., Leo, C. H., Parry, L. J. and Ritchie, R. H. (2018). Relaxing as a therapeutic target for the cardiovascular complications of diabetes, *Frontiers in Pharmacology*, 9, p. 501. doi: 510.3389/fphar.2018.00501.

[53] Ong, E. S., Oh, C. L. Y., Tan, J. C. W., Foo, S. Y. and Leo, C. H. (2021). Pressurized hot water extraction of okra seeds reveals antioxidant, anti-diabetic and vasoprotective activities, *Plants*, 10, p. 1645. https://doi.org/1610.3390/plants10081645.

[54] Langston-Cox, A., Leo, C. H., Tare, M., Wallace, E. M. and Marshall, S. A. (2020). Sulforaphane improves vascular reactivity in mouse and human arteries after "preeclamptic-like" injury, *Placenta*, 101, pp. 242–250.

[55] Leo, C. H., Hart, J. L. and Woodman, O. L. (2011). 3',4'-dihydroxyfla-vonol reduces superoxide and improves nitric oxide function in diabetic rat mesenteric arteries, *PLOS One*, 6, p. e20813. doi: 20810.21371/journal.pone.0020813.

[56] Leo, C. H. and Woodman, O. L. (2015). Flavonols in the prevention of diabetes-induced vascular dysfunction, *Journal of Cardiovascular Pharmacology*, 65, pp. 532–544.

[57] Ng, H. H., Leo, C. H., O'Sullivan, K., Alexander, S. A., Davies, M. J., Schiesser, C. H. and Parry, L. J. (2017). 1,4-Anhydro-4-seleno-d-talitol (SeTal) protects endothelial function in the mouse aorta by scavenging superoxide radicals under conditions of acute oxidative stress, *Biochemical Pharmacology*, 128, pp. 34–45.

[58] Warner, C. R., Brumley, W. C., Daniels, D. H., Joe, F. L. J. and Fazio, T. (1986). Reactions of antioxidants in foods, *Food and Chemical Toxicology*, 24, pp. 1015–1019.

[59] Pawlowska, E., Szczepanska, J. and Blasiak, J. (2019). Pro- and antioxidant effects of vitamin C in cancer in correspondence to its dietary and pharmacological concentrations, *OxiMed and Cellular Longevity*, 2019, p. 7286737.

[60] Leo, C. H., Hart, J. L. and Woodman, O. L. (2011). 3',4'-dihydroxyfla-vonol restores endothelium dependent relaxation in small mesenteric artery from rats with type 1 and type 2 diabetes, *European Journal of Pharmacology*, 659, p. 193–198.

[61] Yao, L. H., Jiang, Y. M., Shi, J., Tomasbarberan, F. A., Datta, N., Singanusong, R. and Chen, S. S. (2004). Flavonoids in food and their health benefits, *Plant Foods for Human Nutrition*, 59, pp. 113–122.

[62] Goulds, G. W. (1995). Biodeterioration of foods and an overview of preservation in the food and dairy industries, *International Biodeterioration and Biodegradation.*, 36, pp. 267–277.

[63] Popp, J., Peto, K. and Nagy, J. (2013). Pesticide productivity and food security: a review, *Agronomy for Sustainable Development*, 33, pp. 243–255.

[64] Oerke, E. C. (2006). Crop losses to pests, *Journal of Agricultural Science*, 144, pp. 31–43.

[65] Fisher, M. C., Henk, D. A., Briggs, C. J., Brownstein, J. S., Madoff, L. C., McCraw, S. L. and Gurr, S. J. (2011). Emerging fungal threats to animal, plant and ecosystem health, *Nature*, 484, pp. 186–194.

[66] Damalas, C. A. and Eleftherohorinos, I. G. (2011). Pesticide exposure, safety issues, and risk assessment indicators, *International Journal of Environmental Research and Public Health*, 8, pp. 1402–1419.

[67] Delcour, I., Spanoghe, P. and Uyttendaele, M. (2015). Literature review: impact of climate change on pesticide use, *Food Research International*, 68, pp. 7–15.

[68] Cooper, J. and Dobson, H. (2007). The benefits of pesticides to mankind and the environment, *Crop Protection*, 26, pp. 1337–1348.

[69] Bolognesi, C. (2003). Genotoxicity of pesticides: a review of human bio-monitoring studies, *Mutation Research*, 543, pp. 251–272.

[70] Andreu, V. and Pico, Y. (2004). Determination of pesticides and their degradation products in soil: critical review and comparison of methods, *Trends in Analytical Chemistry*, 23, pp. 772–789.

[71] Li, Z. H., Velisek, J., Zlabek, V., Grabic, R., Machova, J., Kolarova, J., Li, P. and Randak, T. (2011). Chronic toxicity of verapamil on juvenile rainbow trout (Oncorhynchus mykiss): effects on morphological indices, hematological parameters and antioxidant responses, *Journal of Hazardous Materials*, 185, pp. 870–880.

[72] Cooper, R. L., Stoker, T. E., Tyrey, L., Goldman, J. M. and McElroy, W. K. (2000). Atrazine disrupts hypothalamic control of pituitary-ovarian function, *Toxicological Sciences*, 53, pp. 297–307.

[73] Abarikwu, S. O., Adesiyan, A. C., Oyeloja, T. O., Oyeyemi, M. O. and Farombi, E. O. (2009). Changes in sperm characteristics and induction of oxidative stress in the testis and epididymis of experimental rats by a herbicide, atrazine, *Archives of Environmental Contamination and Toxicology*, 58, pp. 874–882.

[74] Franco, R., Lia, S., Rodriguez-Rocha, H., Burns, M. and Panayiotidis, M. I. (2010). Molecular mechanisms of pesticide induced neurotoxicity: relevance to Parkinson's disease, *Chemico-Biological Interactions*, 88, pp. 289–300.

[75] Zhang, M., Wilhoit, L. and Geiger, C. (2005). Assessing dormant season organophosphate use in California almonds, *Agriculture, Ecosystems and Environment*, 105, pp. 41–58.

[76] Athiel, P., Alfizar Mercadier, C., Vega, D., Bastide, J., Davet, P. and Brunel, B. (1995). Degradation of iprodione by a soil Arthrobacter-like strain, *Applied and Environmental Microbiology*, 61, pp. 3216–3220.

[77] Costa, L. G. (1997). Basic toxicology of pesticides, *Occupational Medicine: State of the Art Reviews*, 12, pp. 251–268.

[78] Hrelia, R. (1996). The genetic and non-genetic toxicity of the fungicide, *Vinclozolin Mutagenesis,* 11, pp. 445–453.

[79] Richardson, J. R., Fitsanakis, V., Westerink, R. H. S. and Kanthasamy, A. G. (2019). Neurotoxicity of pesticides, *Acta Neuropathologica,* 138, pp. 343–362.

[80] Draper, A., Cullinan, P., Jones, M. and Newman Taylor, A. (2003). Occupational asthma from fungicides fluazinam and chlorothalonil, *Occupational and Environmental Medicine*, 60, pp. 76–77.

[81] Mukherjee, I., Gopal, M. and Chatterjee, S. C. (2003). Persistence and effectiveness of iprodione against Alternaria blight in mustard, *Bulletin of Environmental Contamination and Toxicology*, 70, pp. 586–591.

[82] Tleuova, A. B., Wielogorska, E., Talluri, V. S. S. L. P., Štěpánek, F., Elliott, C. T. and Grigoriev, D. O. (2020). Recent advances and remaining barriers to producing novel formulations of fungicides for safe and sustainable agriculture, *Journal of Controlled Release*, 326, pp. 468–481.

[83] Choudri, B. S., Charabi, Y., Al-Nasiri, N. and Al-Awadhi, T. (2020). Pesticides and herbicides, *Water Environment Research*, 92, pp. 1425–1432.

[84] Incledon, B. J. and Hall, J. C. (1997). Acetyl-coenzyme a carboxylase: quarternary structure and inhibition by graminicidal herbicides, *Pesticide Biochemistry and Physiology*, 57, pp. 74–83.

[85] Tchounwou, P. B., Yedjou, C. G., Patlolla, A. K. and Sutton, D. J. (2012). Heavy metal toxicity and the environment, *Experientia Supplementum*, 101, pp. 133–164.

[86] Garba, Z. N., Ubam, S., Babando, A. A. and Galadima, A. (2015). Quantitative assessment of heavy metals from selected tea brands marketed in Zaria, Nigeria, *Journal of Physical Science*, 26, pp. 43–51.

[87] Al Othman, Z. A. (2010). Lead contamination in selected foods from Riyadh city market and estimation of the daily intake, *Molecules*, 15, pp. 7482–7497.

[88] Woyessa, G. W., Kassa, S. B., Demissie, E. G. and Srivastava, B. B. L. (2015). Determination of the level of some trace and heavy metals in some soft drinks of Ethiopia, *International Journal of Current Pharmaceutical Research*, 2, pp. 84–88.

[89] Bottone, E. J. (2010). Bacillus cereus, a volatile human pathogen, *Clinical Microbiology Reviews*, 23, pp. 382–398.

[90] Peck, M. W. and van Vliet, A. H. (2016). Impact of Clostridium botulinum genomic diversity on food safety, *Current Opinion in Food Science*, 10, pp. 52–59.

[91] Dahlsten, E., Lindström, M. and Korkeala, H. (2015). Mechanisms of food processing and storage-related stress tolerance in *Clostridium botulinum*, *Microbiological Research*, 166, pp. 344–352.

[92] Bintsis, T. (2017). Foodborne pathogens, *AIMS Microbiology*, 3, pp. 529–563.

[93] Argudín, M. Á., Mendoza, M. C. and Rodicio, M. R. (2010). Food poisoning and Staphylococcus aureus enterotoxins, *Toxins (Basel)*, 2, pp. 1751–1773.

[94] Hennekinne, J. A., De Buyser, M. L. and Dragacci, S. (2012). Staphylococcus aureus and its food poisoning toxins: characterization and outbreak investigation, *FEMS Microbiology Reviews*, 36, pp. 815–836.

[95] Threlfall, E. J. (2002). Antimicrobial drug resistance in Salmonella: problems and perspectives in food- and water-borne infections, *FEMS Microbiology Reviews*, 26, pp. 141–148.

[96] Croxen, M. A., Law, R. J., Scholz, R., Keeney, K. M., Wlodarska, M. and Finlay, B. B. (2013). Recent advances in understanding enteric pathogenic *Escherichia coli*, *Clinical Microbiology Reviews*, 26, pp. 822–880.

[97] Ram, S., Bajpai, P., Singh, R. L. and Shanker, R. (2009). Surface water of a perennial river exhibits multi-antimicrobial resistant shiga toxin and enterotoxin producing *Escherichia coli*, *Ecotoxicology and Environmental Safety*, 72, pp. 490–495.

[98] Baker-Austin, C., Oliver, J. D., Alam, M., Ali, A., Waldor, M. K., Qadri, F. and Martinez-Urtaza, J. (2018). Vibrio spp. infections, *Nature Reviews Disease Primers*, 4, p. 8. doi: 10.1038/s41572-41018-40005-41578.

[99] Warren, B. R., Parish, M. E. and Schneider, K. R. (2006). Shigella as a foodborne pathogen and current methods for detection in food, *Critical Reviews in Food Science and Nutrition*, 46, pp. 551–567.

[100] Alshannaq, A. and Yu, J. H. (2017). Occurrence, toxicity, and analysis of major mycotoxins in food, *International Journal of Environmental Research and Public Health*, 14, p. 632.

[101] Bennett, J. W. and Klich, M. (2003). Mycotoxins, *Clinical Microbiology Reviews*, 16, pp. 497–516.

[102] Rushing, B. R. and Selim, M. I. (2019). Aflatoxin B1: a review on metabolism, toxicity, occurrence in food, occupational exposure, and detoxification methods, *Food and Chemical Toxicology*, 124, pp. 81–100.

[103] Kumar, P., Mahato, D. K., Kamle, M., Mohanta, T. K. and Kang, S. G. (2017). Aflatoxins: a global concern for food safety, human health and their management, *Frontiers in Microbiology*, 7, p. 2170. doi: 2110.3389/fmicb.2016.02170.

[104] Polak-Śliwińska, M. and Paszczyk, B. (2021). Trichothecenes in food and feed, relevance to human and animal health and methods of

detection: a systematic review, *Molecules*, 26, p. 454. doi: 410.3390/molecules26020454.

[105]   Pellett, P. E., Mitra, S. and Holland, T. C. (2014). Basics of virology, *Handbook of Clinical Neurology*, 123, pp. 45–66.

[106]   Bosch, A., Gkogka, E., Le Guyader, F. S., Loisy-Hamon, F., Lee, A., van Lieshout, L., Marthi, B., Myrmel, M., Sansom, A., Schultz, A. C., Winkler, A., Zuber, S. and Phister, T. (2018). Foodborne viruses: Detection, risk assessment, and control options in food processing, *International Journal of Food Microbiology*, 285, pp. 110–128.

[107]   Shukla, S., Cho, H., Kwon, O. J., Chung, S. H. and Kim, M. (2018). Prevalence and evaluation strategies for viral contamination in food products: risk to human health — a review, *Critical Reviews in Food Science and Nutrition*, 58, pp. 405–419.

[108]   Robertson, L. J. (2018). Parasites in food: from a neglected position to an emerging issue, *Advances in Food and Nutrition Research.*, 86, pp. 71–113.

[109]   Li, J., Wang, Z., Karim, M. R. and Zhang, L. (2020). Detection of human intestinal protozoan parasites in vegetables and fruits: a review, *Parasites and Vectors* 13, p. 380.

[110]   Lind, T., Sundqvist, A., Hu, L., Pejler, G., Andersson, G., Jacobson, A. and Melhus, H. (2013). Vitamin A is a negative regulator of osteoblast mineralization, *PLOS One,* 8, p. e82388.

[111]   Grillo, A., Salvi, L., Coruzzi, P., Salvi, P. and Parati, G. (2019). Sodium intake and hypertension, *Nutrients*, 11, p. 1970.

[112]   van Dam, R. M. and Seidell, J. C. (2007). Carbohydrate intake and obesity, *European Journal of Clinical Nutrition*, 61 Suppl, pp. S75–S99.

[113]   Mortimore, G. E., Pösö, A. R. and Lardeux, B. R. (1989). Mechanism and regulation of protein degradation in liver, *Diabetes/Metabolism Research and Reviews*, 5, pp. 49–70.

[114]   Code, F. F. (2019). Available online: https://www.fda.gov/food/retail-food-protection/fda-food-code. Accessed on: 9 Dec. 2021.

[115]   SFA. (2019). Available online: https://www.sfa.gov.sg/food-retail/food-safety-management-system/food-safety-management-system. Accessed on: 9 Dec. 2021.

[116]   FDA. (2019). Available online: https://www.fda.gov/food/hazard-analysis-critical-control-point-haccp/haccp-principles-application-guidelines. Accessed on: 9 Dec. 2021.

[117]   ISO22000. (2021). Available online: https://www.iso.org/iso-22000-food-safety-management.html. Accessed on: 9 Dec. 2021.

[118]   Liu, L., Meng, Y., Dai, X., Chen, K. and Zhu, Y. (2019). 3D printing complex egg white protein objects: properties and optimization, *Food and Bioprocess Technology*, 12, pp. 267–279.

[119] Singh, P. and Raghav, A. (2018). 3D food printing: a revolution in food technology, *Acta Scientific Nutritional Health*, 2, pp. 1–2.

[120] Liu, Z., Zhang, M., Bhandari, B. and Wang, Y. (2017). 3D printing: printing precision and application in food sector, *Trends in Food Science and Technology*, 69, pp. 83–94.

[121] Dankar, I., Pujola, M., Omar, F. E., Sepulcre, F. and Haddarah, A. (2018). Impact of mechanical and microstructural properties of potato puree-food additive complexes on extrusion-based 3D printing, *Food and Bioprocess Technology*, 11, pp. 2021–2031.

[122] Jin, Y. A., Li, H., He, Y. and Fu, J. Z. (2015). Quantitative analysis of surface profile in fused deposition modelling, *Additive Manufacturing*, 8, pp. 142–148.

[123] Turunen, S. M., Melchels, F. P. W. and Kellomaki, M. (2008). A review of rapid prototyping techniques for tissue engineering purposes, *Annals of Medicine*, 40, pp. 268–280.

[124] Lee, Y. (2021). A 3D food printing process for the new normal era: a review, *Processes*, 9, p. 1495. https://doi.org/1410.3390/pr9091495.

[125] Hertafeld, E., Zhang, C., Jin, Z., Jakub, A., Russell, K., Lakehal, Y., Andreyeva, K., Nagaraj Bangalore, S., Mezquita, J., Blutinger, J. and Lipson, H. (2019). Multi-material three-dimensional food printing with simultaneous infrared cooking, *3D Printing and Additive Manufacturing*, 6, pp. 13–19.

[126] Blutinger, J. D., Tsai, A., Storvick, E., Seymour, G., Liu, E., Samarelli, N., Karthik, S., Meijers, Y. and Lipson, H. (2021). Precision cooking for printed foods via multiwavelength lasers, *npj Science of Food*, 5, p. 24. https://doi.org/10.1038/s41538-41021-00107-41531.

[127] FDA. (2021). Available online: https://www.fda.gov/radiation-emitting-products/home-business-and-entertainment-products/laser-products-and-instruments. Accessed on: 9 Dec. 2021.

[128] Dogan, A., Siyakus, G. and Severcan, F. (2007). FTIR spectroscopic characterization of irradiated hazelnut (*Corylus avellana* L), *Food Chemistry*, 100, pp. 1106–1114.

[129] Pan, M., Yang, J., Liu, K., Xie, X., Hong, L., Wang, S. and Wang, S. (2021). Irradiation technology: an effective and promising strategy for eliminating food allergens, *Food Research International*, 148, p. 110578.

[130] O'Bryan, C. A., Crandall, P. G., Ricke, S. C. and Olson, D. G. (2008). Impact of irradiation on the safety and quality of poultry and meat products: a review, *Critical Reviews in Food Science and Nutrition*, 48, pp. 442–457.

[131] Stefanova, R., Vasilev, N. V. and Spassov, S. L. (2010). Irradiation of food, current legislation framework, and detection of irradiated foods, *Food Analytical Methods*, 3, pp. 225–252.

[132]  da Trindade, R. A., Mancini, J. and Villavicencio, A. (2009). Effects of natural antioxidants on the lipid profile of electron beam-irradiated beef burgers, *European Journal of Lipid Science and Technology*, 111, pp. 1161–1168.

[133]  Diehl, J. F. (1991). Nutritional effects of combining irradiation with other treatments, *Food Control*, 2, pp. 21–25.

[134]  Zhu, M. J., Mendonca, A., Ismail, H. A., Du, M., Lee, E. J. and Ahn, D. U. (2005). Impact of antimicrobial ingredients and irradiation on the survival of Listeria monocytogenes and the quality of ready-to-eat turkey ham, *Poultry Science*, 84, pp. 613–620.

[135]  Anese, M., Manzocco, L., Calligaris, S. and Nicoli, M. C. (2013). Industrially applicable strategies for mitigating acrylamide, furan, and 5-hydroxymethylfurfural in food, *Journal of Agricultural and Food Chemistry*, 60, pp. 10209–10214.

[136]  Fan, X. T. (2005). Formation of furan from carbohydrates and ascorbic acid following exposure to ionizing radiation and thermal processing, *Journal of Agricultural and Food Chemistry*, 53, pp. 7826–7831.

[137]  Marchioni, E., Raul, F., Burnouf, D., Miesch, M., Delince'e, H. and Hartwig, A. (2004). Toxicological study on 2-alkycyclobutanones — results of a collaborative study, *Radiation Physics and Chemistry*, 71, pp. 145–148.

[138]  Cross, A. J. and Sinha, R. (2004). Meat-related mutagens/carcinogens in the etiology of colorectal cancer, *Environmental and Molecular Mutagenesis*, 44, pp. 44–55.

[139]  Hamidi, E. N., Hajeb, P., Selamat, J. and Abdull Razis, A. F. (2016). Polycyclic aromatic hydrocarbons (PAHs) and their bioaccessibility in meat: a tool for assessing human cancer risk, *Asian Pacific Journal of Cancer Prevention*, 17, pp. 15–23.

[140]  Domingo, J. L. and Nadal, M. (2016). Carcinogenicity of consumption of red and processed meat: what about environmental contaminants? *Environmental Research*, 145, pp. 109–115.

[141]  Mordukhovich, I., Rossner Jr., P., Terry, M. B., Santella, R., Zhang, Y. J., Hibshoosh, H., Memeo, L., Mansukhani, M., Long, C. M., Garbowski, G., Agrawal, M., Gaudet, M. M., Steck, S. E., Sagiv, S. K., Eng, S. M., Teitelbaum, S. L., Neugut, A. I., Conway-Dorsey, K. and Gammon, M. D. (2010). Associations between polycyclic aromatic hydrocarbon-related exposures and p53 mutations in breast tumors, *Environmental Health Perspectives*, 118, pp. 511–518.

[142]  Daniel, C. R., Cross, A. J., Graubard, B. I., Park, Y., Ward, M. H., Rothman, N., Hollenbeck, A. R., Chow, W. H. and Sinha, R. (2012). Large

prospective investigation of meat intake, related mutagens, and risk of renal cell carcinoma, *The American Journal of Clinical Nutrition*, 95, pp. 155–162.

[143] Diggs, D. L., Huderson, A. C., Harris, K. L., Myers, J. N., Banks, L. D., Rekhadevi, P. V., Niaz, M. S. and Ramesh, A. (2011). Polycyclic aromatic hydrocarbons and digestive tract cancers: a perspective, *Journal of Environmental Science and Health, Part C*, 29, pp. 324–357.

[144] Tareke, E., Rydberg, P., Karlsson, P., Eriksson, S. and Toernqvist, M. (2000). Acrylamide: a cooking carcinogen? *Chemical Research in Toxicology*, 13, pp. 517–522.

[145] Besaratinia, A. and Pfeifer, G. P. (2007). A review of mechanisms of acrylamide carcinogenicity, *Carcinogenesis*, 28, pp. 519–528.

[146] Erkekoglu, P. and Baydar, T. (2014). Acrylamide neurotoxicity, *Nutritional Neuroscience*, 17, pp. 49–57.

[147] Soares, N. F. (2021). 3D food printing and food safety. Available online: https://nunofsoares.com/3dprinting_foodsafety/. Accessed on: 9 Dec. 2021.

[148] Foamlabs. (2021). The essential guide to food safe 3D printing: regulations, technologies, materials, and more. Available online: https://formlabs.com/blog/guide-to-food-safe-3d-printing/. Accessed on: 9 Dec. 2021.

[149] FSIS. (2021). Available online: https://www.fsis.usda.gov/food-safety. Accessed on: 9 Dec. 2021.

[150] IDDSI. (2019). The IDDSI framework. Available online: https://iddsi.org/Framework. Accessed on: 9 Dec. 2021.

[151] Pant, A., Lee, A. Y., Karyappa, R., Lee, C. P., An, J., Hashimoto, M., Tan, U. X., Wong, G., Chua, C. K. and Zhang, Y. (2021). 3D Food printing of fresh vegetables using food hydrocolloids for dysphagic patients, *Food Hydrocolloids*, 114, p. 106546.

# Problems

1. What are the different types of food hazards?
2. When you are consuming a chicken pie, you find a piece of bone inside the chicken pie. Identify this type of food hazard and explain why?
3. Explain and describe the differences between chemical hazards and biological hazards.
4. What are the different types of chemical hazards and how can they cause harm to human health?

5. What are the different types of biological hazards, and which type of food is commonly associated with those biological hazards?
6. Why is it important to develop a HACCP plan for food safety management?
7. When developing a HACCP plan, once critical control points are determined, what is the next step? Explain your answer.
8. A moist food ink is printed under warm conditions and takes a much longer duration to print. What are the potential food hazards that can occur? Explain your answer.
9. During the 3D food printing of real meat, a laser is also used to cook the food at the same time. What are the potential food hazards that can occur? Explain your answer.
10. Describe the food safety considerations associated with 3D food printing.
11. What are the different tests required to ensure that 3D-printed food is suitable for dysphagia patients?

# Chapter 8

# Business Development for 3D Food Printing

## 8.1 Introduction

Two main areas have been explored in this chapter on business development for 3D food printing (3DFP): chocolates and alternative proteins (APs). The chapter describes how chocolate evolved from a bitter beverage to the sweetened, delectable bar that we know today, and the form it may manifest into with the help of 3DFP. A brief account is provided from a historical perspective, including the uses, health benefits, and potential adverse effects of chocolate consumption. How chocolate is manufactured traditionally, which regions are the biggest producers of cocoa beans, the source material, and the top manufacturers and companies involved in the business are also covered. Comparisons are drawn from other industries where 3D printing has disrupted well-established industry practices and gained traction. Also, explored in this section are the industry players already involved in 3D printing of chocolates and their future plans in this space. Most of the research is still experimental, but a definite advancement has been made in the field of chocolate printers. More and more companies are launching useful 3D printers at reasonable cost for chocolate 3D printing enthusiasts, hobbyists, chefs, and consumers. Most of the research is also directed towards the development and market entry of these printers.

For centuries, APs have been consumed as part of the conventional diets in many countries of the world; however, this term has gained popularity in the current decade as a source of non-traditional dietary proteins as a way of moving away from conventional animal proteins in view of climate change, increased greenhouse gas (GHG) emissions, animal welfare, and an increase in consumers' flexitarian eating habits due to increased health consciousness. This chapter covers AP sources featuring plants, insects, fungi, cyanobacteria, and cell-based meat alternatives. Most of the commercialized products belong to plant-based food APs, which have been popularized recently by companies like Beyond Meat and Impossible Foods as fast-food alternatives to ground meat, typically used in burgers, patties, hot dogs, etc. Cultured meat has also come into fruition based on advanced tissue culture techniques. Although APs are already commercialized and sold in the market, 3D-printed food proteins are still in the exploratory phase. Research paves a way to look at the protein properties essential for 3D printing. With the advancement in APs, there has been pushback from dairy, meat, and other traditional food industries that has fuelled counternarratives on the authenticity and safety of these novel food alternatives. Beyond the exploratory phase that 3DFP has been in over the last decade, advancements must be made to introduce this technology to the market in order for it to be truly viable and disruptive to the existing food industry. Irrespective of the types of these industries, there are currently hardly any viable business models for 3D-printed chocolates and AP products. Here, we have attempted to look into these issues and proposed a few ways in which 3D-printed products can find their niche markets.

## 8.2 Chocolate Additive Manufacturing

### 8.2.1 *Traditional Chocolate Manufacturing*

#### 8.2.1.1 *History and uses*

Chocolate, the beloved treat, is made from cocoa beans. Cocoa grows in the tropics, and its fruit, called cabosside, bears 20 to 40 cocoa beans. Mayans were the first people to cultivate the cacao plant to make xocolatl, a bitter invigorating drink with cinnamon and pepper. During the Aztec period, cacao beans were given a currency-like status. Upon the introduction of the drink into Europe, chilli and spices were added, and finally with the addition of sugar and vanilla, the drink took on its characteristic sweet flavour. From 1600 onwards, the drinking of hot sweet chocolate spread across Europe after the Spanish conquistador Hernan Cortes brought cocoa to the court of King Charles of Spain [1, 2].

Cocoa plants are grown in the tropics, with high humidity levels. The cacao tree starts to bear fruit after 5 years, but the maximum yield may occur only after about 10 years. Out of the three genetic varieties, the Criollo type has superior flavour and is used to make premium fine chocolates, accounting for 5%–10% of the global chocolate production. However, 80% of the total production comes from the Forastero type, as this breed is resistant to diseases, gives higher yield, and is relatively less costly [3]. Cacao bean taste depends on the plant variety as well as the soil type, environmental temperature, rain, and sunlight exposure. According to the International Cocoa Organization (ICCO), a global record production of 5.024 million tonnes has been forecast for the 2020/21 season [4].

Chocolates can be classified into a few different varieties based on specific ingredients and characteristics [1]:

Dark: It is made of cocoa bean solids (80% of total weight) and cocoa butter. It is recognizable for its bitter aftertaste and melt-in-the-mouth feeling. Most of the health benefits can be attributed to this variety of chocolate.

Gianduja: It is made by combining hazelnut, cocoa, and sugar, sometimes with the addition of milk, almonds, or walnuts. It is brown in colour.

Milky: It contains cocoa (not less than 20–25%), cocoa butter, sugar, lecithin, and milk powder. It tastes sweet with a slightly bitter taste of cocoa.

White: Sweet and pleasant in flavour, it contains no cocoa solids and is made of cocoa butter, milk powder, vanilla, and sugar. It is used for the preparation of edibles like cream, mousse, and desserts.

### 8.2.1.1.1 Process of chocolate making

The process of making a chocolate bar from cocoa bean (popularly referred to as Bean to Bar) is elaborate, involving many transformation steps that have been summarized in a simplified form in Figure 8.1. It starts with cacao seed harvesting and processing, followed by the refinement of beans into chocolates, which also is a multi-step process.

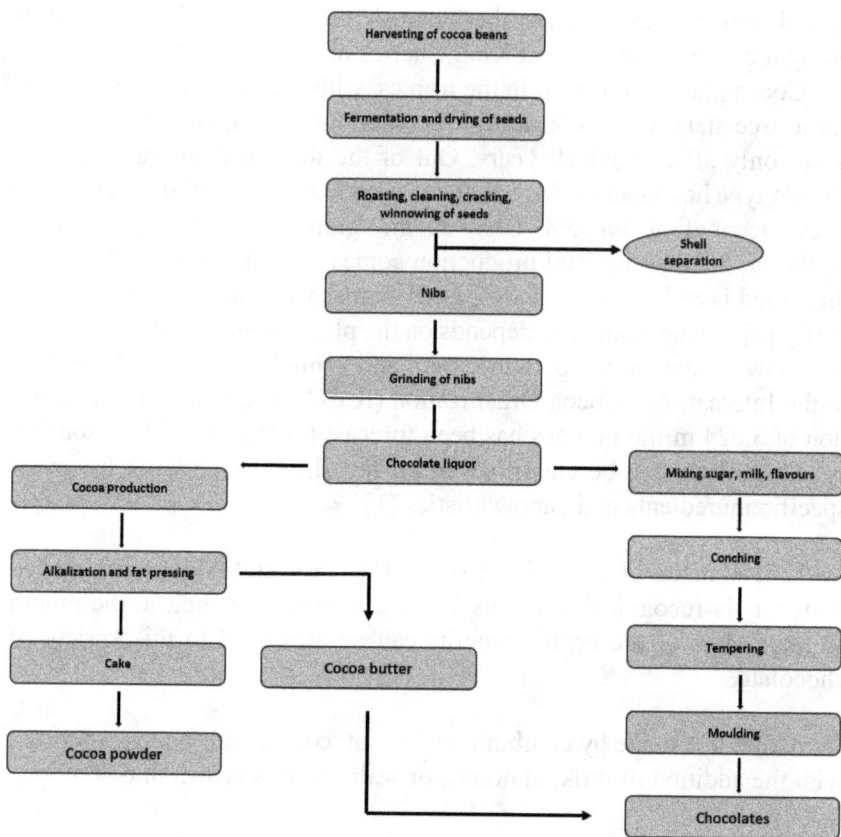

Figure 8.1.   Chocolate manufacturing process.

A description of the chocolate manufacturing process can be found in Section 8.2.1.3.

Chocolate, referred to as the "food of the gods" in many cultures, has permeated important aspects of our lives, be it in the form of a satisfying standalone dessert option in various forms and shapes or as a decoration on top of cakes, pastries, and other types of confectioneries. It also plays a big role in imagination because of the power to transform molten state into various eye-pleasing forms. However, apart from the joy that its consumption brings to the masses, it is important to look at its potential health benefits and risks. Nutritionally, fats are present in a significant proportion in cocoa (40–50% as cocoa butter with approximately 33% oleic acid, 25% palmitic acid, and 33% stearic acid). High concentrations of polyphenols, which have antioxidant properties, are present in the beans [5]. Cocoa is also a source of proteins, theobromine, caffeine, and minerals like potassium, phosphorus, iron, magnesium, zinc, and copper [6]. The health benefits of chocolate can be primarily attributed to the consumption of dark chocolate rather than white or milky chocolate, because cocoa is rich in polyphenols and low in sugar content. During processing, cocoa beans are depleted of polyphenols, which cause the bitter taste of chocolates. Addition of milk and sugar to offset the bitter taste of chocolate has led to negative impressions regarding the health benefits of chocolate. Popular media has often pointed out that highly processed chocolates are high-calorie and high-fat treats with the potential to cause dental problems, obesity, and heart diseases. In spite of these concerns, recent literature presents us with a number of studies linking cocoa consumption to potential health benefits. Cacao polyphenols have been shown to help with cancer prevention and progression delay [7], possibly due to increased vasodilation, which lowers blood pressure by promoting the production of nitric oxide [8]. Chocolate consumption of ≥3–4 servings/week has been associated with a lower risk of heart attack [9]. Increased cocoa uptake is associated with reduced stroke risk, probably due to increased cerebral blood flow [10, 11]. A few studies point to the therapeutic benefits of cocoa consumption against obesity. One study reported an increase in high-density lipoprotein (HDL) cholesterol levels and a decrease in abdominal circumference on consuming dark chocolate (100 g/day) for just a week in normal-weight obese women [12]. Cocoa and flavanols decrease pancreatic enzymes in a dose-dependent manner, resulting in beneficial effects on glucose homeostasis by reducing the rate of carbohydrate digestion and absorption [13]. Immune regulation is also modulated by cocoa; in some clinical trials, reduced cytokines, white blood cells, and other immune markers indicate anti-inflammatory response of dark chocolate consumption [14]. Methylxanthines of cocoa, including theobromine and caffeine, work as central nervous system stimulants, diuretics, and muscle relaxants [6]. Chocolate also exerts its beneficial effects in the form of an antidepressant and as an aphrodisiac. However, most of the studies link flavanols, especially epicatechin, with the associated health benefits, but these phenolic compounds are significantly depleted by about 10-fold during the processing stages. Hence, more studies are still needed to ascertain the potential health effects of different types of chocolates, as many studies have focused on cocoa consumption rather than the finished whole product.

## 8.2.1.2 *Business and revenue of the chocolate industry*

Being a major agricultural crop, *Theobroma cacao* (cacao) propels a large industry with both big and small stakeholders, ranging from farmers to distributors and manufacturers of various types of chocolates. Smallholder farmers contribute about 80–90% of the global cacao production, which is grown by about 6 million farmers worldwide, with about 40 million people deriving their livelihoods from it [15]. Countries like Holland, Switzerland, and France pioneered the development of methods for producing the first forms of chocolate. The British company Cadbury was among the first to manufacture chocolate boxes and bars for worldwide consumption. The global chocolate market is expected to grow at a compound annual growth rate (CAGR) of 4.6% from 2020 to 2027 and is pegged at USD 130.56 billion in 2019. In total, 99.4% of the share is captured by traditional chocolates, with milk chocolate dominating due to the first-mover strategy. Most of the distribution is channelled through supermarkets and hypermarkets, followed by convenience stores, and finally, online retails. Online retail shows significant expansion as more and more retailers are focusing on this channel due to the popularity and ease among today's generation in using the internet. As a medium offering additional services like discounts, customization, and paybacks, online retail is set to expand the customer base [16]. The COVID-19 pandemic led to an increase in the consumption of chocolates, as per a survey from the Cargill North America Cocoa and Chocolate research team [17]. Chocolate is consumed not only in various forms on its own but also in other confectionery products, such as coating in ice creams and biscuits, cake toppings, etc. According to the ICCO, a record production of 5.024 million tonnes of cocoa beans has been forecast for 2020/21, with Côte d'Ivoire being the largest producer. With 77% of global cocoa output, the African continent remains the largest cocoa-producing area, with the shares of Asia and Oceania being 17% and 6%, respectively. The global chocolate market has risen rapidly by about 13% from 2010 to 2015. The manufacture of specially branded chocolate is geographically dominant in North America and Europe, with 65% of apparent cocoa consumption in 2014–2015 [4]. Many of the big chocolate companies have been progressively incorporated as units into multinational food processing firms, one example being Mondelēz through its acquisition of Cadbury [18].

Table 8.1. World's top ten confectionery companies.

| Company | Key Brands | Net Sales in 2018 (USD million)* |
|---|---|---|
| Mars Wrigley Confectionery, a division of Mars Inc. (USA) | Snickers, Mars, M&Ms, Dove | 20,000 |
| Ferrero Group (Luxembourg/Italy) | Ferrero Roche, Nutella | 13,566 |
| Mondelēz International (USA) | Cadbury, Milka, Toblerone | 11,467 |
| Meiji Co. Ltd. (Japan) | Meiji Milk Chocolate | 10,075 |
| Hershey Co. (USA) | Hershey's, Reese's | 8,066 |
| Nestlé SA (Switzerland) | Kit Kat, Crunch | 7,636 |
| Pladis (UK) | United Biscuits, Godiva Chocolatier | 4,655 |
| Chocoladefabriken Lindt & Sprüngli AG (Switzerland) | Lindt | 4,331 |
| Ezaki Glico Co. Ltd. (Japan) | Pocky | 3,311 |
| Haribo GmbH & Co. K.G. (Germany) | Starmix, Goldbears | 3,300 |

*Source*: [19].
* Includes non-chocolate confectioneries.

Table 8.1 presents the top ten global confectionery companies manufacturing different forms of candies and chocolates as per net confectionery sales [19].

Emerging markets in Asia, Latin America, and the Middle East have contributed to the ten fastest-growing chocolate markets [20]. Outsourcing of industrial chocolate manufacturing to non-branded firms is also taking precedence, with leading companies like Mondelēz, Nestlé, and Hershey signing contracts with companies like Barry Callebaut for industrial chocolate supplies [18]. Technological advancements like cocoa butter substitution have enabled Cadbury to produce compound chocolate in China [21].

Retail distribution of heat-sensitive products like chocolates requires close and tight co-ordination from chilled points to the points of sale, which is solved by meaningful partnership of chocolate manufacturers with distributors. The success of Mars in China was due to its entering into various partnerships with local distribution channels to control this process [21]. In 2015, the McLane company, as the sole distributor of

Hershey, contributed to 26% of consolidated net sales across North America.

### 8.2.1.3 *Traditional model of chocolate manufacturing*

Traditional chocolate making is a lengthy and technical process requiring physical and chemical transformations complemented with various ingredients to achieve the desired quality and sensory attributes (Figure 8.1). The manufacturing of chocolate depends on company practices and the preferences of consumers and, as such, differs considerably. An important factor underlying chocolate manufacturing is its rheological property that helps achieve well-defined textures.

Industrial manufacturing of chocolates involves the following steps [22], as shown in Figure 8.2.

Mixing of ingredients can be continuous or in batch over time–temperature combinations. Nestlé and Cadbury employ continuous mixing using automated kneaders to produce tough texture mixtures. This mixture is further refined to a particle size of <30 $\mu$m by using rollers, which not only help to reduce the size but also break down and distribute particles coated with lipid. During the conching process, certain volatile acids are taken out, moisture is reduced, and fats and emulsifiers are added to develop flavour, viscosity, and final texture. Tempering is essential to produce a product with a well-formed shape, colour, and gloss that can be easily demoulded and is harder and heat-resistant. Tempering leads to pre-crystallization of triglycerides to appropriate crystals. It is followed by moulding and casting with the help of automated equipment. Moulds may be of different types, aimed at having a precise product. Moulds were first made of metal with stamped impressions, but these were quite heavy. They have now been replaced with plastic or silicone moulds, which are much lighter and efficient. The viscosity of the tempered chocolate is crucial to obtain uniform weight and size of the final product. Inappropriate temperatures of the moulds may lead to de-tempering, poor gloss, sticking, and markings on the finished product. Following this, moulds are passed through a cooling plant to reduce the temperature to 12–15 °C and take the shape of solidified bars. Small-scale chocolate manufactures carry out this step in refrigerators. Specialized equipment is also available to add nuts, raisins, and other edible

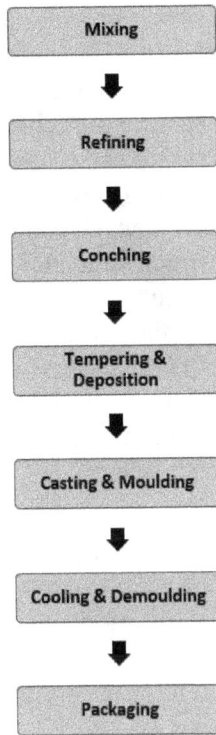

Figure 8.2. Industrial chocolate manufacturing steps.

ingredients. Demoulding is one of the last steps whereby bars are taken out of moulds using hammer pressure and twisting the moulds. Finally, the chocolate bars are sent to the wrapping and packaging facilities before being sent out to the market. Different chocolate production machines are required for separate mouldings [23]. Shell moulding is used to make specialized "Easter Eggs", but the shell thickness is difficult to control in this process. The market is dominated by standard products involving large-scale manufacturing of chocolates in set designs and shapes by leading companies. The traditional way of chocolate manufacturing does not fully satisfy the need for customization of designs and ingredients [24]. The current breed of consumers expects and demands personalized products according to their aesthetic, taste, and health preferences. Most of the personalized chocolate business is through handcrafted chocolates designed and produced by artisans on a small scale, which is difficult to sustain profitably in the market over a long term due to longer and costlier production requirements. To achieve this, chocolate manufacturers would require different moulds for different customers, seasons (e.g., during Easter or Christmas), or occasions (birthdays), necessitating the production of specialized moulds each time, slowing down production.

This in turn would mean significant additional costs and time, which may not be easily accepted by the big chocolate manufacturers. The need for moulds can be obviated by the use of additive manufacturing (AM) technology, where chocolate 3D printers can offer customization benefits to chocolate companies and also to designer chocolate producers and chefs, helping them to personalize and play around with fun designs targeting the more experimental consumers.

## 8.2.2 Impact of Additive Manufacturing on the Chocolate Industry

### 8.2.2.1 Disruptions of additive manufacturing in other industries and parallels

Conventional manufacturing processes, also known as subtractive manufacturing, have been around for a long time. With the advent of additive manufacturing (AM), conventional processes have been threatened and disrupted [25]. When rapid prototyping was introduced, its initial goals were to reduce the lead time and cost of developing prototype, but after almost 30 years, AM is also impacting the production end of manufacturing.

Disruptive innovation theory, introduced by the Harvard professor Clayton M. Christensen, explains the transformation of an existing market by the advent of a simpler, more convenient, less costly, and accessible technology. Targeted initially at a niche market, this new idea has the capability to change and challenge the existing industry [26]. Mainstream market segments do not often consider these disruptions valuable as per standard performance, but with more research and development (R&D), the ideas and technologies will mature, succeeding in satisfying the requirements of a traditional kind. These technologies usually replace the existing industry practices [27]. The real success of AM will depend on cost and productivity improvements with further technological gains in material properties and printing accuracy [25].

The popularity of AM is due to reduced costs involving transport, distribution, packaging, and production, as well as the idea of being able to manufacture anything, as long as a computer, a printer, and materials are available [28]. The spread of AM to consumers and hobbyists is being facilitated by the affordability of printers with desktop 3D printers. Printers that were earlier limited to large manufacturing and design firms are now available for as low as USD 1,000 [29]. Campbell explained AM as having two paths of growth; the high-end path of developing functional technology that matches industry standards and acceptance, and the low-end path of reducing the costs and complexity of AM for wider public acceptance. AM is capable of either improving or disrupting the existing industry processes as well as creating new markets altogether. Manufacturing parts and devices closer to the intended use will have an impact on the transport industry. For example, the Dutch carrier Maersk examined the making of a spare or broken part for an oil rig and tanker as a cost-saving preventive measure [28]. When productivity efficiency increases and costs go down, AM will become more pervasive. Transformative technologies like AM tend to have more assumptions surrounding them and certain facts to be asserted, which can be revisited on a later day [30].

The United States (US) and Europe are at the forefront of R&D in AM. Examples of industries utilizing AM to effectively produce low-cost customized parts at high efficiency are in consumer electronics, aviation, and automobile manufacturing. Consumer products, automotive, medical, and aerospace are the top four industry applications utilizing AM [31]. Two US companies, 3D Systems and Stratasys, have the majority market share [32]. The aerospace industry was among the first to bring 3D printing to its realm. Difficulties in the processing of materials, complex structures with difficult geometries, and thin-walled aircraft engine components made adoption of AM easier in comparison to traditional manufacturing. The use of AM resulted in low material consumption, and complex consolidated designs of high strength aided in the overall weight reduction, making it one of the most desirable attributes. Repair of damaged parts through laser metal deposition is also undertaken to this end [33]. Conventional manufacturing practices that generate much material waste are preventable through AM, thus saving costs. The medical industry too finds itself a significant beneficiary of advanced biocompatible and high-strength printing materials, specifically orthopaedic implant devices, dental devices, prosthetics moulds, anatomical models, custom saw and drill guides, and pre-surgical tools. Reduction in operation room times by using surgical guides for maxillofacial surgery and models for spinal surgery is one of the many reported benefits of 3D printing [34]. Surgical guides have now replaced the anatomical models as the most common medical application of AM. Reduced surgical times, improved medical outcomes, and decreased radiation exposure are very attractive uses, but the cost-effectiveness of 3D printing needs to be looked into with great detail. One Australian study identified the five main barriers plaguing major opportunity areas of AM in MedTech (material science, technology, business models, regulation, and quality) as manufacturing process and post-process approval, medical and professional endorsement, medical device reimbursement, material issues, and staff training [35]. Automotive companies have utilized AM to build parts that survive extreme temperatures and harsh conditions and yet are light-weight. The global automotive 3D printing market size is projected to grow to USD 2,730 million by 2023 from USD 762 million in 2016 [36]. BMW has utilized 3D printing technology to build hand tools for vehicle assembly and testing. Working with Stratasys, BMW manufactured a tool which was 72% lighter and hence easier to use, as well as convenient for reaching hard-to-reach spots. Ford, the earliest to adopt AM, has five 3D prototyping centres, with one of them producing 20,000 prototype products per annum. In 2014, Local Motors achieved the distinction to 3D-print an electric car from an ABS carbon filter blend called "Strati" [37]. The consumer product industry thrives on quick turnover times from the conceptualization of an idea to market delivery. Consumer electronics benefits from AM designs that are both functional and beautiful. 3D printing has also brought special focus to personalized product marketing. Personalization draws consumers' attention by keeping them engaged in the process, resulting in an unforgettable experience. The launch of mini-sized bottles by Coca-Cola Israel is an example of personalized marketing using AM. The brand established a unique way by introducing a competition to recreate users' physical appearances using an app, and selected winners were invited and presented with 3D-printed versions of their mini-me's by having their bodies scanned [38].

### 8.2.2.2 *Companies utilizing additive manufacturing: 3D printing of chocolates*

Chocolate belongs to natively extrudable food ink types and has been researched much more as compared to any other food for extrusion 3D printing. First printed by Schaal in 2007 [39], the year 2010 saw the design of the Choc ALM prototype [40], which was developed further into Choc Creator V1. Choc Edge's Choc Creator 3D printer, first launched in 2012, was the first commercially available chocolate printer [41]. Since then, they have released another version, Choc Creator V2 Plus. CocoJet, a chocolate 3D printer which was created through a collaboration between Hershey, the leading chocolate manufacturer, and 3D Systems, the leading company in 3D-printed applications. The printer was used to print milk, dark, and white chocolate in various patterns, as well as to personalize Hershey's products. CocoJet was introduced at the Chocolate World attraction at Hershey, PA, in 2014 [42]. Further developments need to be seen on the part of Hershey and 3D Systems as both of them are in partnership with the Culinary Institute of America (CIA). One aim of this partnership was to make the students and chefs explore the possibility of creating new ideas and finding applications of 3D printing in the culinary world. Barry Callebaut, with its global decoration brand has opened a design studio, called the Mona Lisa 3D studio. The studio aims to help chefs to innovate novel and bespoke chocolate designs via 3D printing using Belgian chocolate. This studio is offering personalized 3D-printed chocolate at scale. They have teamed up with a renowned pastry chef, Jordi Roca, to innovate a unique signature piece, "Flor de Cacao", whose inspiration came from the cacao flower [43]. Academic research into chocolate 3D printing has also made significant progress. An important study in understanding the various parameters and additives affecting chocolate 3D printing came out in 2017 [44]. Recent studies have focused on the incorporation of additives like vitamin C, cranberry powder, and methylcellulose while printing milk chocolate [45]. Dark chocolate printing also has been experimented by the addition of plant sterols or magnesium stearate. Additives tend to supplement the nutritional profile of the chocolates being printed and also to modify the viscoelastic properties needed for good printing. Most of the chocolate 3D printing work utilizes the hot-melt extrusion technique, whereby chocolates are melted and printed at temperatures ranging from 31 to 36 °C. This is a simple, adaptable technique requiring precise temperature control [44, 46]. There is one report on cold extrusion of chocolate-based inks by modifying the rheological properties at room temperature. The authors termed it as chocolate-based ink 3D printing by combining readily available chocolate pastes and syrups with cocoa powders at 10 to 25% concentrations [47]. An exciting recent study demonstrated 3D printing of low-sugar chocolates with similar sweetness to that of chocolates with 19% higher sugar content. It was achieved by layering the chocolates with different amounts of sugar without affecting the sweetness perception of study participants. This study can serve as a base to carry further experiments to create multi-layered 3D-printed chocolates with different sugar concentrations in order to reduce the overall sugar content without affecting the sweetness and temporal sensory profile of consumers [48]. This will also help to dispel the negative narrative in the mainstream media against chocolates as sugar-laden hedonist delights contributing to adverse health effects.

Apart from these initiatives, the market is flooded with chocolate 3D printers at different price ranges, giving cheer to individuals and chefs interested in customizing and personalizing new food designs. Mass production of chocolates by 3D printing is still in a nascent stage, as one of the limiting factors is the speed of printing. Since chocolate is influenced by temperature, molten chocolate does not set very rapidly at room temperature, so the print speed is kept slow to ensure that the layers do not deform under gravity before they are set. This limits the production capacity, as very few chocolates can be printed in a given time. As stated above, the temperature needs to be controlled carefully, as any drastic variations may lead to the chocolate losing its viscoelastic properties and becoming too liquid or hardening within the extruder. Scaling up is a problem too, so mass production is still not realistic. The use of high-quality, machine-tempered dark chocolates containing high amounts of cocoa solids is recommended. For instance, Choc Edge printers work well on dark chocolates with higher cocoa percentage, and the company recommends using couverture chocolates rather than compound chocolates for easier printing function [49].

Most of the research on chocolate 3D printing has been in the development of printers. Some printers in the market for chocolate printing are Wiiboox Sweetin food printer, FoodBot S2, Zmorph VX, Mmuse Touchscreen printer, byFlow chocolate printer, and Mycusini printer. Most of these printers work well with low-volume ingredients and can produce complex designs and customized offerings not scalable to industry requirements. The printers range in price from USD 1,500 to 6,000; they work on the direct-ink-writing principle, depositing melted chocolate layer by layer. Even though they print chocolate, the printers may also be used to print other foods. For instance, FoodBot S2 can print biscuits, toffees, and jams. Foodini from Natural Machines has been adopted by chefs to print not just chocolate designs but also soft foods. Wiiboox Sweetin is at the low end of the cost spectrum, but it provides reliable and accurate printing. The Zmorph Fab printer has high precision due to its design as a FDM 3D printer. Besides the efforts of 3D printing companies in advancing more sophisticated printers, traditional chocolate manufacturers are beginning to see the market potential of investing in AM for chocolate production, as demonstrated by the collaboration between Hershey and 3D Systems and the recent opening of the Mona Lisa design studio by Barry Callebaut. Customized chocolates can serve as a great marketing tool for small and big businesses to further their own branding as well as pleasing their customers or clients with personalized chocolates.

## 8.3 3D Printing of Alternative Proteins

### 8.3.1 *Need for Alternative Proteins*

Due to an increase in per capita protein consumption in Asia and North Africa, it is estimated that 40–75% more protein will be needed for the rising population [50]. Protein-rich sources based on plants, insects, fungi, algae, and lab-grown meat through tissue culture to replace traditional animal-origin proteins have come to be known as alternative proteins (APs). Though a lot of APs have been consumed traditionally (soy, pea, seitan, cricket, etc.) for centuries, the market is ready for innovation around newer protein products. Consumers are demanding transparency and sustainability in their food choices. Understanding the need for food sustainability, the protein market is looking for healthy, environmentally friendly, and animal cruelty-free options which can be satisfied with APs. Innovative food companies like Beyond Meat and Impossible Foods that create meat alternatives with plant-based proteins have had successful launches in recent times and have been lapped up by consumers. This is also coupled by strong social media campaigns targeting consumer education on the perils of meat consumption and plant-based eating benefits [51, 52]. An outcome of these strategies is the tie-up of various fast-food companies and chains to market vegetarian options of their popular meat products like burgers, sausages, hot dogs, etc. Even though the meat consumption trend continues to rise, especially in the developing world, APs consumption has seen a major growth. Plant-based food sales hit USD 7 billion in 2020, directly replacing animal products. From 2018 to 2020, plant-based food sales increased 2.5 times faster than total food sales [53].

With the global meat market pegged at USD 1.7 trillion, the APs market has grown to USD 2.2 billion. Food manufacturers are slowly rising to this exciting avenue of APs, and a few have focused their efforts on plant-based APs (such as soy and pea protein), insect proteins (cricket, mealworm), and biotechnological innovations (cell-cultured meat). The evolution of APs had several contributing factors like increased health consciousness, prices, and concerns for the environment and animal welfare. Vegan products contributed to the most searched food and beverage options. According to a McKinsey 2018 Daily survey, 73% of the millennials and Generation Z members bought dairy-free alternatives in a year [54]. In the developed markets, protein consumption will remain robust with both traditional and AP sources contributing to it. From 2007 to 2016, new protein products increased from 2% to 5%, with a significant rise in vegan, dairy-free, and ethical labels [55]. Apart from addressing the challenges of food security and climate concerns, the APs market will also help in creating new revenue streams and multiple opportunities for employment generation. The US, China, and Brazil are among the top consumers of beef [56], with extensive meat consumption leading to high GHG emissions, antibiotic resistance, and other adverse environmental issues. One study estimated that a reduction in heavy meat diets and shifting to healthy alternative options will help save USD 1–31 trillion. Half of the GHG emissions from animal agriculture are contributed by beef and lamb foods. GHG emissions involving animal agriculture in the US and their health impacts cost the public about USD 388 billion per year. Direct livestock emissions (methane emission from enteric fermentation and methane and nitrous oxide emissions from manure) contribute to about 4% of total GHG emissions from the US [56]. Land use for meat and dairy production is also one of the biggest problems associated with conventional protein production. Currently, 80% of all agricultural land in the world is used for ruminant meat production [57]. Though APs have their associated pitfalls, they are much less compared to animal agriculture. One claim is that APs can help mitigate this problem, as they are less land-extensive. For example, plant-based proteins compared to beef are 98% less land-intensive per 100 g of protein [58]. Cultivated meats' projected land use will be similar to chicken but less land-intensive than beef or pork. To produce sufficient amounts of products of animal origin, forests and grasslands have been cleared [59]. Land usage for livestock production leads to habitat loss and deforestation in tropical countries [60]. Shifting towards plant-based alternatives will lower the biodiversity loss due to animal agricultural practices. Water use is also lowered when using AP production as shown by plant proteins utilizing 77% less water per 100 g of protein than beef and 89% less water than pork. Another major consideration for switching to the APs diet is due to the effect of red meat on human health. Studies have indicated links between colorectal cancer and the consumption of red and processed meats [61]. Though there are no clear studies establishing a relationship between human health and replacement of animal meat with APs, it has been suggested that reducing red meat consumption with plant-based alternatives might provide more fibre and help with heart health [62]. Antibiotics use is also predominant in animal agriculture, contributing towards antibiotic resistance. Plant-based alternatives have less usage of antibiotics as compared to cultivated meats. Animal welfare also spurred the rise in AP consumption, as plant-based APs do not rely on animal sources and cell-based meat alternatives use cultured animal cells. There is also a possibility of food-borne disease outbreaks like swine flu and mad cow disease [63] with traditional animal products. Ethical practices are in question in rearing animals for feed, as they are sometimes force-fed at systematic times and kept in very small, enclosed spaces. Another push factor for seeking novel protein sources is the allergenicity of existing protein sources, with eggs, dairy, and soy among the top allergens identified by the Food and Drug Administration (FDA). Other than meeting the energy requirements, proteins have a wide variety of functions including flavour and stability enhancement; hence, producers are also looking for natural replacement of synthetic ingredients such as synthetic emulsifiers [64]. All these reasons have led to a market demand for sustainable, healthier, animal cruelty-free, and climate friendly alternatives for getting dietary proteins; hence the need for R&D in APs.

### 8.3.2 *Types of Alternative Proteins*

#### 8.3.2.1 *Plants*

According to Zion Market Research, the global plant-based meat analogue industry is expected to hit USD 21.23 billion by 2025, growing at a CAGR of around 8.6%. Some of the major contributors towards its growth are E. I. du Pont de Nemours and Company, Morningstar Farms LLC, Beyond Meat, Archer Daniels Midland Company, Amy's Kitchen Inc., Garden Protein International Inc., Parabel USA Inc., and Impossible

Foods Inc. [65]. Plant proteins can capture a sizable market portion of animal proteins (dairy, eggs, and meat) if they are produced at economical prices. Among the most well-known and well-established AP sources, plant proteins have been consumed for centuries in many cultures like China and India [66]. Several of the popular plant proteins in the market are soy, pea, chickpea, legumes, rapeseed, and lupin. Plant proteins can be in the form of unprocessed proteins like legumes or highly processed forms like soy or the wheat gluten product seitan. Tempeh, soy, and seitan have been used as protein alternatives in many Buddhist and vegetarian cultures. These can be regarded as the earliest versions of plant-based meat alternatives (PBMAs) [67]. Tofu making can be traced back to the Han dynasty of China followed by its spread to other countries like Japan, Thailand, and Vietnam, with variations coming along [68]. Tempeh, originating in Indonesia, is also prepared by soybeans, with the fermentation aided by Rhizopus forming a solid cake-like state. Seitan also traces its origin to China, prepared from wheat gluten by washing the wheat dough repeatedly to form elastic structures which can be reshaped to imitate mock chicken wings and ducks. Apart from these traditional PBMAs, other plant protein sources (oilseeds) like rapeseed concentrate, glandless cottonseed flour, and defatted peanut flour can also be used to produce meat alternatives. Jackfruit has been purported to produce palatable properties similar to those of meat [69]. Vegetarians accept plant-based protein products that do not taste or look like meat for taste merits and religious reasons, but non-vegetarian consumers are not satisfied entirely in terms of the taste, appearance, and flavour of traditional PBMAs. A few companies have tried in the recent years to produce products that not only taste and look like meat but also have a similar aroma and nutrition profile (such as Beyond Meat, Impossible Foods, Light Life, and many more). The ambition of these new plant product companies is to mimic the taste, texture, and appearance of popular animal products like beef burgers, patties, hotdogs, and other successful products, with special focus on imitating how the finished product compares with the final one. For example, both Beyond Meat and Light Life use beet juice or powder to imitate the bleed of the beef patties. Impossible Foods uses soy leghemoglobin for this purpose, contributing not just the colour but also the heme-influenced flavour. Burger patties, sausages, bacon, and hot dogs are some of the new products introduced by these companies in the market through restaurants, supermarkets, fast-food chains, and online retail. Legumes have been traditionally consumed in a lot of cultures, most prominently in India.

Legume proteins have been investigated (pea, lupine, lentil, chickpea, mung bean, etc.) recently for their potential as meat analogues [70]. Among the plant proteins, soy has been the most commonly used meat analogue. Western countries have been increasingly demanding non-dairy soy milk and soy-based meat analogues as alternatives to animal proteins which are traditionally consumed in Asian countries [71]. Animal proteins have a greater digestibility and higher protein digestibility-corrected amino acid score (PDCAAS) compared to plant proteins. Processed soy protein has an acceptable PDCAAS along with greater essential amino acid bioavailability among all the plant proteins. Proteins in wheat, rice, oats, and barley are classified as cereal proteins, with wheat protein seitan being the most commonly used plant protein. Such proteins have less protein percentage compared to soy, but their structural properties like viscoelasticity and fibrous texture make them good ingredient material for meat analogues. Functional properties of proteins play a major role in them becoming ingredients for various finished products. For example, soy and pea proteins have excellent hydration, fat-absorbing, and emulsifying properties; gluten imparts the cohesiveness and viscoelastic properties found in meat alternatives; oilseed proteins like canola oil provide emulsifying and foaming properties; and legume proteins offer stabilization, emulsification, and gelling properties [64]. Raw plant proteins in general have low PDCAAS compared to animal sources, which can be improved by processing techniques. Among the recent techniques of plant processing, extrusion, shear cell technology, and 3D printing are used to make meat analogues and other protein-rich products.

### 8.3.2.2 *Mycoproteins and algae*

Mycoprotein is the filamentous unprocessed protein produced by the fungal biomass through fermentation. It is known to have about 45% protein, high fibre, and very less carbohydrates, fat, sodium, and cholesterol. Having a chicken breast-like texture, it is made by continuously fermenting *Fusarium venenatum*, a natural microfungus. Since 2002, it has been designated as GRAS (Generally Regarded As Safe) by the FDA in the US. Found in the form of sausages and patties in the supermarkets of the Western world, its global presence is limited by inhibitive prices of production. A savoury pie first debuted as a mycoprotein retail product in the UK in the 1980s with the brand name Quorn. Now, Marlow Foods, a UK-based Quorn company, sells mycoprotein in 17 countries. Since its introduction, around 5 billion servings of Quorn have been consumed, with GBP 205 million sales in 2017. The fungus is first grown in a defined medium, which reduces RNA biomass, and then the addition of egg albumen as a binder and other colours and flavours leads to its characteristic taste and feel, which is like that of meat. Vegan options for Quorn are also being explored [72]. However, these production methods make the cost of mycoproteins similar to that of meat [73]. Mycoprotein has higher values of essential amino acids compared to plant-based proteins with a protein digestibility score of 1.0. The highly insoluble matrix owing to its fibre content of chitin and ß glucan may help in its cholesterol-reducing properties. Research indicates mycoprotein's controlling effect on glucose and insulin, promoting muscle growth and satiety [72, 74].

Asian countries have long had the tradition of using edible algae like seaweed, Spirulina, and Chlorella. Spirulina (blue-green algae) is a microscopic cyanobacterium rich in protein (70%) and contains all the essential amino acids, vitamins, and minerals. Labelled as "superfood", Spirulina supplements have been present in the form of powder, capsules, and tablets in the market and have the distinction of being used by NASA for its space missions [75]. Its dried biomass has been incorporated into daily traditional diets like cereals, beverages, bread, pasta, and cookies as well as in products in the market to enhance the sensory and nutritional properties [76].

### 8.3.2.3 *Insects*

Over 2,000 species of insects are consumed in around 113 countries [77], with beetles, caterpillars, bees, ants, grasshoppers, crickets, and locusts being the common varieties [78]. Insects have been used as edible food sources from ancient times, and in some places, they have become part of the main diet. For instance, in Central Africa, 50% of dietary protein needs are met through insect foods. Most of them are consumed during different life cycle stages and in different forms — raw, roasted, ground, boiled, or fried. Native populations in Africa, South America, and Asia have been consuming insects as part of various dishes. Locusts and grasshoppers are consumed widely in Thailand, and a surge in the demand for insect food has resulted in the establishment of production facilities of crickets by some farmers to augment their income. However, the Western world's perception of insect-eating is associated with disgust and fear, and negative media portrayal has caused a decline in their popularity [79]. Insects are a rich source of energy, proteins, fats, and minerals with varying degrees, depending on insect species and even within species based on their diets, developmental stages, sex, and environmental factors. Proteins constitute about 30–65% of the total dry matter, followed by fats, in which between 46% and 96% of all amino acids are present with limited amounts of tryptophan and lysine. Grasshoppers and mealworms have also been evaluated as potential protein sources. Besides human consumption, insects have been used as feed source, including black soldier fly larvae for their excellent conversion of organic waste into biomass [80]. Protein and lipid contents of commercially reared insects are quite high and carbohydrate content is low. There are studies which have led to different reproductive rates in crickets based on dissimilar diets, raising possibilities of alteration of nutritional profiles via their diets [81]. Most of the companies are utilizing whole insects for producing insect flours as protein sources because isolating proteins from the insects is still cost-prohibitive and labour-intensive. Though insects can be consumed whole (cultural practices in many countries), innovations are targeting to produce more acceptable forms such as flour for consumption by global consumers. Most of the companies utilizing insects as AP sources rear them in considerable numbers and roast and mill them as flours to be incorporated into doughs to make cookies, snacks, bars, etc. In this context, food safety of these insects needs to be studied in detail, in view of potential anti-nutrient and anti-allergenic components. Insects feeding on sources unfit for livestock consumption also contribute to the circular agricultural economy [82]. Insect companies in the US such as Chirps, Bitty Foods, and Tiny Farms have sprung up on the promise of reduced GHG emissions, faster reproducibility, reduced water usage, and energy-efficient methods of protein production [83]. A Finnish company, Fazer, created the first insect flour-based bread in 2017 [84]. Insect-based burgers that have been introduced in the market are from Coop (a Swiss retailer) and Bugfoundation (a German food company) [67].

### 8.3.2.4 *Cell-cultured alternative proteins*

Cell-based lab-grown meat uses tissue culture technology to propagate single animal cell *in vitro* using growth media and culture containing nutrients. Advantages of such a system are manifold — (1) no animal is harmed in the process, (2) since tissue regeneration occurs under sterile conditions in a laboratory, the product is safe, and (3) the protein value mimics that of an original meat product. Presently, due to the technologies involved, this is certainly a costly innovation. In 2013, the first lab-grown meat burger was introduced in London at the price of EUR 250,000 by Mosa Meats. Cultured meat is referred to as clean meat, lab meat, or cell-based meat [85]. Despite the frenzy surrounding cell-cultured meats, it is often very difficult to produce them due to the complex nature and arrangement of different components like cells, extracellular matrix (ECM), proteins, etc., and hence intensive research is still required. An important angle in promoting these foods to the consumers is how these proteins are manufactured and what benefits they offer compared to conventional animal proteins. According to a recent survey, 58% of the respondents were ready to pay a premium amount for cultured meats [86]. With respect to conventional ways of obtaining animal protein which may sometimes require several weeks to years, cultured meat harvest requires considerably shorter times [69], saving precious resources and reducing the externalities associated with animal proteins. Another advantage of these proteins is the avoidance of contamination from conventional production processes as well as the risks of zoonotic diseases, as production is performed under sterile conditions with quality controls in place. MeaTech's recent acquisition of the cultivated meat company Peace of Meat (POM, Belgium), involved in cultivating chicken cells into fat cells, was announced in December 2020. The technology aims to come up with hybrid plant-based products incorporating cultured meat to imitate real meat properties [87]. Changing public perception through advertising and correct branding is needed for cultured meats to lay claim to the "naturalness" of their content. Eat Just, a US start-up, recently served cultured chicken in the Singapore restaurant 1880, the world's first commercially sold cultured meat approved by the Singapore Food Agency (SFA) called "Good Meat" [88]. Still, to make cultured meat commercially viable owing to prohibitive costs and consumer reluctance, more innovation and awareness is required for its wide-scale entry into the markets.

### 8.3.3  *3D Printing of Alternative Proteins*

#### 8.3.3.1  *Research and development*

In terms of 3D printing of APs, the field is still limited to academic research in laboratories. A few research papers are cited in the literature on 3D printing of APs. In terms of 3DFP, soy, cricket, and pea proteins have been experimented in some capacity in 3DFP. Soy protein isolates were 3D-printed with sodium alginate and gelatin for extrusion printing of food mixtures [89]. Good fidelity of 3D-printed designs was achieved by mixing cricket and pea protein in different ratios with mashed potatoes [90]. No adverse clinical outcome was reported after consuming 25 g of whole cricket powder for 14 days. According to a recent study, there are some benefits to gut health. Wheat flour has been enriched with mealworm beetle (Tenebrio molitor) to 3D-print snacks [91]. Microalgae proteins *Arthrospira platensis* and *Chlorella vulgaris* were mixed into dough to make 3D-printed cookies with stable structures and greater mechanical resistance [92]. Novameat has an offering of vegan protein resembling a steak in both appearance and taste. Made with pea protein, seaweed, rapeseed oil, rice protein, and beetroot juice, the texture is made to resemble the actual meat. Based on the patented micro-extrusion technique, it is still in the experimental phase, attracting funding from New Crop Capital, which has invested in Beyond Meat, Memphis Meat, and Mosa Meat. Another company experimenting with 3D printing of plant-based steaks is Redefine Meat [93]. The company has recently launched 3D-printed plant-based beef cuts that consist of soy, pea protein, chickpeas, beetroot, yeasts, and coconut fat mix in Europe. It has been introduced in Israel, Germany, the Netherlands, and the United Kingdom [94]. Among the other alternatives offered by the company in the "New Meat" series are ground beef, cigars, burgers, kebabs, and sausages [95]. Mushrooms, known to have a texture like that of meat, has been 3D-printed as a snack with wheat flour [96]. 3D printing of cell-cultured meat is an advancement to cultured meat technology as it allows the bioprinting of animal cells along with fat cells, ECM within the scaffold to support growth and proliferation. It differs from 3D printing of meat pastes as the muscle cells are inherently grown into the scaffold, and complex and structured meats like steaks can be produced. There are a few reports on the 3D printing of meat pastes. Pork and chicken meat pastes have been 3D-printed successfully [97, 98]. 3D printing offers the promise of personalization, a possibility that was explored by 3D Bioprinting Solutions and an Israeli company, Alephs Farms, by printing cultured meat (beef) in space [99].

### 8.3.3.2 *Economics of alternative proteins*

Time is required for protein product evolution in the market and consumer acceptance, which involves creating a following, further building of interest and use, followed by supply chain build-ups, and finally diversifying the consumer base [100]. PBMAs attempt to create and mimic the mouth feel of meat with plant proteins. This has found success in products introduced by Beyond Meat and Impossible Foods. In the US, APs have been majorly funded by private ventures [56]. One of the reports suggests that public funding in this area will be more effective than private investments in creating a domestic economy in the US. This is because research in this area requires collaboration between multiple disciplines involving higher risks [66]. Beyond Meat and Impossible Foods backed by private funding and venture capitalists have successfully launched their plant-based products in restaurants, supermarkets, and grocery stores. Much funding still needs to focus on R&D and be matched by the public sector for APs to truly make a difference in the protein segment of the US industry and compete with animal proteins. One route to AP acceptability is via fast-food chains, which has already been tested and proven successful. This may be due to easier integration of novel products by these chains due to similar sensory attributes of these proteins and well-established supply chains and infrastructures. Another way of introduction may be at specialized dining places, niche restaurants, and places like schools, hospitals, and canteens to replace cheap meat sources like chicken nuggets, beef burgers, patties, etc. Bioprinting of cultured meat may be an important technique for the customization of nutritional profile and palatability for different consumer groups. For the cell-cultured lab meat growth, high-risk research is essential to answer basic scientific questions to make the process successful.

Soy protein has recently seen some decline in market share by a CAGR of 6%, partly due to the interest in other AP sources such as pea protein. Concerns over soybean's allergenic and estrogenic effects are, however, limited to a smaller populace. Cultured meat saw a market share of 16% CAGR and pea protein was at a CAGR of 30% from 2004 to 2019. According to a McKinsey report, pea and cultured meat will see the major market share though producers need to supplement these plant proteins with other amino acids to make them as protein-rich as the insect proteins [55]. Plant proteins are also cost-effective compared to cell-cultured proteins, which are still cost-prohibitive. Beyond Meat raised a massive IPO of USD 760 million, attracting much interest from private investors. Public funding is critical in realizing the success of APs as is evidenced by the investment of governments in R&D capabilities in their respective countries — for example, a USD 100 million investment in the AP company

Merit Functional Foods by the Canadian government [101]. Singapore, with its "30 by 30" slogan aimed at achieving food security by 2030, has signalled R&D in APs through the Agency for Science, Technology and Research [102]. Asian countries like Singapore, Hong Kong SAR, and Macau SAR have reported six times more sales of Impossible Foods' products, with the company eyeing the Chinese market, especially after the coronavirus pandemic where consumers are more willing to try PBMAs. Competition is building up due to the presence of other plant-based alternatives in China like Beyond Meat, Eat JUSY Inc., a vegan egg company, and Oatly AB [103]. There are now more than 60 start-ups in the cultivated meat industry. Most of the insect trade comes from Southeast Asia, with Thailand contributing the most with a USD 1.14 million per year import market. Insect farming looks like a promising revenue source for farmers as the market value for insects is greater than for other protein sources. A yearly income of USD 4,270–9,970 can be obtained by the production of 500–750 kg of crickets four to five times a year [104]. By 2023, the global edible insect market will outgrow USD 522 million, with the European market seeing projected sales of USD 46 million. Energy bars and insect flours comprised about 54% of the total market share in 2015 [105].

That being said, the rise of the AP industry may also upset the already well-established industry practices in agriculture and livestock, causing significant economic effects. To allay the fears plaguing the traditional meat industry, the narrative should not be towards complete replacement of meat-based foods with APs. Focus should remain on the creation of a new parallel industry, leading to job creation as well as raising the demand for different crops like peas, providing farmers with unique opportunities. Food narratives are increasingly inclined towards consumer awareness to help them make better and informed choices on their daily diets. One way has been to introduce the various players and processes involved in food production so that the consumers feel a direct connection with the producers. A dominant role in the demand for APs in the market will be played by public perception. To make APs safe, sustainable, scalable, and nutritionally and functionally optimized for commercial purposes, continued R&D studies need to be done. Nutritional values linked to AP narratives could be better bone health, healthy cardiovascular properties, and a low glycaemic index, and functional characterization affecting the appearance, taste, and texture can also be modified by using these proteins [100]. Food and commodity companies are also investing in AP start-ups (Tyson Food in Beyond Meat, Cargill in Puris, Maple Leaf Food in Entomo Farms and LightLife Foods, Nestlé in Sweet Earth), indicating confidence in their potential growth [66]. The narrative should shift from the consumption of AP products as fast-food replacements to genuine newer products that share less of the unhealthy characteristics associated with fast foods.

### 8.3.4 *Challenges and Counternarratives*

#### 8.3.4.1 *Safety concerns and regulations*

Safety assessment of proteins investigates two key issues: (a) toxicity or allergenicity potential, and (b) ingredient safety and process-related contaminations [100]. Just like any nascent industry, APs too face regulatory challenges, especially the ones involving insects and cell-based proteins. Different countries have different regulations regarding cultured meat production. The European Union has banned the use of recombinant bovine somatotropin (rBST), whereas it is approved in the US for meat products [106]. The main safety issue facing APs is the question of allergenicity, which may be elicited by any novel dietary food product. It is important to look at a few predictors of a food's allergenicity — is it allergenic in small amounts (cottonseed protein) or allergenic in the country where it is mass consumed (e.g., buckwheat in Japan and Korea)? Does it share a genetic relation with other known allergenic foods (e.g., canola and mustard similarity), and is it is easily digestible [72]? For insect alternatives, regulations are not much of a problem in places where insects are a part of traditional diets; however, Western countries are trying to have regulations around insect foods for human consumption. The European Food Safety Authority (EFSA) has categorized insect products for human consumption as a "novel food" requiring due approval. Likewise, in the US and Canada, approvals must be sought from the respective regulatory institutes [107]. Another area to focus on is the production practices of insect rearing; unlike livestock farming where there are established processes, there are hardly any good practices for insect farming as yet. Data has to be generated on the effect of different techniques of processing and production on the safety and nutritional profile of insects. One narrative used by detractors of APs is that their origin source is not "natural". In 2017, the European Court of Justice made it illegal to use labels like milk and associated words, such as butter and cheese, for non-dairy products [108]. Missouri became the first US state to pass a bill defining meat, making it illegal to misrepresent a product as meat if it is not harvested from livestock or poultry [109].

### 8.3.4.2 *Consumer acceptance*

Meat consumption has been advocated based on four Ns: natural, normal, necessary, and nice [110]. Promissory narratives have been built around APs based on their ability to deliver the said promises. What the concept does is to raise consumer interest and venture capital if these food innovations are successful. APs have a few barriers to cross before widespread acceptability, such as scientific research, production capability, costs, and consumer support, can happen. APs are viewed as proper food by their supporters and non-food by their detractors [111]. Acceptance of APs is a great source of concern for the companies investing in these products as some of the APs may not appeal to the consumers through their sensory profiles. For consumers, convenience, experience, and naturalness are the key drivers for uptaking any food product. In the same way that insect proteins can elicit feelings of disgust, neophobia is associated with them [100]. Certain plant proteins like soy are characterized by a beany flavour, which is not particularly liked by certain consumer groups. Others may have concerns over the authenticity of cultured meat. There has been news about the debate around the realness of both plant- and lab-based proteins, as some may see them as non-natural or fake. The meat industry has been advocating against cultured proteins, claiming them to be non-natural, fake, and engineered, along with a variety of ingredients composing them. Some plant-based proteins have been criticized by activists on their genetically modified organism (GMO) tags. In a bid to counteract these narratives, the AP industry has relied on consumer awareness campaigns. Surveys have shown that many consumers are unaware of recent cell-cultured meat products. Plant-based APs' success depends on the ability to improve taste and texture and replicate the texture and fibrous nature of animal proteins. One challenge plaguing the 3D printing of cultured meat is problematic scaffolding, providing mechanical support to cells so that the structure can be printable, flexible, and cost-effective to be scalable [112]. Replicating the complex vasculature of muscle tissues is quite difficult, as is the isolation and separation of stem cells from animals for growth, differentiation, and proliferation. Problems with scaling also need to be tested as these technologies are still in a nascent phase. Much of the pushback against APs has come from stakeholders in conventional meat, livestock, dairy, and egg industries through media interviews as well as social media campaigns. Studies have suggested a few counternarratives from the opposition groups, the first being "Not a serious threat" [66]. This narrative came into existence during the initial stages of the AP story, in part due to the doubts revolving around technological capabilities to produce quality end products, and in part due to a reliance on consumers to reject these innovative techniques producing APs. A case in point was a dairy giant in New Zealand dismissing the competition introduced by the non-dairy milk production company Perfect Day Foods (milk from yeast) and questioning the technical viability and nutritional content of the yeast milk, calling it "artificial" [109]. Another term used by various anti-AP groups is "fake", or "not real food", to counteract the clean and safe claims of APs like cultured meat. They highlight the use of biomedical techniques and sterile laboratories to build a narrative of not being real or natural, and not being true to the consumers. APs are also labelled as ultra-processed foods as most of the plant-based meat products have been introduced via fast-food chains. From a legal point of view, the question arises in terms of labelling these proteins and whether to classify them as meat, dairy, eggs, etc. Livestock industry groups have asked the regulatory bodies to focus on the regulatory definition of meat as those harvested in a traditional manner. The dairy industry has been contesting the use of the labels milk and butter to the proclaimed healthier options like soymilk, almond butter, etc. Another trend in labelling food products is "free from" (e.g., gluten free, GMO free, and BPO free), inherently implying that products that do not contain these ingredients are superior in product comparisons. Plant-based proteins tend to have sodium in high amounts and are available in highly processed forms like burgers, sausages, and patties, defeating the purpose of healthier and sustainable choices. Many of the plant ingredients used for meat analogues make these products ultra-processed foods despite offering a nutritional profile similar to the meat product they mimic [100]. Despite these facts, consumers in North America are willing to pay a premium for these products. Many modern meat analogues satisfy the macronutrient requirements of the traditional meat diet, yet increased consumption of these ultra-processed foods may have unintentional and yet unknown health effects [70]. Even though plant-based meat producers claim high similarities in the taste, texture, and appearance of the meat analogues compared to real meat, it has been difficult to achieve a satisfactory sensory profile.

For example, the algal proteins Spirulina and Chlorella impart a distinctive green colour, making the products less appealing.

A few concerns with human consumption of insects remain, one among them being the possible anti-nutrient effects of chitin, the main component of insects' tough exoskeleton, on digestibility. Though there are research studies promoting chitin as a good source for fibre and having a cholesterol-lowering effect, long-term ingestion effects are still unknown [107]. Microbial risks associated with bacteria living in the gut of insects could be addressed by applying processing techniques, such as washing and heating, as in other products of animal origin [113]. A study has identified a novel cricket allergen, hexamerin 1B, as well as cross reactivity of cricket in people known to have prawn allergies [114]. There are also reported cases of contact allergy due to frequent handling of insects by breeding farm workers [115]. In Western societies, the issue is an increased acceptance of entomophagy, as recent surveys have reported a low consumer willingness to consume insects as food. For example, one study showed that only 12.8% males and 6.3% females were willing to adopt insects as a meat substitute [116], and another study reported this percentage to be 19% [117]. Plant proteins such as legumes and soy have been characterized by their "off-flavours", hindering wider consumer acceptance. Though there is a strong co-relationship between consumption of plant-based diets and good health as opposed to consuming animal-dependent diets, there is still no study that directly assesses the health impact of substituting meat with PBMAs, insects, and other AP diets over the long term. In terms of nutritional profile, many AP sources are marketed as a replacement of fast-food options like burgers, patties, and sausages, again falling into the trap of a higher degree of processing, thus losing their nutrient-dense appeal. Looking at the sales data, burger and sausage products as PBMAs emerge as the top choices [53]. Hence, one conclusion is that the consumers are led to believe that they are consuming healthy nutrient-dense foods, which may not be true. A major driver of consumer acceptance is cost, even though some surveys suggest that consumers are willing to pay a premium for PBMAs, supporting the healthy narrative, but the same is not true for insect- or culture-based food products. Even though traditionally used plant proteins are cheap (soy, pea protein), the protein ingredients currently used by companies drive up the costs. For example, due to the use of pea protein isolate in Beyond Sausages, its cost is 70% higher than that of conventional pork sausages [118]. Looking at all the claims and counterclaims, more studies and research need to be done in order to arrive at any decisive conclusion regarding the future of APs.

### 8.3.5  *Business Models of 3D Printing of Chocolate and Alternative Proteins*

Business models are essential for any new and existing businesses as they help a company map out its target market and related expenses and contribute to competitiveness, survivability, and sustainability. They involve planning strategies to estimate a company's profit plan based on the products and services it intends to sell. Business models answer questions on four different aspects: (a) market — identifying the various customer segments and their needs, (b) potential — value proposition of the product, (c) technical aspects — features contributing to the product's value, and (d) profitability — profit via revenue streams [119]. Though the 3D printing industry is hailed as a disruptor, there is limited research on its business models. Classification of its services has been categorized into three categories: generative — generation of 3D models, facilitative — 3D printing without model generation, and selective — services aimed at creating a database of 3D models [120]. One recent study demonstrated the existence of five distinct model design patterns revolving around a host of printing services [121], which are as follows:

1) The digital marketplace model: Exclusively functioning online, eight companies use this model which connects customers to other users of the technology and focuses on the entrepreneurial skills of the users. They offer their customers an array of 3D printing facilities and engage social media channels for communication through blogs and press articles. Profits are generated via selling 3D-printed objects or hardware and earning commission fees utilizing digital services.
2) The single-tech specialist model: Used by 30 companies, it enables value generation through different services like pre-print services and training services and revenue generation via sales, ensuring good 3D printing outcomes. Distribution is via online web shops and offline stores. Half of them also perform individual contract work.
3) The local service centre model: Featuring 32 firms, it enables value proposition through an assortment of services like 3D training and

pre-print service. Extrusion and jetting processes are chosen as 3D printing types. Partnering with professionals' focus is mostly offline stores with limited web shops. Revenue follows sales and contract work with the rental option as a minor stream.

4) The multi-tech champion model: Involving 23 companies, it enables value generation through their industry experience, offering high expertise along with associated services with quality and swift ordering process. Most of them provide online services, and the majority also have offline stores. Contract work is the most important revenue source, putting emphasis on individual printing jobs.

5) The all-rounder model: Used by the maximum number of companies (48), it utilizes different business model characteristics. Value creation comprises several factors like speed, quality, expertise, confidentiality, variety, flexibility, and customization. Web shops and physical stores provide solutions through professional partners. Contract work and sales represent the revenue streams. Dominant yet limited communication is through social media.

In terms of 3DFP, even less are known vis-à-vis business models. For value proposition, emphasis is laid more on personalization, i.e., customization of food products specifically geared towards people with special needs like dysphagia patients (patients with swallowing difficulties), babies, malnourished people, pregnant women, and overweight people by modifying the nutritional profile of the 3D-printed foods. One research paper presented three business value propositions for 3D-printed foods after conducting workshops based on interviews, surveys, and consumer focus groups. These were 3D-printed designer chocolates, personalized snacks at semi-public places dispensed through vending machines integrated with 3D printers, and catering for hospital canteens equipped with digitized 3D kitchens. Apart from benefits by personalization, convenience was rated high among the attractive attributes of these business models, though the study underlined significant risk taking during developmental phase [122].

In this section, we provide some examples of utilizing the value proposition canvas (VPC) and business model canvas (BMC) to identify particular customer segments and the value that 3D-printed products bring. VPC is an established tool in helping businesses to understand their customers' needs and position their products and services around them for profit generation. Initially developed by Dr Alexander Osterwalder, it

**Value Proposition**                    **Customer segment: F&B patrons**

**Gain Creators**

Fun aesthetic 3D designs

Good feeling after purchase

**Products & services**

Easy access

Personalization through design and nutrition

Immediate accomplishment feeling

Unique dining experience

**Pain Relievers**

Involvement with the finished product either through design choice or ingredient selection

**Gains**

Making a difference through food

Making own choices in nutrition through diet

**Customer Jobs**

Immediate Design selection

Novel experiences

Diet customization

Contributing towards food sustainability

**Pains**

Convenience

Excess cost for customization, Overpaying

No control over nutrition profile

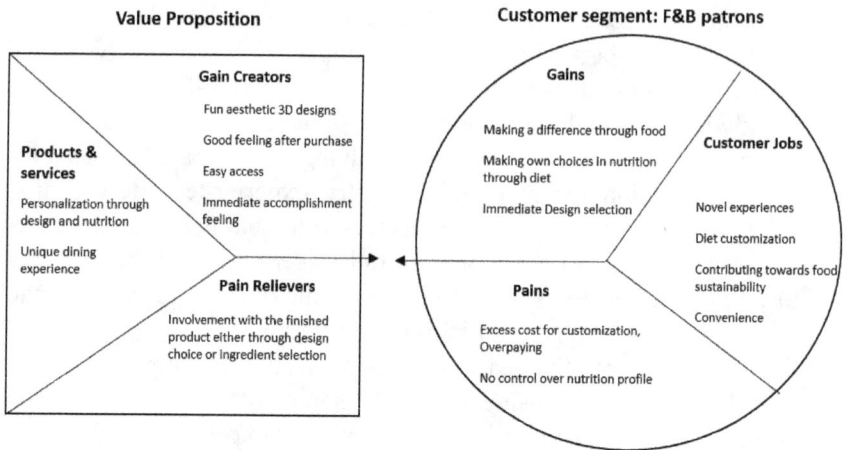

Figure 8.3.   Value proposition canvas (VPC) for 3D-printed foods for the food and beverage industry.

examines the relationship between customer segments and value propositions [123]. Figure 8.3 depicts one example of VPC for 3D-printed foods focusing specially on the food and beverage consumer segment. Likewise, businesses can generate independent models by doing consumer research surveys and interviews to identify the needs, gaps, and opportunities in their areas. Also important for the businesses is to study their competitor's practices in terms of the market and how to position their products more favourably and design better products that offer more value to the consumers. Since 3DFP is still a novel and niche area, both for chocolates and APs, first-mover advantage will be accorded to the companies initiating their products using AM. Understanding the VPC (Figure 8.3), for a consumer interested in 3D-printed food, the customer profile is represented by the circle. Assuming the customer's role in choosing a specific food outlet is to have a "novel experience" around food, to achieve "personalization" by having a say in the 3D-designed meal as well as keeping in mind the wish to have a "balanced meal" which contributes to less "food wastage". What the value map does is to try to fit and address most of the customers' expectations of achieving gains and relieving pains. Here, the prospect of 3D printing of customers' meals is to add the fun and entertainment experience simultaneously, providing ease of access to the prospect of 3D dining. A consumer who has no access to 3D printers or 3D technology can easily choose to patronize a restaurant with these facilities. The pain of not having a choice in ingredients selection can also be overcome through this proposition.

VPCs are mapped out for each customer segment as the individual's needs and expectations may vary. After identifying top customer segments for the intended technology, a BMC should be mapped out for a structured and visual assessment of the value proposition offered and the channels through which revenues might be generated [124]. There are nine building blocks that need to be defined on this canvas. The "customer segments" could be identified as supermarket shoppers representing the mass market, 3D-printed designer chocolate consumers, F&B patrons, and institution managers (hospitals, canteens, nursing centres looking for catered foods) representing niche market areas. For instance, a nursing home for the elderly might be interested in soft food 3D printed services. Efficient "channels" will be needed to reach the target consumers, which may be through partner distributors or through web shops of manufacturers, or both. The business also needs to maintain "customer relationship", which can be through before and after sales surveys to gauge customer satisfaction with the products and through blogs, articles, and social media to attract new customers. The business must also look at the "key infrastructure needs" comprising resources, partners, and activities to set up and sustain, which in the case of 3D-printed foods may be investing in the R&D of source ingredients (chocolates and APs) by collaborating with research institutes, 3D printer development, and manpower training. The value proposition for these types of food can be novelty, personalization, automation, and sustainable eating. Finally, the finances, which include "cost structures" with fixed and variable costs such as manpower salary, production costs, and maintenance costs, need to be discussed. "Revenue" can come by the selling or renting of 3D printers and selling of specific ingredients for formulation of 3D food inks, design ideas, software licences, manpower training, etc. Now let us take a look at value chains specific for two different business models for 3D-printed foods.

### 8.3.5.1 *Model for 3D-printed chocolates: Combined AM processes at the manufacturer and the retailer*

For this model (Figure 8.4), the main printing of chocolates can take place at the established manufacturer's factory, as they already possess the required ingredients, know-how on handling chocolates, and necessary

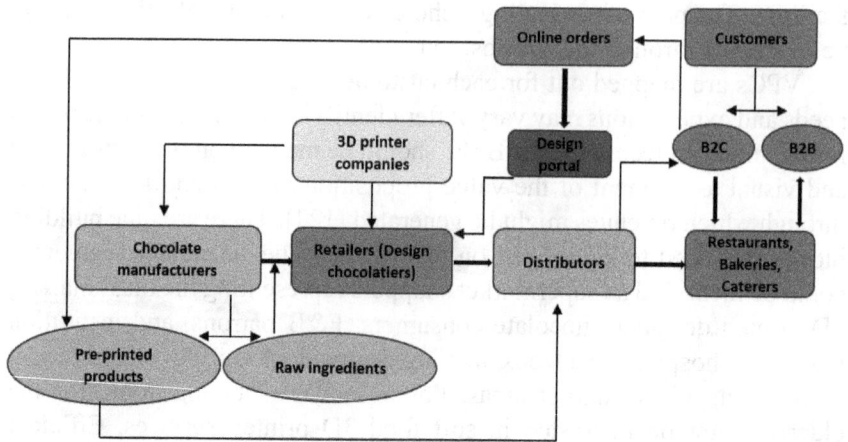

Figure 8.4.   Proposed business model for 3D-printed chocolate.

storage and equipment. This model also utilizes an intermediary designer retailer facility for customized 3D printing. One way to make this happen is to have trained personnel at the retailer's setting as chocolate printing can be challenging. The customers comprise B2C (business-to-consumer) clients who can place orders directly through the internet at the manufacturer's web shop by selecting the design from the design library maintained by the chocolate manufacturer as well as ingredients to go into the 3D-printed chocolates. Another way of ordering will be from the design portal maintained by the retailer stores that feature higher levels of customization and customers can submit their own 3D designs for their printing. The portal will also contain recipes for chocolate ink formulations for consumers who are looking for low-sugar or other dietary options. Other clients might be bakeries and confectionery shops specializing in customized chocolates and toppings on cakes. These specialized bakeries and shops will then be designated as B2B clients as they will source their 3D chocolates from the chocolate manufacturers and design retailers for bulk orders. The clients can also directly purchase from these places (B2C). The manufacturer and retailers will engage 3D printing service providers to help them with setting up the printers as well as the requisite manpower training and the upkeep of 3D designs gallery. Value proposition for this type of model is the unique eating experience offered through customized 3D-printed chocolates. This involves both a manufacturer and a retailer-led model. Since chocolate 3D printing still does not have a mainstream market, combining both the business models will assure better revenue generation. With their technical know-how, manufacturers can produce 3D-printed chocolates at a fast rate, whereas the retailers can help drive profits by giving competition to handcrafted chocolate producers. The advantage of this model will be reduced set-up costs as the manufacturing and printing products will be in the same location, reducing the need for transporting raw materials. The chocolatiers can source from the manufacturer based on the orders received, eliminating the need to maintain a ready inventory of raw materials and produced goods. The manufacturers can utilize pre-existing distribution and supply channels to deliver their finished 3D-printed products, reducing costs further. Additional costs incurred include the initial investment in 3D-printed machinery, maintenance, the setting up of a design library, and manpower training. Customer outreach can be through social media posts, press releases and blogs, catering events, and competitions on 3D chocolate printing. This business model is viable because there is already enough research in chocolate printing and a host of printers in the market for this purpose.

### 8.3.5.2 *Model for 3D-printed alternative proteins: Centralized facility*

A centralized facility will have the 3D printers installed in-house, with maintenance and upkeep provided by the 3D printer suppliers. In this model (see Figure 8.5), AP producers will supply raw materials (e.g., peas, cricket protein, soy protein) to the centralized facility. The recipes for the 3D-printed products will be determined by the businesses (restaurants — meat analogues; supermarkets — snacks, desserts; fast-food

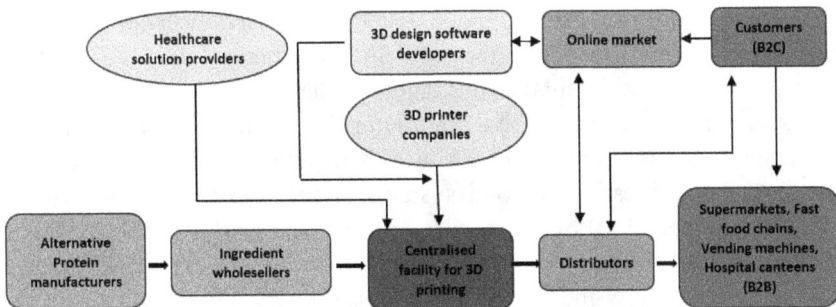

Figure 8.5.    Proposed model for 3D-printed alternative proteins — centralized facility.

chains — burgers, hot dogs; hospital canteens — fortified foods, etc.) and finalized after consultation with chefs, nutritionists, and dieticians. The centralized facility will be responsible for formulating the food inks for 3D printing as per special customized recipes provided by different businesses. They will work with distribution channels to supply the finished products to customers. The consumers will have an option of walking into these supermarkets, food chains, or vending machines and picking up the products, or using the internet to order personalized AP products provided by the distributors. The value propositions are healthy ingredient choices, protein fortification, sustainable eating through food choices, and experiential dining. The costs incurred by the consumers may be more than for other options, but it may target specific demographics looking to change their diets and being a part of conscious eating choices. This model has the potential to generate higher revenues as it involves large-scale consumption and distribution through supermarkets, F&B industry players, as well as vending kiosks.

## 8.4  Summary and Future Prospects

Traditional food chains are less transparent due to their complex nature, and food products are mass standardized to withstand long journeys, with added preservatives to maintain flavour and appearances. Although it is still in nascent stages, 3DFP is capable of addressing some of these issues [122]. APs will have to co-exist with conventional animal proteins as consumers may prefer to replace a certain part of their diets with APs but not entirely for the foreseeable future, as established eating habits are hard to change or let go despite good intentions. Historically and culturally, Asian countries have greater acceptability of meat alternatives — PBMAs and insects. For instance, in India, a vegetarian lifestyle is a common and prevalent practice. The acceptance of PBMAs is increasing in the Western world, but there still remains hesitancy in the adoption of insect-based foods. APs have received global attention riding on promissory narratives of reduced environmental impact, health benefits, animal welfare, and sustainability. Investing in food systems is sensible to meet the rising dietary demands of 10 billion people by 2050 and also to cut the costs of treating diseases by adopting healthier diets, because food influences human health. In low-income countries, meat is very important for providing nutrients and energy, as there are hardly any alternative food sources for these needs. For middle- or high-income countries, consumption of a varied diet fulfils these energy and nutrient requirements without overt reliance on meat [66]. Much of the scientific research into new market products and investment in APs has come from the US and Europe. Instead of replacing the conventional meat products entirely (or even partially) with alternative protein diets, options can also include a mixed or combined product. Some studies have suggested that hybrid products, beef burgers, and sausages containing real meat with APs are quite similar in sensory profile to full-meat products. For example, even a 10% replacement in beef patty with tempeh produced a juicier and tender product similar in flavour [125]. Consumer education and awareness are critical, along with the necessary information from AP producers on the source, nutritional profiles, and processing techniques in simple labelling for transparent information. Incorporation of supplementary nutrients should be considered for both chocolate- and AP-based 3D-printed foods. Consumers may see APs in a new light, not just as fast-food alternatives, and chocolates may move beyond the negative consumer perception of fattening products. In terms of cost, as production facilities scale up and more R&D happens, the pricing of APs and chocolate may come down. APs might see a shift from niche to mainstream food category in high- and middle-income countries, riding on quality novel ingredients and government interventions in the next decade [66]. Again, more needs to be done to understand the business models of companies with 3D-printed products in the market. Though PBMAs are hailed as meat alternatives, the form in which they are currently available leaves much to be desired for a good health narrative. There is a need to modify the processing treatments of APs to lower the sodium and fat content while preserving taste and texture. Moreover, novel ingredients such as soy leghemoglobin in burgers, which are introduced in these products, must be assessed for long-term effects. Joint public–private investment opportunities, as well as consumer awareness studies, should pave the way for higher assimilation of these novel products and technologies such as 3DFP.

# References

[1] Verna, R. (2013). The history and science of chocolate, *Malaysian Journal of Pathology*, 35, pp. 111–121.
[2] Lippi, D. (2013). Chocolate in history: food, medicine, medi-food, *Nutrients*, 5, pp. 1573–1584.

[3]  Rusconi, M. and Conti, A. (2010). *Theobroma cacao* L., the food of the gods: a scientific approach beyond myths and claims, *Pharmacology Research*, 61, pp. 5–13.

[4]  ICCO. (2021). Retrieved from https://www.icco.org/may-2021-quarterly-bulletin-of-cocoa-statistics/

[5]  Montagna, M. T., Diella, G., Triggiano, F., Caponio, G. R., De Giglio, O., Caggiano, G., Di Ciaula, A. and Portincasa, P. (2019). Chocolate, "food of the gods": history, science, and human health, *International Journal of Environmental Research and Public Health*, 16.

[6]  Latif, R. (2013). Chocolate/cocoa and human health: a review, *Netherlands Journal of Medicine*, 71, pp. 63–68.

[7]  Martin, M. A., Goya, L. and Ramos, S. (2013). Potential for preventive effects of cocoa and cocoa polyphenols in cancer, *Food Chemical Toxicology*, 56, pp. 336–351.

[8]  Fisher, N. D. L. and Hollenberg, N. K. (2006). Aging and vascular responses to flavanol-rich cocoa, *Journal of Hypertension*, 24, pp. 1471–1474.

[9]  Larsson, S. C., Akesson, A., Gigante, B. and Wolk, A. (2016). Chocolate consumption and risk of myocardial infarction: a prospective study and meta-analysis, *Heart*, 102, pp. 1017–1022.

[10]  Francis, S. T., Head, K., Morri, P. G. and Macdonald, I. A. (2006). The effect of flavanol-rich cocoa on the fMRI response to a cognitive task in healthy young people, *Journal of Cardiovascular Pharmocology*, 47, pp. 215–220.

[11]  Walters, M. R., Williamson, C., Lunn, K. and Munteanu, A. (2013). Chocolate consumption and risk of stroke: a prospective cohort of men and meta-analysis, *Neurology*, 80, pp. 1173–1174.

[12]  Renzo, L. D., Rizzo, M., Sarlo, F., Colica, C., Iacopino, L., Domino, E., Sergi, D. and Lorenzo, A. D. (2013). Effects of dark chocolate in a population of normal weight obese women: a pilot study, *European Review for Medical and Pharmacological Sciences*, 17, pp. 2257–2266.

[13]  Ramos, S., Martín, M. A. and Goya, L. E. (2017). Effects of cocoa antioxidants in type 2 diabetes mellitus, *Antioxidants*, 6, p. 84.

[14]  Esser, D., Mars, M., Oosterink, E., Stalmach, A., Muller, M. and Afman, L. A. (2014). Dark chocolate consumption improves leukocyte adhesion factors and vascular function in overweight men, *FASEB Journal*, 28, pp. 1464–1473.

[15]  Beg, M. S., Ahmad, S., Jan, K. and Bashir, K. (2017). Status, supply chain and processing of Cocoa — a review, *Trends in Food Science and Technology*, 66, pp. 108–116.

[16]  Grand View Research. (2020). Chocolate market size, share and trends analysis report by product (traditional, artificial), by distribution channel

(supermarket & hypermarket, convenience store, online), by region, and segment forecasts, 2020–2027.

[17] Shoup, M. E. (2021). Chocolate consumption on the rise, survey finds. Retrieved from https://www.foodnavigator-usa.com/Article/2021/06/16/Chocolate-consumption-on-the-rise-survey-finds

[18] Neilson, J., Pritchard, B., Fold, N. and Dwiartama, A. (2018). Lead firms in the cocoa–chocolate global production network: an assessment of the deductive capabilities of GPN 2.0, *Economic Geography*, 94, pp. 400–424.

[19] Candy Industry. (2021). 2021 Global top 100 candy companies. Retrieved from https://www.candyindustry.com/2021/global-top-100-candy-companies

[20] KPMG. (2014). A taste of the future: the trends that could transform the chocolate industry.

[21] Allen, L. L. (2010). Chocolate fortunes: the battle for the hearts, minds and wallets of China's consumers, *Thunderbird International Business Review*, 52, pp. 13–20.

[22] Afoakwa, E. O. (2016). *Chocolate Science and Technology*, 2nd edition (John Wiley & Sons).

[23] Jeffery, M. S., Glynn, P. A. and Khan, M. M. U. (1977). Method of manufacturing a chocolate product. No. 4,045,583, US Patent and Trademark Office, United States.

[24] Beckett, S. T. (2008). *The Science of Chocolate*, 2nd edition (Royal Society of Chemistry).

[25] Winston, S. (2012). Additive manufacturing as a disruptive technology: how to avoid the pitfall, *American Journal of Engineering and Technology Research*, 12, pp. 86–93.

[26] Christensen, C. M. (2006). The ongoing process of building a theory of disruption, *Journal of Product Innovation Management*, 23, pp. 39–55.

[27] Danneels, E. (2004). Disruptive technology reconsidered: a critique and research agenda, *Journal of Product Innovation Management*, 21, pp. 246–258.

[28] Stapleton, D. and Pande, V. (2016). Evaluating additive manufacturing as a disruptive technology in transportation and logistics. *AM POMS*.

[29] Campbell, T., Williams, C. B., Ivanova, O. and Garrett, B. (2011). Could 3D printing change the world? Technologies, Potential, and Implications of Additive Manufacturing. Atlantic Council.

[30] Barnes, J. E. and Slattery, K. (2021). Additive manufacturing: incremental improvements to a disruptive technology, *Accounts of Materials Research*, 2, pp. 574–576.

[31] Abe, T. and Sasahara, H. (2016). Dissimilar metal deposition with a stainless steel and nickel-based alloy using wire and arc-based additive manufacturing, *Precision Engineering*, 45, pp. 387–395.

[32] Hao, B. and Lin, G. (2020). Additive manufacturing (3D printing): a review of materials, methods, applications and challenges. Presented at the IOP Conference Series: Materials Science and Engineering.

[33] Kalender, M., Bozkurt, Y., Ersoy, Y. and Salman, S. (2020). Product development by additive manufacturing and 3D printer technology in aerospace industry, *Journal of Aeronautics and Space Technologies*, 13, pp. 129–139.

[34] Tack, P., Victor, J., Gemmel, P. and Annemans, L. (2016). 3D-printing techniques in a medical setting: a systematic literature review, *BioMedical Engineering OnLine*, 15, p. 115.

[35] Tavassoli, S., Brandt, M., Ma, Q., Arenius, P., Kianian, B., Diegel, O., Mention, A.-L., Cole, I., Elghitany, A. and Pope, L. (2020). Adoption and diffusion of disruptive technologies: the case of additive manufacturing in medical technology industry in Australia, *Procedia Manufacturing*, 43, pp. 18–24.

[36] Bajaj, S. (2017). Automotive 3D printing market by component (technology, material, and services) and application (prototyping & tooling, R&D and innovation, and manufacturing complex products) — global opportunity analysis and industry forecast, 2017–2023.

[37] Sreehitha, V. (2017). Impact of 3D printing in automotive industries, *International Journal of Mechanical and Production Engineering*, 5, pp. 91–94.

[38] Wu, W. and Beng, T. S. (2014). 3D printing for marketing and advertisement industry. In *Proceedings of the 1st International Conference on Progress in Additive Manufacturing*, Research Publishing Services, pp. 303–308.

[39] Periard, D., Schaal, N., Schaal, M., Malone, E. and Lipson, H. (2007). Printing food. In *Proceedings of the 18th Solid Freeform Fabrication Symposium*, pp. 564–574.

[40] Hao, L., Mellor, S., Seaman, O., Henderson, J., Sewell, N. and Sloan, M. (2010). Material characterisation and process development for chocolate additive layer manufacturing, *Virtual and Physical Prototyping*, 5, pp. 57–64.

[41] Choc Edge Ltd. (2019). About Choc Edge. Retrieved from http://chocedge.com/about.html

[42] Buhr, S. (2015). Hershey's chocolate 3D printer whips up any sweet design you want. Retrieved from https://techcrunch.com/2015/09/16/likeachocolateselfie/

[43] Barry Callebaut. (2020). Barry Callebaut opens world's first 3D Printing Studio to craft unseen chocolate experiences. Retrieved from https://www.barry-callebaut.com/en/group/media/news-stories/barry-callebaut-opens-worlds-first-3d-printing-studio-craft-unseen

[44]  Mantihal, S., Prakash, S., Godoi, F. C. and Bhandari, B. (2017). Optimization of chocolate 3D printing by correlating thermal and flow properties with 3D structure modeling, *Innovative Food Science and Emerging Technologies*, 44, pp. 21–29.

[45]  Hao, L., Li, Y., Gong, P. and Xiong, W. (2019). Fundamentals of 3D food printing and applications, *Material, Process and Business Development for 3D Chocolate Printing*, pp. 207–255.

[46]  Lanaro, M., Forrestal, D. P., Scheurer, S., Slinger, D. J., Liao, S., Powell, S. K. and Woodruff, M. A. (2017). 3D printing complex chocolate objects: platform design, optimization and evaluation, *Journal of Food Engineering*, 215, pp. 13–22.

[47]  Karyappa, R. and Hashimoto, M. (2019). Chocolate-based ink three-dimensional printing (Ci3DP), *Scientific Reports*, 9, p. 14178.

[48]  Khemacheevakul, K., Wolodko, J., Nguyen, H. and Wismer, W. (2021). Temporal sensory perceptions of sugar-reduced 3D printed chocolates, *Foods*, 10.

[49]  Choc Edge. (2021). 3D chocolate printing guide. Retrieved from http://chocedge.com/faq.html

[50]  United Nations. (2017). World population prospects: the 2017 revision.

[51]  Impossible Foods. (2021). Retrieved from https://impossiblefoods.com/

[52]  Beyond Meat. (2021). Retrieved from https://www.beyondmeat.com/

[53]  Good Food Institute. (2021). US retail market data for the plant-based industry. Retrieved from https://gfi.org/marketresearch/

[54]  Adams, C., Maluf, I. T., Meilhac, L., Ramirez, M. and Paula, R. U. d. (2019). A winning growth formula for dairy. Retrieved from https://www.mckinsey.com/industries/consumer-packaged-goods/our-insights/a-winning-growth-formula-for-dairy

[55]  Bashi, Z., McCullough, R., Ong, L. and Ramirez, M. (2019). Alternative proteins — the race for market share is on. McKinsey & Company.

[56]  Smith, A., Shah, S. and Blaustein-Rejto, D. (2021). The case for public investment in alternative proteins. Breakthrough Institute. Retrieved from https://s3.us-east-2.amazonaws.com/uploads.thebreakthrough.org/Alternative-Protein-Report_v6.pdf

[57]  Ritchie, H. (2019). Half of the world's habitable land is used for agriculture. Retrieved from https://ourworldindata.org/global-land-for-agriculture

[58]  Bigelow, D. (2017). A primer on land use in the United States. Economic Research Service, US Department of Agriculture. Retrieved from https://www.ers.usda.gov/amber-waves/2017/december/a-primer-on-land-use-in-the-united-states/

[59]  Bonnedahl, K. J. and Heikkurinen, P. (2018). The case for strong sustainability. In *Strongly Sustainable Societies: Organising Human Activities on a Hot and Full Earth* (Routledge).

[60] Machovina, B., Feeley, K. J. and Ripple, W. J. (2015). Biodiversity conservation: the key is reducing meat consumption, *Science of the Total Environment*, 536, pp. 419–431.

[61] Godfray, H. C. J., Aveyard, P., Garnett, T., Hall, J. W., Key, T. J., Lorimer, J., Pierrehumbert, R. T., Scarborough, P., Springmann, M. and Jebb, S. A. (2018). Meat consumption, health, and the environment, *Science*, 361.

[62] Hu, F. B., Otis, B. O. and McCarthy, G. (2019). Can plant-based meat alternatives be part of a healthy and sustainable diet? *The Journal of the American Medical Association*, 322, pp. 547–1548.

[63] Sharma, S., Thind, S. S. and Kaur, A. (2015). *In vitro* meat production system: why and how? *Journal of Food Science and Technology*, 52(12), pp. 7599–7607.

[64] Ismail, B. P., Senaratne-Lenagala, L., Stube, A. and Brackenridge, A. (2020). Protein demand: review of plant and animal proteins used in alternative protein product development and production, *Animal Frontiers*, 10, pp. 53–63.

[65] Zion Market Research. (2019). Zion market report, global plant based meat market will reach USD 21.23 billion by 2025. Retrieved from https://www.zionmarketresearch.com/news/plant-based-meat-market

[66] Godfray, H. C. J. (2019). Meat: the future series — alternative proteins. White Paper, World Economic Forum. Retrieved from https://www3.weforum.org/docs/WEF_White_Paper_Alternative_Proteins.pdf

[67] He, J., Evans, N. M., Liu, H. and Shao, S. (2020). A review of research on plant-based meat alternatives: driving forces, history, manufacturing, and consumer attitudes, *Comprehensive Reviews in Food Science and Food Safety*, 19, pp. 2639–2656.

[68] Shurtleff, W. and Aoyagi, A. (2013). *History of Tofu and Tofu Products (965 CE to 2013)* (SoyInfo Center).

[69] Gaydhane, M. K., Mahanta, U., Sharma, C. S., Khandelwal, M. and Ramakrishna, S. (2018). Cultured meat: state of the art and future, *Biomanufacturing Reviews*, 3.

[70] Bohrer, B. M. (2019). An investigation of the formulation and nutritional composition of modern meat analogue products, *Food Science and Human Wellness*, 8, pp. 320–329.

[71] Hoogenkamp, H. W. (2004). *Soy Protein and Formulated Meat Products* (CABI Publishing).

[72] Finnigan, T. J., Wall, B. T., Wilde, P. J., Stephens, F. B., Taylor, S. L. and Freedman, M. R. (2019). Mycoprotein: the future of nutritious nonmeat protein, a symposium review, *Current Developments in Nutrition*, pp. 1–5.

[73] Hannah, R., Jim, L. and David, R. (2017). 3f bio: halving the cost of mycoprotein through integrated fermentation processes, *Industrial Biotechnology*, 13, pp. 29–31.

[74] Souza Filho, P. F., Andersson, D., Ferreira, J. A. and Taherzadeh, M. J. (2019). Mycoprotein: environmental impact and health aspects, *World Journal of Microbiology and Biotechnology*, 35, p. 147.

[75] Karkos, P. D., Leong, S. C., Karkos, C. D., Sivaji, N. and Assimakopoulos, D. A. (2011). Spirulina in clinical practice: evidence-based human applications, *Evidence-Based Complementary and Alternative Medicine*, 2011, p. 531053.

[76] Lafarga, T. (2019). Effect of microalgal biomass incorporation into foods: nutritional and sensorial attributes of the end products, *Algal Research*, 41.

[77] Yen, A. L. (2009). Edible insects: traditional knowledge or western phobia? *Entomological Research*, 39, pp. 289–298.

[78] Raheem, D., Carrascosa, C., Oluwole, O. B., Nieuwland, M., Saraiva, A., Millán, R. and Raposo, A. (2018). Traditional consumption of and rearing edible insects in Africa, Asia and Europe, *Critical Reviews in Food Science and Nutrition*, 59, pp. 2169–2188.

[79] Durst, P. B. and Hanboonsong, Y. (2015). Small-scale production of edible insects for enhanced food security and rural livelihoods: experience from Thailand and Lao People's Democratic Republic, *Journal of Insects as Food and Feed*, 1, pp. 25–31.

[80] Bessa, L. W., Pieterse, E., Marais, J. and Hoffman, L. C. (2020). Why for feed and not for human consumption? The black soldier fly larvae, *Comprehensive Reviews in Food Science and Food Safety*, 19, pp. 2747–2763.

[81] Morales-Ramos, J. A., Rojas, M. G. and Dossey, A. T. (2018). Age-dependent food utilization of *Acheta domesticus* (Orthoptera: Gryllidae) in small groups at two temperatures, *Journal of Insects as Food and Feed*, 4(1), pp. 51–60.

[82] van Huis, A. (2013). Potential of insects as food and feed in assuring food security, *Annual Review of Entomology*, 58, pp. 563–583.

[83] Guthman, J. and Biltekoff, C. (2020). Magical disruption? Alternative protein and the promise of de-materialization, *Environment and Planning E: Nature and Space*. 10.1177/2514848620963125.

[84] Tso, R., Lim, A. J. and Forde, C. G. (2020). A critical appraisal of the evidence supporting consumer motivations for alternative proteins, *Foods*, 10, p. 24.

[85] Ong, S., Choudhury, D. and WinNaing, M. (2020). Cell-based meat: current ambiguities with nomenclature, *Trends in Food Science & Technology*, 102, pp. 223–231.

[86] Rolland, N. C. M., Markus, C. R. and Post, M. J. (2020). The effect of information content on acceptance of cultured meat in a tasting context, *PLOS One*, 15, pp. 1–10.

[87]  Meatech. (2021). Our cultivated chicken fat: extraordinarily ordinary. Meatech Blog. Retrieved from https://meatech3d.com/our-cultivated-chicken-fat-extraordinarily-ordinary/

[88]  Bennett, D. (2021). Diners enjoy world's first restaurant meal made from lab-grown meat, *Science Focus*. Retrieved from https://www.sciencefocus.com/news/diners-enjoy-worlds-first-restaurant-meal-made-from-lab-grown-meat/

[89]  Chen, J., Mu, T., Goffin, D., Blecker, C., Richard, G., Richel, A. and Haubruge, E. (2019). Application of soy protein isolate and hydrocolloids based mixtures as promising food material in 3D food printing, *Journal of Food Engineering*, 261, pp. 76–86.

[90]  Scheele, S. C., Hoque, M. N., Christopher, G. and Egan, P. F. (2021). Printability and fidelity of protein-enriched 3D printed foods: a case study using cricket and pea protein powder. In *Proceedings of the ASME 2021 International Design Engineering Technical Conferences and Computers and Information in Engineering Conference*.

[91]  Severini, C., A. DerossiI, Ricci, I., Caporizzi, R. and A. Fiore (2018). Printing a blend of fruit and vegetables: new advances on critical variables and shelf life of 3D edible objects, *Journal of Food Engineering*, 220, pp. 89–100.

[92]  Uribe-Wandurraga, Z. N., Igual, M., Reino-Moyón, J., García-Segovia, P. and Martínez-Monzó, J. (2021). Effect of microalgae (*Arthrospira platensis* and *Chlorella vulgaris*) addition on 3D printed cookies, *Food Biophysics*, 16, pp. 27–39.

[93]  Carrington, D. (2020). "Most realistic" plant-based steak revealed. Retrieved from https://www.theguardian.com/food/2020/jan/10/most-realistic-plant-based-steak-revealed

[94]  Reuters. (2021). Israel's redefine meat serves up plant-based whole cuts of beef. Retrieved from https://www.ynetnews.com/business/article/sy8t2jbdt

[95]  Redefine Meat. (2021). Products. Retrieved from https://www.redefinemeat.com/products/

[96]  Keerthana, K., Anukiruthika, T., Moses, J. A. and Anandharamakrishnan, C. (2020). Development of fiber-enriched 3D printed snacks from alternative foods: a study on button mushroom, *Journal of Food Engineering*, 287.

[97]  Dick, A., Bhandari, B., Dong, X. and Prakash, S. (2020). Feasibility study of hydrocolloid incorporated 3D printed pork as dysphagia food, *Food Hydrocolloids*, 107.

[98]  Yang, G., Tao, Y., Wang, P., Xu, X. and Zhu, X. (2022). Optimizing 3D printing of chicken meat by response surface methodology and genetic algorithm: feasibility study of 3D printed chicken product, *LWT*, 154.

[99]   Carlotta V. (2019). Aleph Farms and 3D Bioprinting Solutions collaborate to create slaughter-free meat. Retrieved from https://www.3dnatives.com/en/aleph-farms-3d-printed-meat-181120194/#!

[100]   Karmaus, A. L. and Jones, W. (2021). Future foods symposium on alternative proteins: workshop proceedings, *Trends in Food Science & Technology*, 107, pp. 124–129.

[101]   Enjoli, A. (2020). Canada just invested $100 million into vegan meat. Retrieved from https://www.livekindly.co/canada-invested-100-million-vegan-meat/

[102]   Liu, V. (2021). A*Star launches new research institute to drive innovation to tackle food security challenges. *The Straits Times*. Retrieved from https://www.straitstimes.com/singapore/astar-launches-new-research-institute-to-drive-innovation-to-tackle-food-security

[103]   The Business Times. (2020). Impossible Foods launches plant-based sausage product in HK. Retrieved from https://www.businesstimes.com.sg/consumer/impossible-foods-launches-plant-based-sausage-product-in-hk

[104]   Hanboonsong, Y., Jamjanya, T. and Durst, P. (2013). Six-legged livestock: edible insect farming, collecting and marketing in Thailand. Food and Agriculture Organization. Retrieved from https://www.fao.org/publications/card/en/c/76e0a383-3ca0-5a7c-8d5a-b3a4262b857f/

[105]   Global Market Insights Inc. (2015). Edible insects market size set to exceed USD 520 million by 2023, with over 40% growth from 2016 to 2023. Global Market Insights Inc.

[106]   Petetin, L. (2014). Frankenburgers, risks and approval, *European Journal of Risk Regulation*, 5, pp. 168–186.

[107]   Dobermann, D., Swift, J. A. and Field, L. M. (2017). Opportunities and hurdles of edible insects for food and feed, *Nutrition Bulletin*, 42, pp. 293–308.

[108]   BBC. (2017). EU court bans dairy-style names for soya and tofu. *BBC News*. Retrieved from https://www.bbc.com/news/business-40274645

[109]   Sexton, A. E., Garnett, T. and Lorimer, J. (2019). Framing the future of food: the contested promises of alternative proteins, *Environment and Planning E: Nature and Space*, 2, pp. 47–72.

[110]   Piazza, J., Ruby, M. B., Loughnan, S., Luong, M., Kulik, J., Watkins, H. M. and Seigerman, M. (2015). Rationalizing meat consumption. The 4Ns, *Appetite*, 91, pp. 114–128.

[111]   Jonsson, E. (2016). Benevolent technotopias and hitherto unimaginable meats: tracing the promises of *in vitro* meat, *Social Studies of Science*, 46, pp. 725–748.

[112]   Gungor-Ozkerim, P. S., Inci, I., Zhang, Y. S., Khademhosseini, A. and Dokmeci, M. R. (2018). Bioinks for 3D bioprinting: an overview, *Biomaterials Science*, 6, pp. 915–946.

[113] Rumpold, B. A., Fröhling, A., Reineke, K., Knorr, D., Boguslawski, S., Ehlbeck, J. and Schlüter, O. (2014). Comparison of volumetric and surface decontamination techniques for innovative processing of mealworm larvae (*Tenebrio molitor*), *Innovative Food Science and Emerging Technologies*, 26, pp. 232–241.

[114] Srinroch, C., Srisomsap, C., Chokchaichamnankit, D., Punyarit, P. and Phiriyangkul, P. (2015). Identification of novel allergen in edible insect, *Gryllus bimaculatus*, and its cross-reactivity with *Macrobrachium* spp. allergens, *Food Chemistry*, 184, pp. 160–166.

[115] Jensen-Jarolim, E., Pali-Schöll, I., Jensen, S. A. F., Robibaro, B. and Kinaciyan, T. (2015). Caution: reptile pets shuttle grasshopper allergy and asthma into homes, *World Allergy Organization Journal*, 8, pp. 1–5.

[116] Verbeke, W. (2015). Profiling consumers who are ready to adopt insects as a meat substitute in a Western society, *Food Quality and Preference*, 39, pp. 147–155.

[117] Hartmann, C. and Siegrist, M. (2017). Insects as food: perception and acceptance. Findings from current research, *Ernahrungs Umschau*, 64, pp. 44–50.

[118] Root, A. (2019). Beyond meat and the limits of a Tesla strategy at the grocery store. Barron's. Retrieved from https://www.barrons.com/articles/beyond-meat-price-comparison-51559339044

[119] Gassmann, O., Frankenberger, K. and Csik, M. (2015). *The Business Model Navigator: 55 Models That Will Revolutionise Your Business* (Pearson Education Limited, London, UK).

[120] Rogers, H., Baricz, N. and Pawar, K. S. (2016). 3D printing services: classification, supply chain implications and research agenda, *International Journal of Physical Distribution & Logistics Management*, 46, pp. 886–907.

[121] Holzmann, P., Breitenecker, R. J., Schwarz, E. J. and Gregori, P. (2020). Business model design for novel technologies in nascent industries: an investigation of 3D printing service providers, *Technological Forecasting and Social Change*, 159.

[122] Jayaprakash, S., Paasi, J., Pennanen, K., Flores Ituarte, I., Lille, M., Partanen, J. and Sozer, N. (2020). Techno-economic prospects and desirability of 3D food printing: perspectives of industrial experts, researchers and consumers, *Foods*, 9.

[123] Osterwalder, A., Pigneur, Y., Smith, G. B. A. and Papadakos, T. (2014). *Value Proposition Design: How to Create Products and Services Customers Want* (John Wiley & Sons).

[124] Osterwalder, A. (2022). Building blocks of business model canvas. Retrieved from https://www.strategyzer.com/business-model-canvas/building-blocks

[125]　Taylor, J., Ahmed, I. A. M., Al-Juhaimi, F. Y. and Bekhit, A. E.-D. A. (2020). Consumers' perceptions and sensory properties of beef patty analogues, *Foods*, 9, p. 63.

# Problems

1.　What is a value proposition canvas (VPC)? Provide a hypothetical VPC for any chosen consumer segment for 3DFP.
2.　What are the building blocks of a business model canvas (BMC)? Draw a BMC for soft foods targeting nursing homes housing people with swallowing difficulties.
3.　What are the challenges in the acceptance of alternative proteins by consumers?
4.　Discuss a few companies engaged in alternative proteins' market space.
5.　What are the advantages and disadvantages of chocolate 3D printing? Give examples of companies developing chocolate printers.

# Chapter 9

# Emerging Technologies and Future Outlook of 3D Food Printing

## 9.1 Sustainability and 3D Food Printing

Sustainability is built on this basic principle: "Everything needed for existence and well-being is either directly or indirectly reliant on the natural environment". Sustainability is defined as "the combination of environmental health, social fairness, and economic vitality in order to produce vibrant, healthy, diversified, and resilient communities for this generation and generations to come," according to the University of California, Los Angeles (UCLA) Sustainability Committee's charter. Sustainability demands a systems perception and an appreciation of complexity since it recognizes how these challenges are interconnected. Environmental human and economic health and vitality are all guided by sustainable practices. Sustainability assumes that resources are finite and should be used carefully and intelligently, with a long-term perspective on priorities and implications of resource usage. Hence, practicing sustainability entails establishing and preserving conditions that allow the environment and living organisms to coexist in productive harmony for the benefit of current and future generations.

In the context of making 3D-printed alternative meat that mirrors the physical features of real meat products, it is crucial to think about sustainability in order to determine if this process can benefit humanity in the long run. However, numerous consideration factors, including energy use, trash creation, and pollution, influence sustainability [1]. For this purpose, a structure for competitive environmental strategies has been proposed.

This structure comprises affordable tactics such as minimizing investment and operating expenses and diversifying strategies such as generating sustainability-based value plans. Additionally, efforts might be directed toward capable governance processes/activities and the design of superior consumer products [2]. However, previous research found that 3D printing can be cost-effective in low-volume, high-value industries like medical and aerospace component manufacturing [3]. The equipment (machines), materials, pre- and post-processing procedures, and skilled personnel all influence the cost of manufacturing in 3D printing [1]. In high-value sectors, requiring fewer inputs and outputs can result in decreased energy consumption and $CO_2$ emissions [3]. In other industries, such as 3D-printed high-speed gears, the lack of a beneficial influence on the environment has been documented. Nonetheless, 3D food printing has yet to be subjected to a life cycle assessment (LCA). The LCA is a method for analyzing a product's environmental impact across its full life cycle. LCA incorporates 3D printing and post-processing stages independent of the industrial industry [1]. Some of the factors to be considered in the LCA for 3D food printing include: (1) energy consumption, (2) carbon footprint, and (3) sidestream valorization of by-products within the food value chain.

### 9.1.1 *Energy Consumption*

Even though the technology is still in the nascent phases of research, 3D food printing has been reported to offer a number of advantages, including decreased raw material and energy costs, as well as reduced waste [3]. The quantity of energy consumed by 3D food printing, particularly electrical energy, has a significant influence on the environment. This type of energy is mostly needed to pre-heat the machine and print and cool the finished product. While energy consumption is comparable for samples made using various 3D printing technologies (e.g., additive manufacturing or injection molding), it differs when large-scale production is involved [1]. Food manufacturing is a high capacity (number/type of goods) industry that necessitates a wide range of components in huge amounts. As a result, 3D printers' economic viability is determined by their capacity (speed) to print huge quantities of things in a short amount of time. Hence, the design of 3D printers, aimed at mass producing food, would be more energy-efficient, improving the sustainability of food manufacturing. In addition to the food printer itself, the type of materials used has an impact on energy use. Energy is consumed more by materials that demand higher temperatures and vice versa. Tuna-shaped constructs, for example, are 3D printed using tuna purée and a pressure-controlled extrusion machine at 20 °C [4]. In addition, post-processing treatments such as oven drying for 20–30 min at 100 °C are necessary for 3D-printed, fibre-rich goods [5]. Similarly, larger-mass goods need a greater quantity of energy. At room temperature, hummus or other printed materials are often extruded [6]. Many food products can also be printed at room temperature. Depending on the variety of circumstances, determining energy consumption is a difficult task. In addition, there is a paucity of LCA studies on the 3D printing of numerous types of food items, which warrants further investigation to demonstrate the sustainability of 3D food printing.

The input of raw materials and resources for 3D printing of meat alternatives (manufacturing, distribution, and storage) also has an environmental impact. For example, LCA studies on lab-based muscle cell culturing have found that, despite the lower land and input requirements, the total energy needed for the procedure is much higher than the production of cattle meat itself [7, 8]. Cellular cultivation of meat in the laboratory has been linked to the use of power and fuel. Although lab-grown meat's global warming potential was lower than that of bovine meat in one research, it was not found to be lower than that of real meat. Even though several assumptions were made in this work, large-scale cell culture production at the industrial level was shown to be energy-intensive [7]. Hence, given that it is debatable if the production of cultivated meat is more sustainable than cattle meat production, the additional input of energy to 3D print cultivated meat to create a steak-like physical appearance and texture will likely be a less sustainable approach.

Farming or collecting insects from pastures has been observed to have a lesser environmental effect [9]. While some research suggests that insect-derived food is more environmentally friendly than cattle meat, data on their impact on the environment is scarce. Although the energy usage for worms (Tenebrio molitor or Zophobas morio) was equivalent to porcine meat (per kg of edible protein), it was higher than chicken but lower than bovine meat, according to an LCA-based analysis. The greater energy usage was mostly due to the need for heating in order for the worms to develop. To solve this problem, large-sized larvae were added to the mix, along with smaller larvae, to produce more flesh metabolic heat. The increased rate of breeding, effective feed transformation, and absence of methane output are all important benefits of insect farming over animal farming [10]. Similarly, it is thought that repurposing meat by-products and trash has a good impact on the environment. Bovine slaughterhouses produce the most solid waste, followed by sheep slaughterhouses, and then pig slaughterhouses (ranging from 27% to 4% of the animal weight), according to a previous assessment. As a result, it is critical to repurpose these wastes, which would otherwise be disposed to landfills [11]. Commercial and residential unwanted fish remains are frequently dumped in landfills or dumped in the sea [12]. Similarly, unutilized plant/agricultural remains may be used for the creation of 3D meat alternatives in a more environmentally friendly manner. Extraction of useful biomaterials from these forms of animal/seafood industrial wastes, on the other hand, is a time-consuming and energy-intensive procedure. In this context, a recent study used non-thermal pulsed electric fields or mechanical pressing to recover proteins from discarded poultry meat. When non-thermal therapy using a low voltage long pulse was applied on discarded muscles of chicken breast, the extraction of liquid was observed to be enhanced. Without the use of any chemicals, extraction of discarded chicken biomass was accomplished with a 12% yield of liquid fraction. Some of the observed advantages [10] include increased revenue, decreased waste, and reduced waste-related environmental impact. 3D food printing also produces wastes, much like other food production systems, which must be removed, minimized, and repurposed [1]. However, knowledge of the wastes created and their environmental impact is lacking at this current point in time.

### 9.1.2 *Carbon Footprint*

The impact of additive manufacturing on air pollution has gotten a lot of attention lately because of its rising application in numerous businesses and academia. Benchtop 3D printers were suggested in this area. Consumers will be able to print their own food, transforming the transportation industry. Furthermore, printing items or components closer to where customers are might have a positive impact on the environment. Reduced end-product transit (at least over large distances) is predicted to reduce $CO_2$ emissions, hence improving the air condition. The acquisition of input materials from various sources (bioinks, animal wastes or insects) might result in greenhouse gas emissions. Greenhouse gas emissions during black soldier fly breeding were measured in recent research. Black soldier fly rearing, a viable dietary alternative, resulted in direct greenhouse gas emissions of about 17+/–8.6 g $CO_2$eq per kilogram of dry larval growth. The findings revealed that carbon and energy efficiency improvements are required [13]. However, the kind of input materials has an effect on greenhouse gas emissions from 3D-printed food products. Food goods, for instance, may be 3D printed with low-temperature bio-polymers. Biopolymer-derived filaments are also non-hazardous, eco-friendly, and considerably inexpensive [1]. Printers used for 3D food printing, on the other hand, must be tested in a clean room for emissions of particles (submicron to nano-sized) and gases. Furthermore, 3D food printing uses electricity, which is likely that it has an indirect impact by reducing $CO_2$ emissions, in contrast to standard animal meat, which produces greenhouse gas emissions ($CH_4$, $N_2O$, and $CO_2$). Based on a sustainability assessment, 3D printing technology can reduce $CO_2$ emissions by 130.5–525.5 Mt by 2025 [3]. Furthermore, $CO_2$ emissions induced by the consumption of electricity can be reduced by decarbonization [14]. Another benefit of 3D food printing is lower labour costs, which might be even lower for 3D food manufacturing due to the automation process and does not need highly skilled personnel [1].

### 9.1.3 *Sidestream Valorization of By-products in the Food Value Chain*

As the worldwide need for food rises, so will the demand for food industry by-products. The necessity for efficient side-stream management is critical for a circular food value chain and the valorization of underused biomass, both economically and environmentally. Vegetable-based sidestreams provide a diverse source of renewable biomaterials, with unique features within a wide range of food items. Recognizing food production waste as valuable raw materials is the first step toward resource sustainability and the conversion of agri-food waste into higher-value goods. Hence, the transformation of unutilized primary food products or secondary by-products into higher-value items that provide back to the food supply chain is known as food waste valorization. This adds to the circular economy strategy, in which usable materials that were previously considered waste are upcycled back into the supply chain to generate new goods or products. It is one of the food waste upcycling solutions that can help end the loop on food waste. Food loss and waste is estimated to reach over 1.3 billion tonnes per year worldwide. As a result, reducing waste and loss at various stages of the food value chain is critical [15], which is applicable to the 3D printing of meat alternatives. Real-time instruments or monitors evaluate the quantity of nutrients and wastes in animal cell culture for meat, lowering operating expenses [16]. Plant waste biomass is another possible supply of protein from plant-based meat replacements. Protein obtained from discarded peanut biomass was utilized to make meat replacements via high-moisture extrusion in one study [17]. Plant wastes may be utilized to produce a huge range of proteins and fibres that can be used to make meat alternatives. Waste utilization is both economical and environmentally good [18]. Indeed, bioactive compounds that are rich in polyphenolic compounds that have antioxidant activities can be extracted from unutilized plant biomass, and these products may be suitable as food additives [19, 20]. Furthermore, once these plant or food products are processed for extraction, they can be stored for a longer period of time. Hence, this may also potentially increase the shelf-life of food, particularly those that are unutilized or wasted due to cosmetic filtering [21]. The extraction method that uses the green extraction method has also been shown to be more sustainable and environmental friendly compared to traditional extraction techniques [22]. Meat derivatives and discarded by-products can also be used as key components in the creation of items that resemble typical meat dishes like patties or sausages. For this purpose, 3D printing appears to be a potential method for reducing food waste. For example, okara, which is a waste by-product of soybean, has been formulated into food ink that is suitable for 3D food printing, potentially upcycling the food waste into a novel food product [23]. Even in the procedure of 3D printing, waste reduction is critical. Highly efficient and productive printers can help with this. Nevertheless, efficiency is based on the optimization of all activities in a food value chain, including activity rationalization, material flow, and logistic expenses [24]. When 3D printing is upscaled and marketed, this becomes critical. In the instance of bioprinting, the creation of generic/ universal elements or scaffolds, as well as improvements in cell procurement and construction logistics, would reduce manufacturing costs and ensure long-term viability [25].

## 9.1.4 *Future Outlook*

Food security and sustainability are recognized as a big worldwide concern, especially in the era of the global pandemic (COVID-19), where the food supply chain is massively disrupted. Food security describes the accessibility of food as well as people's capacity to obtain it. It is also described as "having physical, social, and economic access to adequate, safe, and nutritious food that fits one's food choices and dietary needs for an active and healthy life", according to the United Nations' Committee on World Food Security. Food quality, nutritional content, climate change, environmental implications, and having enough food to feed everyone are just a few of the problems associated with food security and sustainability. Thus, the rising worries about global food security and sustainability have prompted the development of innovative technology, such as 3D food printing. One advantage of 3D food printing is the capacity to manufacture 3D structures with complicated geometries, complex textures, increased nutrition, and realistic tastes. While 3D printing has been widely considered a suitable technology for food production to combat food insecurity and promote sustainability, there are several challenges due to the nascent stage of the technology.

For example, livestock meat production is recognized to be associated with extensive land use and resources, as well as a huge carbon footprint. In order to give long-term advantages, academics have bolstered their endeavours to replace normal meat with other options. Although 3D printing has been claimed to minimize waste and reduce energy usage, its environmental impact is yet to be investigated. Materials obtained mostly from meat by-products, insects, plants, and cultured cells are investigated for 3D printing of meat analogs in terms of scientific feasibility, environmental impact, and consumer acceptability. While skeletal muscle bioprinting may be done with a variety of materials, meat analog creation requires only food-safe ingredients. Using generic/universal components or scaffolds, as well as streamlining cell procurement and manufacturing logistics, can help to increase sustainability. Some start-ups declared to have printed meat analogs using a range of plant materials. Plant by-products or unutilized components can be used to increase the sustainability of 3D food printing. Proteins (soluble/insoluble), lipids, and fibres can all be generated in addition to powders created by milling. These compounds have the potential to be employed in meat replacements at appropriate doses and combinations. Meat by-products or discarded parts may be both cost-effective and environmentally benign, making them sustainable. Wastes from the fish industry, such as skin, have been found to provide cost-saving benefits. Because the intake of insects and animal by-products encounters perception-related issues, 3D food printing technology is highly suited for their application. The use of lab-grown cultures and animal by-products in 3D printing meat analogs might be complicated with regulatory issues. Energy usage is linked to environmental effect, which is linked to the kind of printers and materials being used, as well as the bulk of the final products. Although animals and enormous land areas are not necessary for cell cultivation or insect farming, manufacturing these meat substitutes at a large scale is nevertheless energy demanding. While 3D food printing has several drawbacks, including 3D design requirements, modest accuracy, and low efficiency, it has one key advantage — the food production process may be automated and less reliant on skilled workers. The use of wastes in 3D printing can increase the sustainability of plant-derived or animal-based food supply systems. Printer advancements, logical activity organization, material flow optimization, and logistic expenses may all help 3D printers and the printing process to be more efficient. In the present situation, 3D food printing remains in its early phases, and its long-term ramifications for large-scale meat replacement manufacturing and proper LCA-based studies must be determined before concluding that 3D food printing technology can support food security and sustainability.

## 9.2  4D Food Printing

### 9.2.1  *Current Status*

3D-printed foods are static, but eating is a dynamic interaction between humans and food. To introduce dynamics into 3D-printed food, the concept of 4D printing has been brought into food printing. 4D printing is rather vaguely defined; it generally refers to the extension of 3D printing into the time domain, which involves the pre-defined reconfiguration of 3D-printed components over time after printing [26]. Since its emergence from the MIT media lab, 4D printing has attracted much attention from the additive manufacturing (AM) community and developed into a field of its own. The 4D reconfiguration is mainly in the form of shape transformation at the moment, and quite a number of mechanisms have been derived to stimulate these transformation processes. For instance, the 4D transformation can be achieved by heating. Parts printed with shape-memory materials can be mechanically programmed into a secondary shape and reconfigured back to their original shape by heating the part above the glass transition temperature ($T_g$) of the material [27, 28]. One may also control the shape of 3D-printed parts by differential swelling to accomplish 4D printing. Typically, a bilayer structure is tailored to induce a desired curvature upon differential swelling of the two layers while their interface remains fully in contact. This bilayer structure could also be implemented only at the joints to accomplish origami 4D printing [29]. The aforementioned examples are one-way 4D printings triggered by only one external stimulation and require manual programming after each cycle. In contrast, two-way 4D printing transforms the parts between two permanent shapes in a reversible manner with two different external stimulations. For instance, differential swelling and heating are both used to drive the forward and backward shape transformation of 4D-printed polymers [28]. When soaked in ethanol, a bilayer slab would bend towards the side that swelled less. The shape of the curved slab is then fixed by quenching it in ice-cold water. At a later stage, the curved slab could return to its original shape upon heating up in the air, which reduces the swelling by promoting solvent evaporation and heating the materials above their Tg for stress relaxation. Two-way 4D printing can also be accomplished by using magnetic stimulation. By adding ferrous magnetic nanoparticles to silicone rubber, a magnetically responsive soft part is 4D printed [30]. Once permanently magnetized, the shape of the parts can be reversibly controlled by the applied magnetic field. If readers are interested in 4D printing, a more comprehensive discussion can be found in these review articles [31–35].

4D food printing, on the other hand, is still in its infancy. All the studies so far on 4D food printing, in the authors' opinion, are either stretching the concept of food or 4D printing. One of the most cited works on 4D food printing is the shape-morphing "pasta" created by the MIT media lab [36]. The body of the "pasta" was made of gelatin that naturally formed a dense top layer and a porous bottom layer upon cooling, which led to differential swelling due to different water absorption capacities at the top and bottom. In addition, cellulose derivatives were printed into semi-rigid strips on the top surface of the gelatin to regulate the water absorption rate and control the bending direction. As a result, when soaked in water, the flat "pasta" would transform into various curled 3D shapes depending on the configuration of the cellulose strips. The same concept was used to create self-wrapping sushi and self-chopping noodles. However, gelatin and cellulose are not what people commonly have in mind for pasta or sushi, and the gelatin body of these food items is not made by printing. Although the cellulose strips can be printed onto the flat gelatin "dough" to achieve the desired pattern, it is essentially a 2.5D printing process that can be easily replaced by traditional printing techniques such as screen printing, as suggested by the authors. Nonetheless, the gelatin-cellulose system is likely the most successful and well-studied combination for 4D food printing so far. Another highly but wrongly cited work on 4D food printing was also from the MIT media lab [37]. In this work, the authors developed a shape-transformation mechanism for traditional flour-based food. By making a bilayer sheeted dough out of two types of flour with different expansion rates and imprinting groves on the sheeted flour dough, one can regulate the hydration/dehydration rate in the bilayer; hence, the bending angle of the flour sheet. The authors demonstrated making self-wrapping cannoli, self-wrapping tacos, and self-folding cookies based on this mechanism. In addition, they also presented scenarios wherein the shape-changing food could be used to deliver information, such as conveying a love message via a shape-changing pasta that transforms from a straight line into a heart shape upon heating. Admittedly, the accurate, predicted shape transformation of these real pastas is impressive; however, these are not created by 3D printing. The flour dough used in this study is too viscous to be 3D printed by extrusion, and the imprinting of grooves on the sheeted dough can hardly be considered printing. Therefore, this work should not be considered 4D food printing.

In fact, limited by the physical and chemical properties of food inks, which are typically in the form of purée, many stimulation mechanisms used to trigger the shape reconfiguration in typical 4D printing do not apply to 4D food printing. As a result, assistive materials, which are often non-edible, are employed to initiate a strong mechanical response for shape transformation in 4D food printing. Therefore, "assisted 4D food printing" may be a more appropriate term to describe these techniques. For instance, He *et al.* [38] accomplished 4D food printing by extruding food inks made of purple sweet potato onto a piece of food-grade thin PET/PE plastic sheet to form a bilayer structure. The dehydration of the food ink by microwave heating led to differential shrinking of the food-plastic bilayer, while the rough surface of the plastic sheet provided a strong adhesion at the interface. Consequently, the printed food ink, together with the plastic sheet, curled into a different configuration, and the direction of curling could be adjusted by the orientation of the infill strips. In this work, the non-edible plastic sheet was essential for the 4D food transformation. Another example of assisted 4D food printing is the award-winning Neo Fruit (Figure 9.1) [39]. Neo Fruit was a collection of artificially designed edible fruit artwork. The shape of the Neo Fruits was

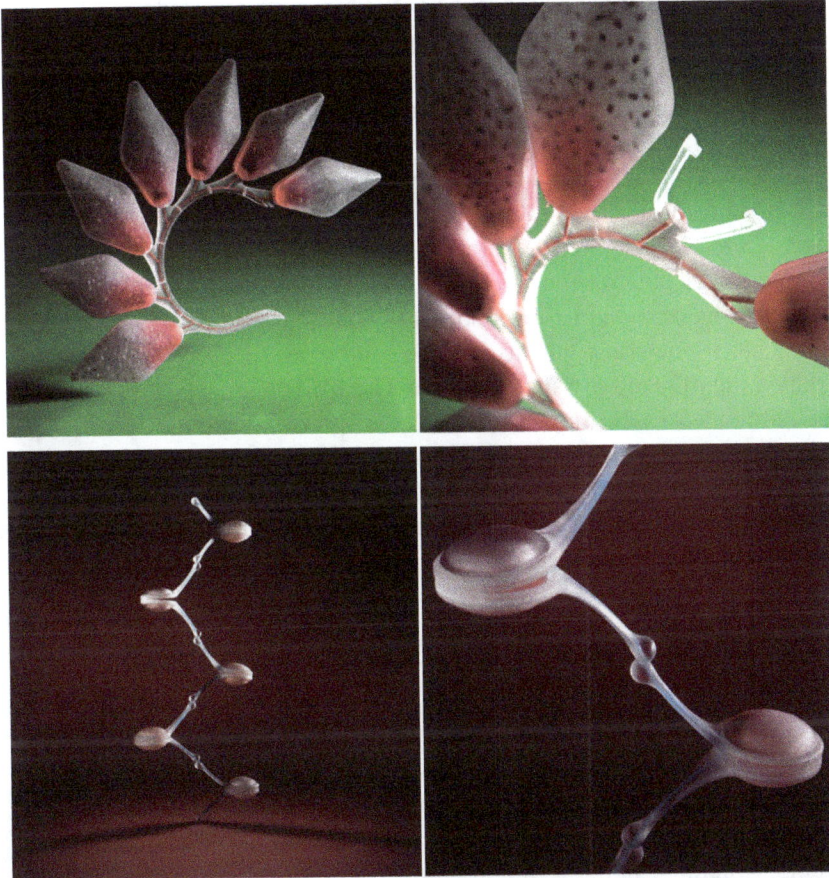

Figure 9.1. Neo Fruit created by Meydan Levy. Credit: Designer: Meydan Levy in Bezalel Academy of Art and Design in 2018. Academic art director: Dov Ganchrow. Photography & photo editing: Bogdan Sokol & Shay Maman. Video shooting & video editing: Bogdan Sokol & Shay Maman. Writing, production & video directing: Esti Baranets. Actress: Emilya Galdon. Makeup & styling: Alona Balona.

set by 3D-printed cellulose peels with embedded microfluidic channels. Fruit purées were injected into the 3D-printed peels to stimulate the 4D transformation of the Neo Fruits to their final forms. The design of Neo Fruits, according to its creator Meydan Levy, "presents an aspiration, a vision, derived of curiosity and thought, a bridge between industry, nature, and human".

The 4D transformation is not limited to shape morphing. The reconfiguration of other food properties over time, such as colour and flavour, is also regarded as 4D food printing. The 4D colour transformation of food is often facilitated by a pH-sensitive, natural pigment anthocyanin. By 3D printing and combining components [40] comprising acidic, neutral, and alkaline mashed potato with components comprising mashed purple sweet potato into a single part, molecules in various components diffuse into each other over time, resulting in a change in pH that in turn led to a change in the colour of anthocyanin present in the mashed purple sweet potato. A more direct way of alternating pH is to simply spread solutions [41] or even soak [42] the 3D-printed food containing pH-sensitive natural pigments into the solution of specified pH in order to induce the 4D colour transformation. The pH change in food could also be stimulated by heating. By heating 3D-printed lotus root powder gel containing curcumin that served as the colourimetric pH indicator and sodium bicarbonate as the pH modifier, the sodium bicarbonate in the food was decomposed into sodium carbonate, resulting in an increase in alkalinity, which caused the colour change.

### 9.2.2 *What is Next for 4D Food Printing?*

#### 9.2.2.1 *4D food printing for visual response*

At present, 4D food printing is more of a novelty than a technology breakthrough. It still lacks the wow factor that can get the public excited and investors interested. The future direction of 4D food printing, in the authors' humble opinion, hinges on a deep understanding of food and eating.

In the brief discussion above, we have reviewed mechanisms currently used to stimulate the 4D transformation of food appearance (shape and colour), which triggers a visual response. We can incorporate many new stimulation mechanisms by learning from culinary techniques. Lava cake is a simple dynamic food; its chocolate filling flows when the cake is heated and cut open. Chocolate is a versatile ingredient in 4D food printing as it can be easily shaped and melted. People have already taken advantage of this property to create dynamic desserts. For example, molten chocolate is poured over a hollow chocolate sphere that melts to reveal the dessert inside. An updated version of lava cake, tsunami cake, takes the "dynamic dessert" to another level. A tsunami cake is made by creating a plastic skirt on the top half of a hemispheric cake and filling the skirt with viscous ingredients such as ganache. Also known as a "pull-me-up" cake, the ganache inside the plastic skirt flows down and covers the entire cake when the consumer pulls the skirt up. This mechanism is now frequently used to create 3D cakes to resemble Disney princesses' dresses. These culinary techniques could certainly be adopted to stimulate the food transformation in 4D food printing.

A raising agent, such as baking powder or yeast, is another material we could explore for 4D food printing. Long associated with food, raising agents are used for better texture and taste in foods. Food with added raising agent often expands during baking or frying. If the distribution and degree of expansion of raising agents in food can be precisely controlled, it is an ideal way of modulating 4D food printing.

Almost all foods comprise biological components, and bio-stimulation presents another possible way of stimulating 4D food transformation. Animal muscles respond to nerve stimulation, and this mechanism would remain active for a relatively long period after the animal is dead. Readers may be familiar with Galvani's frog leg experiment. By stimulating its sciatic nerve, the dead frog's leg kicks as if it is still alive. Applying this concept to food, the muscle response by nerve stimulation has been used to dynamically present food that is to be consumed raw. For example, when pouring soy sauce onto a decapitated octopus, its legs start to wiggle as the headless octopus "dances" on the plate. In a legendary Chinese dish called the "yelling fish", care is taken not to damage the spinal cord and head of the fish during cooking. When presenting the food to customers, sizzling hot sauce is poured on top of the fish, and the fish's mouth starts to open and close as if it is yelling. These mechanisms could certainly be explored for assisted 4D food printing. For instance, biologically active food could serve as a substrate for food printing, and the purée ink printed on top of these substrates could move together with them when stimulated by heat or chemical stimulants in the sauce and seasonings. Even microorganisms could serve as bio-stimulants for 4D food printing. The microorganism is an essential component in our daily diet. Yeast is the key to fermentation, which is an indispensable procedure for bread and alcohol production. Many types of bacteria grow in our food and add unique texture and flavour to it. Natto cells (*Bacillus Subtilis*) secrete a large amount of peptidoglycan, with a large shrinking/expansion coefficient upon dehydration/hydration. When made into a bilayer structure with a layer of natto cells and plastic substrate, the differential swelling and shrinking in these two layers cause the bilayer structure to bend [43]. Therefore, natto cells may be used as an edible bioactuator for shape transformation in 4D food printing.

### 9.2.2.2   *4D food printing for other sensory response*

Food is indeed rather complex. In addition to fulfilling people's basic need for nutrition, food arouses a combination of sensory responses, such as vision (appearance), hearing (sound), smell (aroma), taste (flavour), and touch (texture), into a wholesome eating experience. While current 4D food printing destructures food ingredients and restructures them into dynamic forms that provoke a strong visual response, other sensations triggered by food are largely overlooked.

Aroma and flavour are the two most important aspects of food. These characteristics basically define food as we know it and are the determining factor for consumer acceptance of food. The sensation in response to aroma and flavour contributes to the taste of food, which is often associated with strong emotions [44]. The aroma and flavour of food can be complex, and their dynamic change stimulated by temperature and acidity/alkalinity can be explored for 4D food printing. In fact, pH-sensitive aroma compounds have already been designed for this purpose [45]. The MIT media lab has created a toolbox of pH-sensitive organic primitives to invoke dynamic interactions between humans and food. In this toolbox, an alginate hydrogel containing vanillin shows a pH-dependent aroma strength. At pH 9 or above, this hydrogel has almost no aroma. As the pH decreases, it starts to generate a strong aroma. The food aroma can also be activated by heat. Guo *et al.* 3D printed a buckwheat dough with yellow flesh peach. The 4D transformation of aroma was activated by microwave heating; four aroma compounds showed statistically different concentrations before and after heating. As aroma and flavour are extensively studied in the food industry, a plethora of aroma and flavour compounds are at our disposal, and many of them are readily applicable for 4D food printing.

Sound is an often-ignored aspect of food. We often eat with our ears without realizing it [46]. Sounds made from eating (e.g., chewing a potato chip, biting a crunchy apple, slurping the last sip of Coke through a straw) have a strong influence on our perception of taste. The sound from food is not only generated by eating but also from the food itself when prepared in certain ways. The first sensation aroused by the meat served on a hot cast iron plate is the sizzling sound, and this sound can be further accentuated by spraying a cold sauce over the food. One famous Chinese dish, simply translated to "rice cracker with stir-fried meat", is a dish specifically created for the enjoyment of its sound — gravy is poured onto smoking hot deep-fried rice crackers, resulting in a "thunder-like sound". Many customers order the food simply to enjoy that sound. 4D food printing could certainly learn from this culinary technique and develop novel stimulation mechanisms for 4D sound transformation.

Food texture, specifically referring to the attributes of food perceptible by mechanical and tactile receptors [47], is another sensory factor of food. Food texture, such as the tenderness of meat, the firmness of an apple, and the creaminess of ice cream, are associated with high-quality food. Limited by the current food printing technique, which is primarily based on extrusion, the texture of printed food is significantly different from the conventional forms we are familiar with. On one hand, these 3D-printed purée foods possess textures that could cater to groups with special needs [48, 49]. On the other hand, such a texture might not suit the palate of everyone. Therefore, the 4D transformation of food texture may substantially extend the acceptance of 3D-printed food. Common cooking techniques, such as steaming, baking, frying, or microwave heating, can change the food texture but may also disrupt the structural integrity of the printed food. Other cooking techniques may also be explored for 4D food texture transformation. For instance, surface flaming with a cooking blow torch instantly creates a crunchy surface.

## 9.3 Advanced Fluidic Technologies for 3D Food Printing

### 9.3.1 *Microfluidic Technologies for 3D Food Printing*

Microfluidics is fluid mechanics and technologies at the microscale, typically handling tiny volumes of fluid between $10^{-6}$ and $10^{-9}$ L [50]. The laminar flow of the fluids on a small scale, characterized by the low

Reynolds number, offers unique ways to control the flows in the device [51]. The dimension of fluidic channels ranges from a few micrometres to a few millimetres [52]. Low manufacturing costs, quick processing, and reusability of the microfluidic device mould are just a few of the benefits of such devices [53]. Microfluidic device applications first appeared in the early 1990s for blood rheology [54] and chemical analysis [55]. Biosensors [56], capillary electrophoresis [57], drug screening [58], flow cytometry [59], genetic analyses [60], and multiple component reactions [61] have all become applications since then.

Microfluidic technologies have found their applications in food science, technology, and manufacturing [51]. With the world's rapidly increasing population, there is a growing demand for the improvement of advanced food processing technologies that offer safe and functional foods [51]. Microfluidic technology, one of the latest food processing technologies, has garnered attention recently over the past years, as evident by the number of related patents and publications in the field [51].

Examples of microfluidic techniques applied to food science and engineering include food emulsification, food quality testing, food safety measurements, and the extraction of bioactive compounds from plant-based products [51]. Microfluidic droplet generation has been widely applied for food emulsification with high reproducibility and controllable operational conditions [62]. Additionally, microfluidic techniques enable quicker reactions and produce emulsions of smaller sizes than traditional emulsification techniques like homogenization and high-pressure homogenization [63].

In this section, prominent examples of microfluidic technologies relevant to food engineering are discussed. Droplet generation systems, among other microfluidic technologies, are designed to generate discrete volumes using immiscible phases, and these droplet-based microfluidic devices are utilized for individual analysis, mixing, and transportation of these droplets or fluids [64]. Manipulated droplets with nanolitre or femtolitre volumes can serve as isolated reaction containers for biochemical testing [65]. Most applications requiring very homogenously sized droplets ensure controlled and predictable analyses. In food engineering and processing, food emulsification using microfluidic droplet generators ensures uniform emulsion droplet sizes and reproducibility in their methods under highly controlled operational parameters. The configuration of the microchannels used in microfluidic droplet/emulsion generators can be used to enhance extrusion-based 3D printing of food and biomaterials, which will be illustrated in the later sections of this chapter.

## 9.3.1.1 *Microfluidic emulsion generator*

Emulsions are heterogeneous systems made up of at least two immiscible liquids, such as water and oil, with one phase dispersed in another phase. The dispersed phase comprises small droplets, whereas the continuous phase forms the surrounding liquid. Emulsions are grouped as oil-in-water (O/W) or water-in-oil (W/O) emulsions. Researchers have developed different types of microfluidic emulsion generators characterized by the geometry of the microchannels, which are grouped into two categories: (1) capillaries and (2) planar. A planar device can be divided into a T-junction, a Y-junction, and flow-focusing by the configuration of the channels. The ability to generate droplets is the cornerstone of these device designs and is critical to the property and functionality of the generated emulsions.

- Capillary emulsion generator

Capillary emulsion generators are microfluidic devices that use capillary effects (also called capillary action or capillary force) to manipulate fluids for producing droplets. Droplets are produced in different modes of generation, depending on the flow rates of the continuous phase and dispersed phase. Three modes of droplet generation have been identified — (1) dripping, (2) jetting, and (3) squeezing. At the tip of the capillary, dripping and squeezing happen. The squeezing of the dispersed phase occurs when the interfacial tension exceeds the shear force, causing the droplet to grow until it completely obstructs the channel's cross-section; the continuous phase fills the thin film between the droplet and the channel walls. Detachment of the droplet neck happens when the pressure upstream of the droplet is increased [66]. The droplet is detached in the dripping regime because there is still space for the continuous phase flow [67]. The jetting causes a long stream of inner fluid to form, with a variable point of droplet detachment causing a wider distribution of droplet sizes. Agitations in the jet increase the Laplace pressure, and the droplet is generated by the Rayleigh–Plateau instability. When pressure is sufficiently high, the jet thins out and breaks the stream into droplets [68].

Coaxial capillary microfluidic devices allow for both dripping and jetting and are commonly employed for producing droplets, microparticles, and vesicles [69]. Co-flow and flow-focusing systems are commonly used for generating single emulsions (Figure 9.2). Two fluids flow in the same direction in a co-flow system. The dispersed phase is constrained by

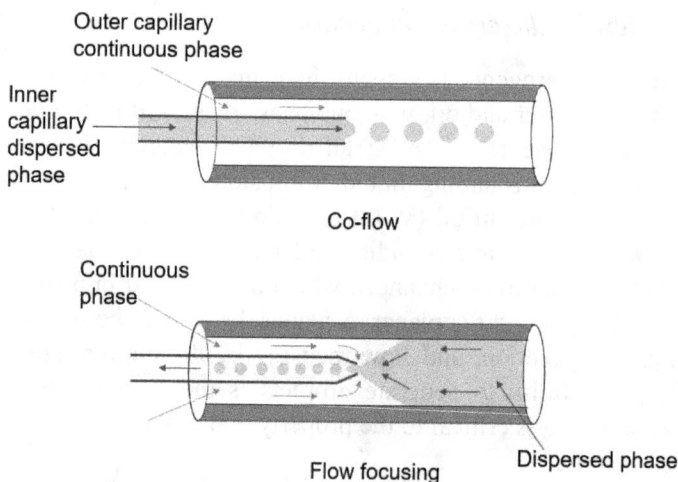

Figure 9.2.  Schematic illustration of the two types of capillary-based emulsion generator: co-flow and flow-focusing [74].

the continuous phase, which enters the outlet channel with an open end on the other side of the external capillary.

The recent development of 3D printing technologies has enabled the fabrication of capillary-type emulsion generators by desktop 3D printers. For example, Vijayan *et al.* assembled 3D-printed fittings, commercially available needles, and low-cost soft tubings as different modules to produce a single droplet generator. Such droplet generators are capable of producing single and complex emulsions such as bi-compartmented Janus particles (dispersed phase) of different sizes or double emulsions (e.g., water-in-oil-in-water (W/O/W)) [70].

Using 3D printing, Martino *et al.* fabricated a screw-nut structure to align the capillaries to generate double emulsions [71], while Meng *et al.* fabricated positioning grooves and connection fasteners for the flexible assembly of flow-control modules [72]. Various flow-control modules can be assembled and disassembled with ease to produce different higher-order multiple emulsions and to scale up the production of droplets. Zhang *et al.* employed a desktop stereolithography apparatus (SLA) printer to create a tubing chamber, which was then inserted with tubing to create a gap channel where droplets could be formed [73]. The droplet size was mostly determined by the tubing inner diameter and gap channel, and so droplets as small as 50 µm could be formed using the small tubing

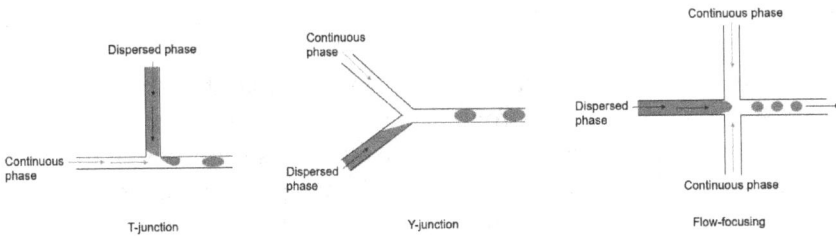

Figure 9.3.  Schematic illustrations of the three types of planar geometry devices: T-junction, Y-junction, and flow-focusing channels.

inner diameter and gap channel, overcoming the problem of generating small droplets using a 3D-printed device.

- Planar emulsion generator

The fabrication of planar microfluidic devices has been demonstrated by soft lithography and replica moulding using polydimethylsiloxane (PDMS) [75]. With the micromold patterned by photolithography, planar microfluidic devices typically possess rectangular cross-sections of channels with uniform height [51]. Commonly employed planar microfluidic emulsion generators have T-, Y-, and flow-focusing channels (Figure 9.3) [51].

In T-junction devices, fluids are delivered through two perpendicular microchannels to produce monodisperse droplets. The capillary number ($Ca$) is a measurement describing the relative effect of the viscous drag forces versus surface tension forces acting across an immiscible liquid interface. $Ca$ is low in the squeezing regime, which makes droplet separation more difficult. Droplet formation happens in two stages in T-junction devices: droplet growth followed by droplet detachment [68]. The angle between the channels in a Y-junction device is frequently greater than 90°, which affects droplet production due to the shear force acted by the continuous phase on the dispersed phase [51].

Unlike the T-junction device, the mechanism of droplet detachment in the Y-junction device occurs in one stage [76]. Droplet formation in a Y-junction is directly dependent on the balance between shear forces and interfacial tension, with no droplets being produced at low $Ca$ [77]. Alternatively, a flow-focusing emulsion generator has four perpendicular channels, two side inlets, a central inlet, and a single outlet. These four channels cross to form a junction, where the flow from the side channels normally constricts the flow of fluid in the central inlet by the process called hydrodynamic focusing [51].

In this process, the dispersed phase flowing from the central channel meets the continuous phase from the side channels at 90° and undergoes a constraint, creating droplets through symmetric shearing from the continuous phase [78, 79]. In flow-focusing droplet production, the flow rate of the continuous phase is usually higher than the flow rate of the dispersed phase, whilst reducing the flow rate of the continuous phase increases the droplet size [64]. The flows in the flow-focusing geometries typically exhibit higher shear than T- and Y-junctions because they are imposed by a continuous phase from the flow at both sides of the dispersed phase. As a result, smaller droplets may form in the flow-focusing junction than T- or Y- junctions [51].

- Microfluidic droplet generations for food manufacturing

The fast-rising global population requires advanced food processing technology to provide more functional and safer foods. Food emulsification can be performed using microfluidic methods [62]. Emulsions are dispersions that can be found in a variety of foods. The most basic emulsion structures are O/W (e.g., mayonnaise and salad dressings) and W/O (e.g., margarine and butter) [80]. Emulsions can also be used in functional applications such as carriers for the stabilization of bioactive substances [81]. In traditional methods, energy is physically provided in mixing equipment, such as a magnetic stirrer and homogenizer, to create food emulsions [82]. These approaches, however, have limited control over the distribution of particle sizes, leading to polydisperse emulsions. The energy efficiency of physical methods is also limited. For instance, in a traditional homogenizer, only ~5% of the energy input is utilized to form the emulsion, with the remainder wasted as heat [62]. Not only is it inefficient in terms of energy, but it also risks denaturing any bioactive substances with the emulsion due to the heat produced during emulsification. Although numerous improvements, such as high-pressure homogenization, ultrasonic homogenization, and membrane homogenization, have been implemented, energy efficiency remains an issue [51]. Microfluidic devices can create emulsions in a regulated manner, allowing to tailor the emulsion properties with improved energy efficiency.

Microfluidic devices provide the ability to regulate the size of emulsion droplets. Additionally, with the control offered by microfluidic devices, more complex emulsions — double-emulsions and multiple emulsions — can be produced. Oil-in-water-in-oil (O/W/O) and water-in-oil-in-water (W/O/W) are double emulsions that provide an improved capability to contain hydrophilic and hydrophobic substances in a single emulsion system. Multiple emulsions have been shown to increase the stability of the food components within these emulsions. For instance, colour retention was observed to remain high for at least 30 days with W/O/W emulsions encapsulating water-soluble food pigments, especially when stored at low temperature (i.e., 4 °C) and low pH (i.e., pH 3) [83]. In another example, W/O/W emulsions containing liposomes encapsulating beta-carotene were found to be physically stable for storage at room temperature for 7 days [84]. Microfluidic droplet generation has also significantly reduced the time, costs, and materials required for food safety testing involving food pathogens such as *Escherichia Coli* [85] and *Salmonella* [86]. They are also applicable for the quantitative detection of bioactive compounds such as polyphenols in wine samples [87].

Although the use of microfluidic droplet generation has found many important applications in food processing, especially in the production of emulsions and food testing, its applications in 3D food printing remain largely unexplored. However, similar microfluidic technology involving the use of microchannels has been integrated with extrusion-based 3D printing for bioprinting and food printing applications. Micromixers can be incorporated to print heads to enable last-minute mixing of several components immediately before ink dispensing. Some microfluidic systems allow for the switching between different inks while printing [88]. These improvements widen the applications in extrusion-based printing. In the subsequent sections, we describe how microfluidic technologies have been integrated with extrusion-based 3D printing for applications related to bioprinting and food printing.

## 9.3.1.2 *Microfluidic improved extrusion-based printing for multi-material printing and 3D food printing*

Pioneering works of microfluidic-enhanced extrusion-based 3D printing have served to overcome the limitations of existing bioprinting techniques, including multi-material printing, precision control, and the handling of living materials (i.e., cells and microorganisms). In comparison to standard printing nozzles, microfluidic printheads have enabled access to several biomaterials or diverse cell-laden bioinks [89]. Additionally, improvements in bioprinting with microfluidic printheads may suggest potential enhancements to the current technology of printing structured cell-based meats, which is a significant development in the field of 3D food printing. Currently, coaxial printheads have been used to produce filaments and fibres within 3D-printed foods to obtain desirable texture, appearance, and shape; they allow for the simultaneous printing of two different food materials into a single 3D-printed food product.

In this section, we explored the use of microfluidic-enhanced extrusion-based 3D printing in two broad areas, namely, bioprinting and 3D food printing. Firstly, we described the three main types of microfluidic enhancements in extrusion-based printing: (1) microfluidic chips, (2) coaxial and triaxial printheads, and (3) printheads that combine the microfluidic chip with the coaxial printhead. Some applications in bioprinting are discussed. Secondly, we described the applications of coaxial printheads in recent works for various applications in 3D food printing.

- Microfluidic systems in 3D bioprinting applications

Microfluidic-based 3D printing is useful in biofabrication since it allows for the creation of complex structures with an improved print resolution. Such extrusion printers can handle a wide spectrum of bioinks with a variety of viscosities [90]. 3D organs and tissues resembling their corresponding biological structures are produced by a layer-by-layer deposition of cell-laden biomaterials (also called bioinks). 3D bioprinting is a relatively new biofabrication technique, with limitations in terms of (i) heterogeneity of cells and biomaterials, (ii) precision of printing, and (iii) fidelity of printing [91]. To address these limitations, microfluidic nozzles with specialized designs have been developed to produce a model with complex intra-structures. Microfluidic printing heads can be used for micro and nano-fabrication; the resultant printed structures may resemble blood arteries, nerves, and muscles [91]. Microfluidic systems developed for current extrusion-based printing are broadly classified into three types: (1) microfluidic chips, (2) multiaxial (i.e., coaxial, triaxial) nozzles, and (3) combined printheads [90]. These printing systems improve the control of extrusion-based printing. These technologies are also transferrable to 3D food printing.

- Microfluidic chip

Microfluidic chips consist of a series of microchannels with varying geometries (Figure 9.4). Such chips can be used to control heterogeneous distributions of materials within single microchannels and

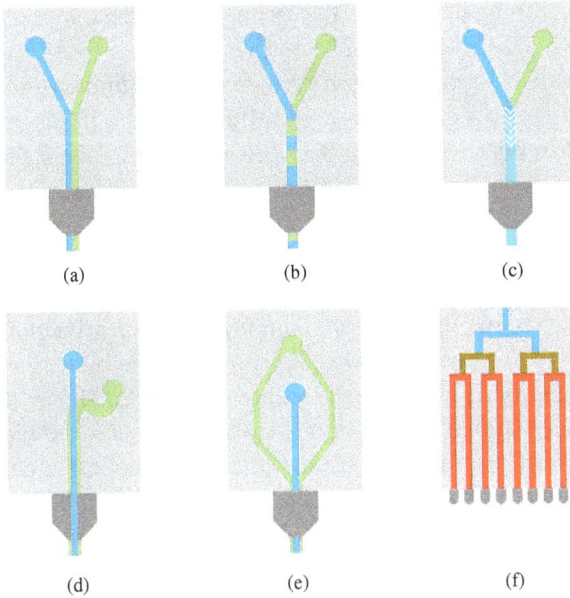

Figure 9.4.   Schematic illustrations of the different types of microfluidic chips employed in biofabrication. (a)–(f) Microfluidic chips having one or more inlets for different functions: (a) Parallel dispensing of two bioinks, (b) switching between two bioinks for alternate dispensing, (c) herringbone micromixer to facilitate mixing of two bioinks, (d) sheath flow maintains the flow of cells in the centre for protection against shear stress against the walls of the nozzle, (e) flow-focusing regulates the diameter of the printed fibre by changing the ratio between the core flow (blue) and sheath flow (green), and (f) multiple outlets in a microfluidic chip to raise scalability by high-throughput bioprinting [90]. Reproduced from [90], which is licensed under CC BY 4.0.

microchambers. Because human tissue is seldom a homogeneous blend of cells and extracellular matrix, heterogeneity is critical for precisely replicating natural tissue complexity and its function [92]. Microfluidic chips enable automatic and seamless switching among different materials in 3D printing. Handling multiple bioinks requires multiple cartridges and print-heads [89]. Instead, the use of branched microfluidic channels allows the rapid switching of multiple materials.

Serex *et al.* demonstrated that material switching could be accomplished without slowing down the printing by employing a device with numerous inlets and channels that converged into a single output [88]. Each inlet was linked to a bioink reservoir controlled by an actuated syringe pump, which enables the printed materials to be changed in <500 ms [88]. This method can be further improved by substituting actuated virtual valves for the actuated syringe pump to improve accuracy [88].

The laminar flow and low rate of diffusion of the fluids in microchannels resulted in minimal mixing between the bioinks. This condition provided a clean and clear transition between the bioinks in multi-material bioprinting [90]. However, some applications require efficient and controlled mixing of different fluids in microchannels. To this aim, a mixing unit can be included inside the printhead [93]. Mixing highly viscous fluids with a passive mixer is challenging; passive mixers are normally suited for low-viscosity fluids [93]. A passive herringbone micromixer has been used to facilitate such mixing in microchannels [88, 94]. Active mixers demand an external source of energy to stir or mix the materials with a continual effort. Because the mixing intensity is not affected by the flow rate and volume, active mixers are often more controlled and efficient than passive mixers despite the higher complexity and working volumes [93, 95]. Lewis *et al.* built an impeller-based active mixer into the extrusion-based bioprinting printhead for efficient homogenization of bioinks to facilitate *in situ* reactions before printing [93].

Extrusion-based bioprinting may result in shear-driven damage to the printed cells that differ in their resilience and susceptibility to environmental stressors [96]. Neuronal cells, for instance, are extremely challenging to print due to their high sensitivity. Willerth *et al.* used a microfluidic printhead to create a protective sheath layer to decrease shear stress on the cells, leading to improved cell survival within the printed structures [97]. Shear stress is high at the edge of the channels, where friction is present between the nozzle wall and the fluid [98]. Sheath flow can be used to hydrodynamically concentrate the cells towards the centre of the flow to improve cell viability [90]. Bioink flow-focusing is important, not just for reducing shear stress but for reducing the diameter of the printed fibre as well [88].

Scalability and throughput are critical in assessing the viability of the general adoption and use of bioprinted constructs in a medical environment. While small constructs should suffice for academic research, patient-oriented tissue engineering requires the production of larger constructs in greater quantities over a shorter period [90]. This requirement should also apply to 3D food printing. To address this problem, Lewis *et al.* developed microfluidic chip-based printheads with multiple outlets for simultaneous production of the same objects [99, 100].

- Multiaxial extrusion printheads

Multiaxial extrusion printheads have been often developed for 3D bio-printing. Coaxial printheads consist of two axially aligned needles, enabling two completely distinct fluids to flow before contacting one another at the point of extrusion. Coaxial and triaxial nozzle systems may be used to fabricate both solid and hollow fibres [101, 102]. To this end, a coaxial printhead may concurrently extrude the bioink and crosslinking agent. The bioink in the core flow swiftly crosslinks upon contact with the crosslinking agent in the sheath flow, causing gelation of the fibre. When the configuration of the fluids is reversed, the bioink in the sheath flow is crosslinked to generate a hollow and perfusable fibre. By moving the printhead in pre-determined toolpaths, the fibre is dispensed with spatial accuracy in layer-by-layer 3D printing [101].

Crosslinking in coaxial extrusion-based bioprinting does not depend on the shear-thinning characteristics of the bioink as much as it does in conventional extrusion-based bioprinting. Instead, it depends exclusively on the interaction of the bioink with the crosslinking agent, which permits the use of low-viscosity bioinks [101, 103]. A popular bioink is alginate-based; it can be physically crosslinked by cell-compatible calcium chloride ($CaCl_2$) solution. Alginate hydrogels show excellent mechanical properties and biocompatibility. Because the natural polysaccharide alginate does not have intrinsic bioactivity, peptide ligands such as the arginine-glycine-aspartic acid motif can be added to alginate to improve cell adhesion and migration [90].

Biomaterials that have intrinsic bioactive properties, for instance, extracellular matrix-like gelatin and collagen, have poor printability on their own because of their poor mechanical properties at lower concentrations. To address this limitation, alginate has often been combined with extracellular matrix-like hydrogels to create mechanically robust yet biocompatible inks [104–106]. For instance, alginate and gelatin methacryloyl (GelMA) has been used to encapsulate human umbilical vein endothelial cells (HUVECs) to create endothelialized networks [105, 107]. In these works, the alginate in the bioink was chemically crosslinked when extruded through the coaxial nozzle, followed by photo-crosslinking of GelMA after printing. The continuous sheath flow prevented the fibre from drying while allowing it to be crosslinked.

Coaxially printed hollow fibres may be utilized to mimic numerous tissues with microtubular structures. Such hollow fibres are especially promising for the vascularization of bigger scaffolds [92]. Vascularization is an important aspect of tissue engineering because the vascular network provides cells with oxygen and nutrients; vasculatures are also essential to eliminate their waste products. This function is increasingly necessary as the construct size increases, as the typical diffusion limit for oxygen and nutrients is in the range of 100–200 μm [108]. There is a great demand for pre-vascularized tissue-engineered constructions; the incorporation of functioning blood artery networks into organ constructs is still one of the biggest challenges. Ozbolat *et al.* were among the first to use a coaxial printhead for direct hollow fibre production [103, 109, 110]. In their work, hollow fibres with a lumen diameter of <200 μm were bioprinted using either cell-laden alginate hydrogels or chitosan hydrogels. Alginate is more advantageous than chitosan in terms of its mechanical and structural integrity. Additionally, the hollow fibre of alginate hydrogels is perfusable with oxygenized medium without any leakage. Human vasculatures do not occur as a straight line but contain numerous bifurcations. Hu *et al.* used a triaxial nozzle method to fabricate a branched hollow fibre in his first attempt at creating such a bifurcated structure [111]. The core flow contained calcium chloride that was enclosed by an alginate shell. This method was utilized to print the Y-shaped hollow fibre, which was then immersed in a calcium chloride bath to completely crosslink the outer layer. Two fused fibres were merged to produce a bifurcated structure. The ungelled outer layer of the hollow fibre was critical for inter-layer adhesion in this fabrication. The third nozzle was employed to quickly clean the clogged nozzle by dispensing a chelating solution to avoid the clogging of the nozzle. Additional research needs to be done to build sophisticated vascular networks with multiple bifurcations.

Vascularized tissues and organs consisting of hollow fibres resembling blood vessels were successfully bioprinted using coaxial extrusion printheads. For instance, Zhang *et al.* printed vessel-like hollow hydrogel filaments that have the potential to be used in the creation of vascularized tissues and organs using a coaxial microfluidic needle; sodium alginate was extruded through the sheath section of the coaxial nozzle, while the calcium chloride solution was dispensed through the core section of the coaxial nozzle [112]. The channels offered mechanical support and permitted fluids to be transported within the 3D cellular environment. Additionally, the cartilage progenitor cells entrapped in the hydrogel wall showed high cell viability during the prolonged culture *in vitro* and demonstrated cartilage-producing functions [112].

Similarly, Gao *et al.* used a coaxial microfluidic nozzle to create a high-strength 3D-printed cell-laden hydrogel structure comprising multi-layers of hollow filaments (produced by the crosslinking of sodium alginate with calcium chloride). This work demonstrated good perfusion of nutrients media and biocompatibility with encapsulated L929 mouse fibroblasts within the filament walls [113]. Gao *et al.* also printed two sets of hollow alginate filaments (i.e., microchannels) using two separate coaxial nozzles, one containing mouse smooth muscle cells and the other L929 mouse fibroblasts. The printed scaffold was seeded with endothelial cells in the inner wall of the two-walled vessel [114]. Various types of vessel-like structures with mm-scale diameter were printed, with applications including vascular circulation, a cell co-culture model, and a cerebral artery surgery simulator [114]. Attalla *et al.* developed a method for creating instantly perfusable vascular networks with cell-seeded gel scaffolds [115]. Vascular-inspired hollow tubes were produced using a coaxial nozzle which permitted the diffusion of calcium ions from the interior into the surrounding annular cell-containing alginate ink. Using 3D printing, these tubes can be assembled into scaffolds or tissue constructs loaded with cells. Their study demonstrated that perfusion of the cell media through these hollow channels increased the cell viability of the embedded HUVECs, as compared to the cells in non-vascularized bulk gels. A similar study was also conducted using coaxial extrusion for the fabrication of cell-laden hydrogel structures assembled from calcium alginate fibres with openings; the exchange of the media supported the viability of loaded human embryonic kidney-293 (HEK-293) cells within the fibres [116].

Overall, these works demonstrated the pioneering uses of multiaxial extrusion printheads for bioengineering applications, such as the creation of perfusion cell culture models and fabrication of vasculature constructs. These works advanced the use of multiaxial nozzles for their subsequent use in 3D food printing, such as the creation of edible fibres.

- Combination of a microfluidic chip with a coaxial extrusion printhead

The advantages of both microfluidic chips and coaxial printheads can be simultaneously utilized. For example, a microfluidic chip can be installed upstream of a coaxial printhead, allowing for the use of low-viscosity bioinks to print complex structures [90]. Microfluidic chips with specific functions can be developed and included in a printhead with a coaxial, triaxial, or multiaxial extruder. Khademhosseini *et al.* created a heterogeneous cell-laden construct constructed of solid fibres, utilizing bioink mixes of alginate (4%) and GelMA (4.5%) containing fluorescent beads using a simple Y-shaped microfluidic chip and a coaxial nozzle (Figure 9.5) [117]. In this approach, the bioinks were first crosslinked with calcium chloride due to the presence of alginate, followed by exposure to UV light for the photo-crosslinking of the GelMA. Because of the low viscosity of the alginate/GelMA ink, a nozzle with a small diameter was used to achieve a higher resolution and faster dispensing. The alginate/GelMA ink displayed Newtonian behaviour, and the dispensing speed was adjusted without sacrificing the print resolution. This method displayed flexible deposition of patterns as well as fast switching between

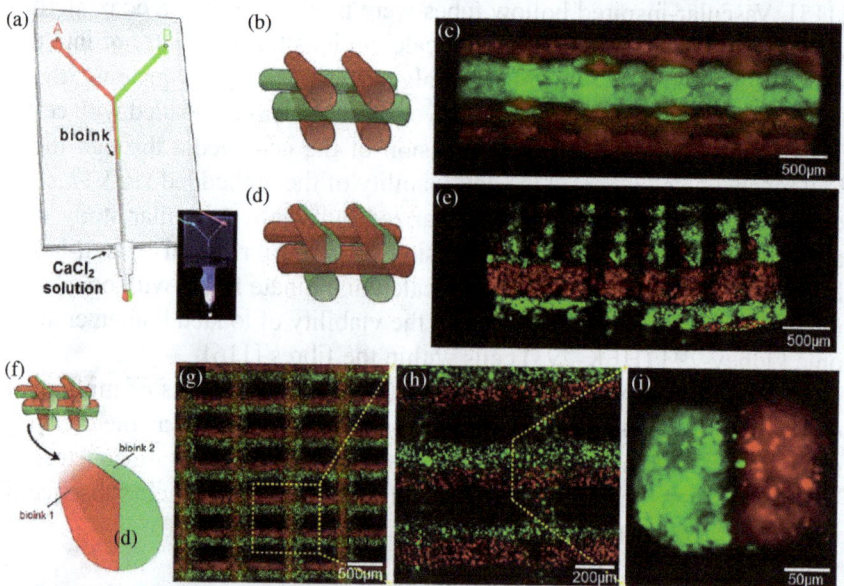

Figure 9.5.    (a) Microfluidic system for two distinct bioinks carrying red and green fluorescent beads flowing out of the device through one extruder. The photograph (insert) displays a coaxial needle system containing a microfluidic chip with a Y-shaped channel. Schematic illustration and fluorescence microscopy cross-sectional images of the 3D construct: (b) and (c) for alternate deposition of two different bioinks, (d) and (e) for alternate/simultaneous deposition of two different bioinks, and (f) to (i) for simultaneous deposition of two different bioinks. Reproduced with permission from [117]. Copyright (2015) from WILEY-VCH Verlag GmbH & Co. KGaA, Weinheim.

two bioinks. By dispensing two bioinks in alternation, longitudinally varied fibres were formed. In contrast, simultaneous deposition of two bioinks produced a variety of fibres in pre-set parallel patterns, resulting in transverse heterogeneity within the fibres. Both techniques were employed to print a single construct with on-demand switching of the materials.

Traditional bioprinting technologies have often faced difficulties to establish a smooth transitory gradient of the various bioinks. To tackle this limitation, Wiszkowski *et al.* created an osteochondral analogue by imitating the zonal cartilage architecture consisting of hyaline articular cartilage layers and calcified cartilage layers separated by an intermediate zone [118]. A passive microfluidic mixer was combined with a coaxial nozzle to establish a continuous transition between bioinks for the osteochondral analogue. This method allowed a continuous gradient to form within the printed construct, which replicated the intermediate zone representing an *in vivo* environment. This method will be especially useful in the fabrication of tissue-engineered analogues with transitory properties.

Hollow fibres may be created using both coaxial dispensers and microfluidic chips to generate tissue-engineered human tubular structures that mimic their native counterparts. Because of their ordered hierarchical architecture and cellular heterogeneity, most human tubular structures are made of multiple layers with diverse biochemical and mechanical characteristics. For example, du Chatinier *et al.* demonstrated the production of multi-material hollow fibre using longitudinal coding (Figure 9.6) [90]. Switching between bioinks was regulated by on-chip pneumatic valves to enable programmed patterning of multiple materials, which were then extruded through a coaxial nozzle to form a hollow structure. These material transitions allowed the continuous bioprinting of hollow fibres with transverse heterogeneity.

Using a triaxial extruder, Pi *et al.* created double- and triple-layered hollow fibres representing blood vessels and urethral tissues (Figure 9.7) [98]. This method enables one to three layers of deposition in a single extrusion to generate multi-layered hollow fibres. The bioink consisted of an eight-arm poly(ethylene glycol) (PEG) acrylate with a tripentaerythritol core in the GelMA/alginate hydrogel, which improved the overall mechanical strength and stability of the crosslinked matrix [98]. Native blood vessels comprising HUVECs and human vascular smooth muscle cells and urethral tissue comprising human urothelial cells and bladder smooth muscle cells were recapitulated using this method [90]. The

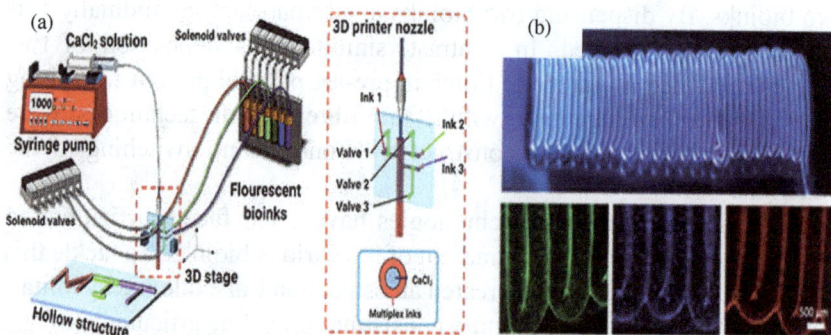

Figure 9.6.   Integration of a microfluidic chip and a coaxial nozzle into one printhead. (a) Schematic representation of the microfluidic system for the longitudinal printing of the multi-material hollow fibres. (b) Hollow fibres are printed longitudinally with a gradual transition between different bioinks. Reproduced with permission from [90]. Copyright (2021) from AIP Publishing.

3D-printed constructs demonstrated high cell viability and functioning for two weeks.

In another demonstration, Costantini *et al.* built a dispensing device to support multi-material extrusion and quick switching between materials by combining a Y-junction microchannel (two inlets and one outlet) with a coaxial needle system [119]. This method enabled the creation of multi-cellular 3D-printed scaffolds, including two distinct cell types, C2C12 myoblasts and BALB/3T3 fibroblasts (Figure 9.8) [119]. Previous literature has reported that myoblast development is aided by fibroblasts secreting growth factors and extracellular matrix components and growth factors [120]. The cell-laden alginate ink (containing PEG-fibrinogen) was extruded through the inner needle and underwent rapid gelation, with calcium ions present in the outer flow. In this way, rapid gelation ensured that the printed hydrogel fibres did not collapse to maintain the shape fidelity of the 3D-bioprinted constructs (Figure 9.9). Additionally, C2C12 myoblasts showed proper spreading and fusing after 21 days of *in vitro* culture, creating well-aligned long-range multinucleated myotubes with high levels of functional expression in laminin and myosin heavy chains. Lastly, 3D-bioprinted myostructures implanted subcutaneously *in vivo* within the backs of immunocompromised mice showed the ability to generate organized muscle-like tissues and their maturation, unlike the implantation to bulk hydrogels [119]. This work allows the scaling up of

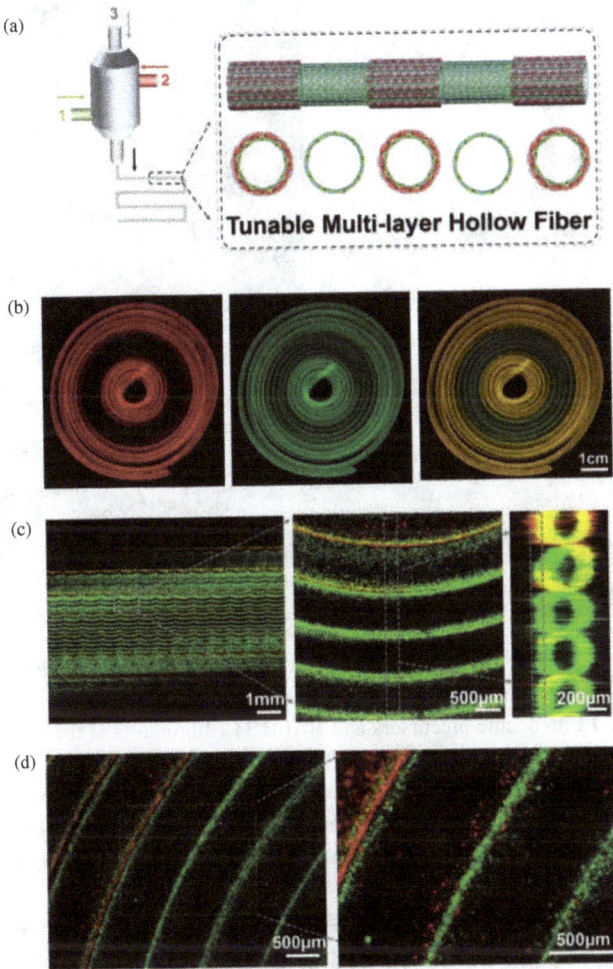

Figure 9.7.  Fabrication of tunable and perfusable double-layered structures using multichannel coaxial extrusion. (a) Schematic representation showing the tuning of layers in a hollow fibre (between double and single layers at fixed intervals) using a multichannel coaxial extrusion system during printing. (b) Fluorescence confocal microscopy images displaying the dynamic conversion between single and double-layered hollow tubes (scale bar = 1 cm). (c) and (d) Simulation of 150 scanned layers exhibiting a dynamic conversion from a double-layered to a single-layered hollow tube with a gradual delineation of single- and double-layered regions of the hollow tube (scale bar = 500 μm). Reproduced with permission from [98]. Copyright (2018) from WILEY-VCH Verlag GmbH & Co. KGaA, Weinheim.

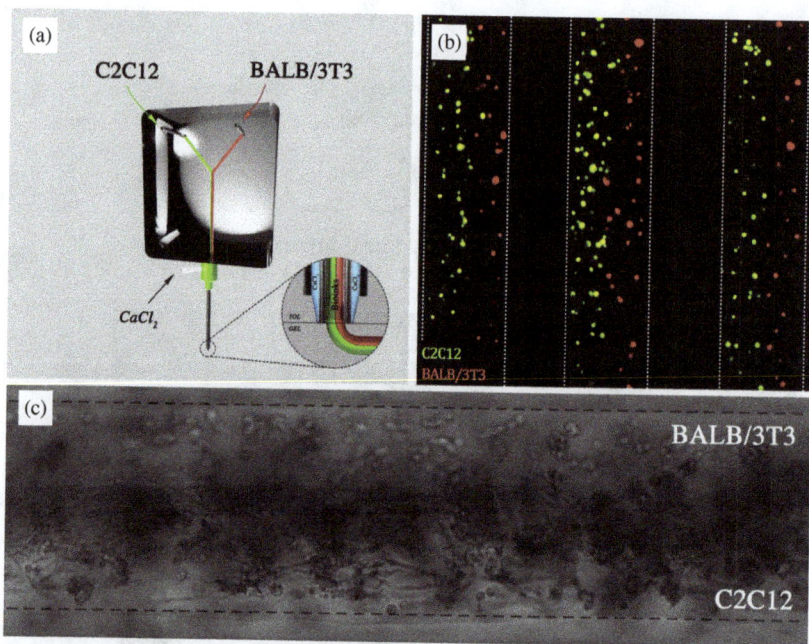

Figure 9.8.    Microfluidic and coaxial combined printhead for multi-cellular 3D bioprinting. (a) Janus-like compartmentalization was obtained in each dispensed hydrogel fibre by extruding C2C12 myogenic precursors and BALB/3T3 fibroblasts at the same flow rate. (b) GFP (green fluorescent) and mOrange (red fluorescent) lentiviruses were used for infecting C2C12 and BALB/3T3, respectively, infected for optimal identification after the 3D bioprinting. (c) Two cell types remained compartmentalized after 5 days of culture. Reproduced with permission from [119]. Copyright (2017) from Elsevier Ltd.

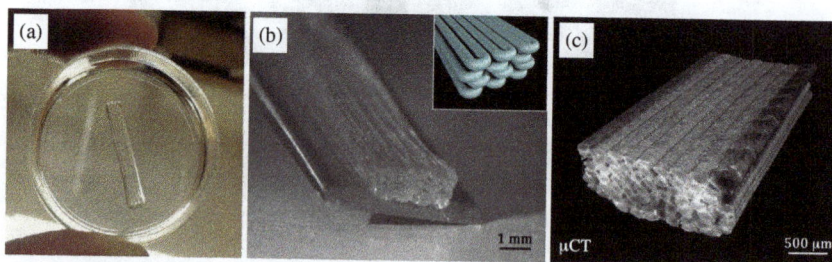

Figure 9.9.    3D-bioprinted structure comprising unidirectionally aligned PEG-fibrinogen hydrogel fibres. (a) Macrograph of a 3D-bioprinted construct inside a 35-mm Petri dish. (b) High magnification optical micrograph of the 3D-bioprinted structure; a schematic illustration showing the fibre arrangement is shown in the inset. (c) X-ray microtomographic scan of the 3D-bioprinted construct in (a). Reproduced with permission from [119]. Copyright (2017) from Elsevier Ltd.

artificial skeletal muscle tissue production and has prospects for muscle repair in the future.

The pioneering works described above have employed the use of microfluidic chips combined with coaxial extrusion for bioengineering applications. These works help pave the way for novel and interesting ideas for 3D food printing in the future, especially for the printing of cell-based meats, whereby biocompatible fibres are printed in unique configurations to achieve various functions and purposes.

• Microfluidic systems in 3D food printing

Among the various microfluidic-enhanced extrusion-based printing systems described in the previous section, the use of coaxial printheads has been researched the most and used in 3D food printing. These coaxial printheads have been employed for the fabrication of filaments [121] and fibres [122] within 3D-printed foods to achieve desirable properties of the food constructs. Tailored properties include their texture, appearance, and shape. In addition, coaxial printheads allowed two different food materials to be printed simultaneously into a 3D food construct.

In an early work of coaxial food printing, Sun *et al.* achieved *surimi*-based (i.e., fish paste) products of various textures by varying their infill patterns [121]. Potato starch was chosen for the coaxial printing as it is a commonly used starch-based binding agent in the food industry [123–125]. The potato starch was dispensed through the outer nozzle, while the *surimi* gel was dispensed through the inner nozzle of the coaxial printhead. Adjacent filaments of the printed *surimi* gel were glued together by the potato starch during printing. They realized that concentric, rectilinear infill patterns resulted in greater hardness than aligned-rectilinear infill patterns. The 3D-printed *surimi* products possessed a hardness comparable to commercial *surimi* products. Previous findings suggested that interlayer bonding of the rectilinear and concentric designs was impervious to deformation because of the alignment of fibres [126]. The study also reported that infill patterns did not influence the cohesiveness, chewiness, resilience, and springiness of the 3D-printed samples, though commercial *surimi* was significantly less springy and chewy than the 3D-printed samples. The higher chewiness of the 3D-printed samples is attributed to the presence of potato starch between layers of *surimi* that acted as a filler [127].

In another work, Ko *et al.* applied coaxial printing to produce artificial meat with muscle-like fibre insertions made of crosslinkable hydrocolloid fibres to achieve a meat substitute analogous to the original meat texture (Figure 9.10) [122]. Soy protein paste (consisting of isolated soy

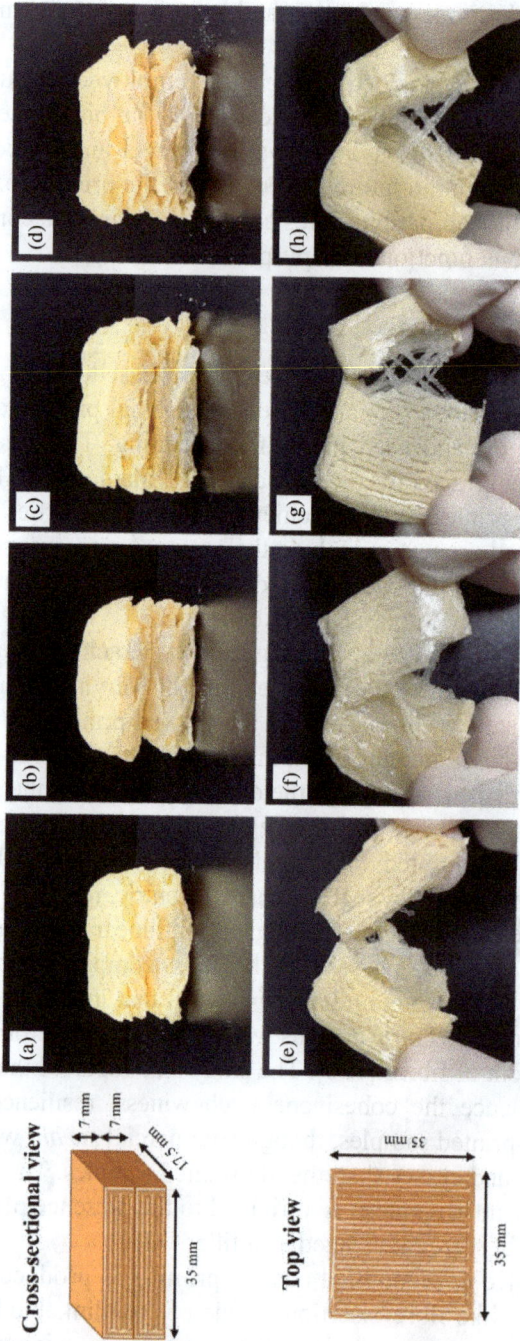

Figure 9.10.  Images displaying the appearance of folded and torn 3D-printed constructs along the fibre direction after the samples have been post-processed; (a) and (e): Samples containing carrageenan (1.5% w/w) glucomannan (0% w/w); (b) and (f): Samples containing carrageenan (2.5% w/w) glucomannan (0% w/w); (c) and (g): Samples containing carrageenan (1.5% w/w) glucomannan (1.5% w/w); (d) and (h): Samples containing carrageenan (2.5% w/w) glucomannan (1.5% w/w). (a) to (d) show the cross-sectional view; (e) to (h) show the top view of the samples. Reproduced with permission from [122]. Copyright (2021) from Elsevier Ltd.

protein (100% hydrolized-defatted soybean), potato starch, xanthan gum (XG), calcium chloride, and potassium chloride) was extruded through the outer nozzle while the fibre solution consisting of carrageenan (containing a mixture of κ-carrageenan and ι-carrageenan), sodium alginate, and glucomannan was extruded through the inner nozzle. The crosslinking of alginate gels formed fibres rapidly upon dispensing both the inner and outer solutions simultaneously. In the absence of glucomannan, fibres were mechanically weak and yielded a break while tearing. However, in the presence of glucomannan, firm and elastic fibre structures were observed within the meat analogues. Fibres were characterized by the breaking time (defined as the time it took for the sample fibre bundles to be detached); the real beef samples exhibited a breaking time similar to the printed sample containing 2.5% w/w of glucomannan. This similarity in breaking time could be attributed to the elasticity of the glucomannan after heating and cooling, which contributed to the fibre strength and elasticity similar to that of beef [128]. However, a similar hardness to beef was not obtained for the tested meat analogues. Future studies aim to increase the hardness of the protein paste to achieve a beef-like texture.

Vancauwenberghe *et al.* performed comparative studies between coaxial extrusion and simple extrusion printing of pectin-based food simulants [129]. In their previous work, they demonstrated that low methoxylated pectin gel could be used as a food ink for 3D printing of customized water-based porous food simulants [130]. Pectin-based sweets are one example of such a food. In this study, the low methoxylated pectin-based food ink was printed as the inner flow, while calcium chloride solution was used as the crosslinking solution in the outer flow for the coaxial 3D printing of cubic structures (Figure 9.11). The gelation of the low methoxylated pectin food ink was facilitated by the crosslinking of calcium ions with the free carboxyl groups within the pectin [131, 132]. Coaxially printed pectin samples did not require any incubation with calcium chloride as the post-treatment because gelation of the pectin food ink took place during printing. In contrast, a conventional simple extrusion printing required at least an additional hour of incubation of the printed pectin samples in 50 mM of calcium chloride solution. Three printed cubic samples having three equivalent Young's modulus (i.e., ~62–65 kPa, 101–106 kPa, and 114–120 kPa) obtained from simple extrusion and coaxial extrusion were selected for structural comparison. The print quality was appropriate for all samples since the cubic shape of the printed pectin structures was respected with adequate layer superposition without

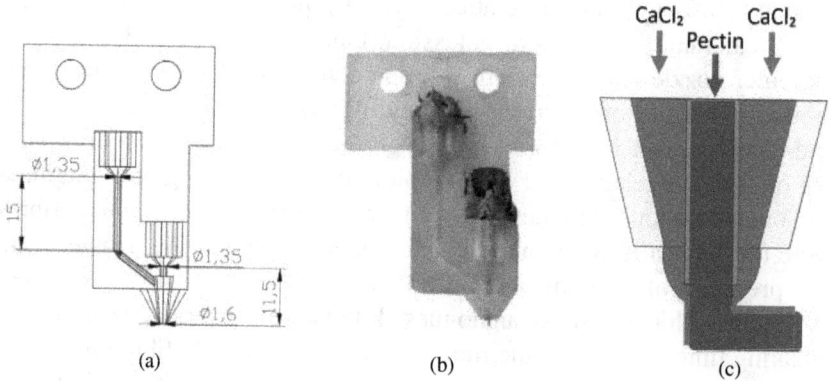

Figure 9.11.   (a) Schematic illustration of the 3D-printed extruder. (b) Photo of the coaxial printhead. (c) Schematic illustration of the junction in between the inner flow of the pectin ink and the outer flow of calcium chloride. Reproduced with permission from [129]. Copyright (2018) from Elsevier Ltd.

internal defects [129]. However, a slight variation in the volume of the printed samples can differentiate coaxial extrusion from simple extrusion printing. The layering ripples of the printed samples by coaxial extrusion had visible reliefs, which were detected by the light contrast of the images. In particular, the difference was evident for samples obtained from simple extrusion and coaxial extrusion containing 15 g/L pectins. Because the pectin food-ink was still semi-liquid while printing, the printed layers collapsed slightly during simple extrusion printing [129]. The X-ray centralized tomography (CT) cross-section scans indicated minor variations from the desired shape for the samples produced from simple extrusion printing; all printed pectin cubes produced by simple extrusion printing had edges with small curvatures, which were induced by the swelling of the printed structures during the incubation, with calcium chloride as a post-treatment (Figure 9.12). As a consequence, samples produced using simple extrusion had higher volumes than the ones printed using the coaxial extrusion method.

The swelling of the materials was influenced by the differences in the ionic concentration between the interior and exterior of a polyion gel [133]; this difference in ionic concentration is greater with pectin concentration, which agrees with the observed results. By raising the concentration of calcium chloride for incubation in the post-printing step, the swelling of the object printed by simple extrusion may be reduced.

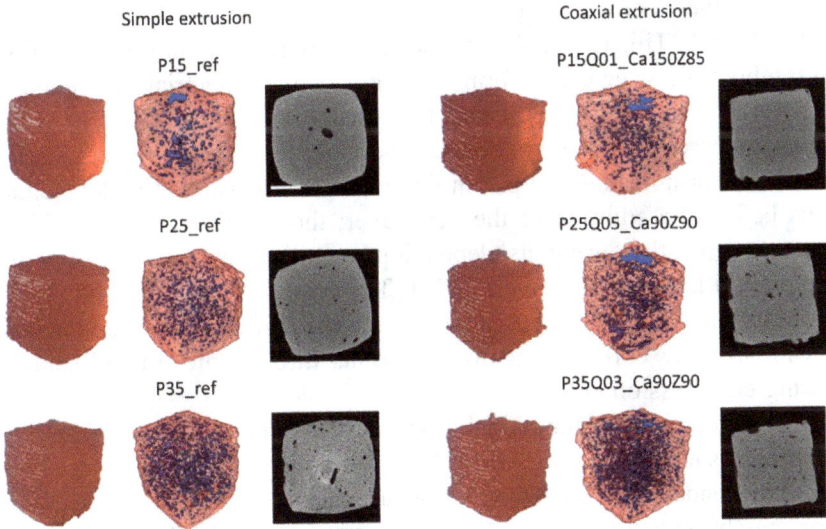

Figure 9.12.　3D-printed samples of pectin cubes at macro- and microscales: Photographic, volume rendering, and horizontal slice X-ray CT image of the 3D-printed samples by simple extrusion printing (left) and coaxial extrusion printing (right). (Scale bar of the micro CT images = 5 mm.) Reproduced with permission from [129]. Copyright (2018) from Elsevier Ltd.

The borders of printed pectin samples via coaxial extrusion were straight. Based on CT cross-sections, small accumulations of materials might occasionally be seen at the boundary of the cross-section surface. During simple extrusion printing, irregular deposition of the food ink can also take place. However, because the food ink did not entirely complete gelation in simple extrusion printing, the printed inks can spread on the deposited layer to form smooth surfaces.

The method of extrusion (i.e., simple or coaxial) can result in different mechanical behaviour of the printed samples at a high compression strain ($\varepsilon > 0.2$), with the rupture point in the stress-strain curve indicating that a sample has raptured under high compression. The rupture point of the samples printed by coaxial extrusion was observed at lower stress and a greater strain than the samples printed by simple extrusion, which demonstrated maximum stress at the rupture point. These findings demonstrated the variation in bonding and adhesion between the superposed layers. Layers of printed samples produced by simple extrusion had excellent adhesion since the food ink was only partially crosslinked during the printing; they completed gelation upon post-treatment in the calcium chloride bath. This protocol of fabrication may result in uniform gelation throughout the structure to form a continuous bulk material that retained the stress until it ruptured. Coaxial extrusion printing resulted in weak adhesion between the layers because the food ink began to undergo gelation at the printhead. Most pectin dimers were cross-linked with calcium ions before the addition of the next layer; the gelation of the food ink happened mostly during its deposition, resulting in a weaker binding interaction between the layers [134]. The printed sample can be represented as a stack of superposed layers, and the adhesion between the layers most likely collapsed at diverse points through the printed sample during compression [129].

The bioactivity of functional foods may be lost due to post-processing processes such as cooking [135]. By formulating functional ingredients into independent food inks through the use of a coaxial nozzle, this limitation may be addressed. Jeon *et al.* demonstrated coaxial extrusion printing to develop a novel nutritional carrier based on nanoemulsion as an independent food ink for 3D food printing [135]. Curcumin nanoemulsion was dispersed in a combination of XG in varying quantities and potato starch to create a nanoemulsion-filled gel (NFG). The outer flow of the coaxial nozzle contained bean paste, while the inner flow contained the NFG (Figure 9.13). Bean paste, which shows high stability in shape and extrudability, was employed as a demonstration ingredient for the base

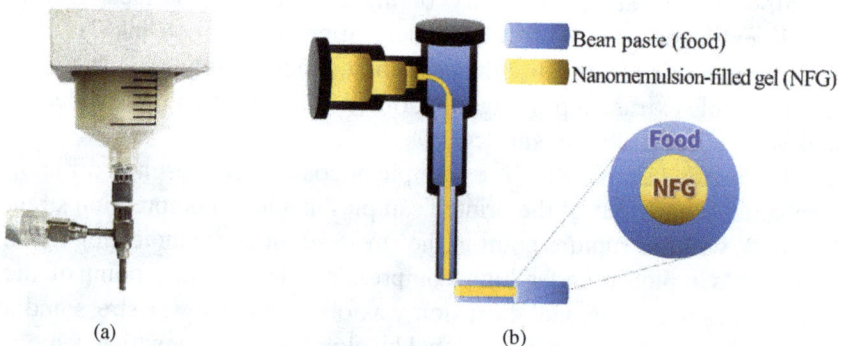

Figure 9.13.   Coaxial 3D food printing system: (a) set-up of the coaxial-assisted 3D food printing system, and (b) schematic illustration showing the cross-sectional view of the coaxial nozzle. Reproduced with permission from [135]. Copyright (2021) from Elsevier Ltd.

material [136]. The fundamental ingredient for the NFG was curcumin, a lipophilic molecule with outstanding anticancer and antioxidant properties [137, 138]. Because of their low cost and extensive availability, polysaccharide-based compounds such as XG and potato starch are used for producing the NFG. The combination of XG and potato starch has been widely employed in delivery systems to alter the texture of the emulsion and improve process convenience [139]. In a liquid-type emulsion, the matrix of XG and potato starch gel can act as a barrier that will help retain the functionality of the encapsulated material [140].

The effect of XG on the rheological property of the NFG was investigated. The values of $G'$ and $G''$ of the NFG gradually increased until 3% w/w of XG was added, demonstrating that the integration of XG could stabilize the NFG matrix. However, a reduction in the values of $G'$ and $G''$ was reported if XG increased above 5% w/w. This phenomenon is likely due to the strong water-holding capacity of the XG, which prevents other polymers from being sufficiently hydrated during the creation of the continuous phase matrix [141]. NFG is a mixture in which the emulsions were dispersed within a continuous phase matrix where XG and starch gel interacted.

The coaxially 3D-printed products containing different amounts of XG in NFG were evaluated (Figure 9.14). The shell matrix (i.e., bean paste) without any NFG demonstrated good print resolution and shape retention after printing, suggesting that the base matrix offered adequate printing performance by itself. However, despite the high mechanical

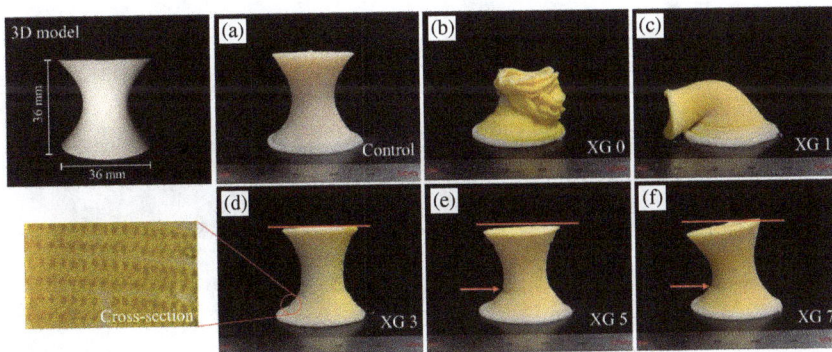

Figure 9.14. Photographs displaying coaxial printed products containing 0, 1, 3, 5, and 7% w/w XG, respectively. Reproduced with permission from [135]. Copyright (2021) from Elsevier.

strength of the base matrix, the supporting force was not sufficient without XG, which resulted in the deformation of the layered structure. Printed samples containing 1% w/w of XG in NFG exhibited improved stability yet collapsed within 30 min post-printing. The printed sample had adequate mechanical strength to sustain the top of its structure when the ink contained 3% w/w of XG. More importantly, a further increase in the contraction of XG resulted in the deformation of the printed samples. The addition of excess XG may induce incompatibility with the starch matrix, resulting in a weak gel [142].

### 9.3.2 *Embedded 3D Food Printing*

Embedded 3D printing is another emerging method in additive manufacturing; the printing is performed in a supporting medium, and the printed ink filaments are embedded within the supporting medium (Figure 9.15) [143]. In embedded 3D printing, additional support materials are not required for the printing of the 3D constructs. The supporting media are usually yield-stress fluids, which are materials that change reversibly from solid to fluid at critical applied stress. When the applied stress surpasses its yield stress, these fluids undergo deformation. The shear stress by the motion of the nozzle can temporarily disrupt the non-covalent and reversible bonds within the supporting medium, enabling direct ink writing in the support medium. After the shear stress is removed, the medium rapidly restores its bonding and wraps the printed ink. The printed ink is

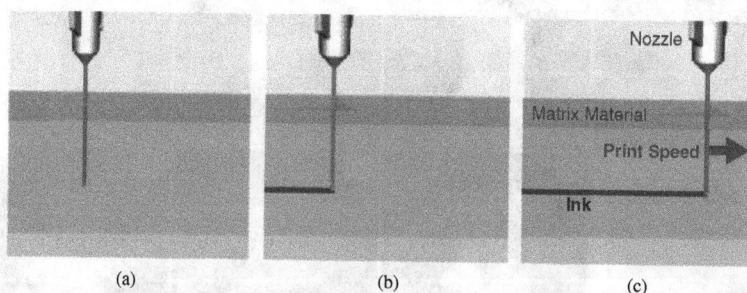

Figure 9.15.   Schematic drawing of embedded 3D printing. (a) Print nozzle enters the viscoplastic matrix, (b) the deposition of the ink into the matrix, and (c) translation of the nozzle in the matrix at a given speed. Reprinted (adapted) with permission from [145]. Copyright (2018) American Chemical Society.

therefore suspended in the supporting medium [144]. In comparison to fused deposition modelling and stereolithography, embedded 3D printing does not require additional support during printing, which helps to enhance the stability of the 3D-printed object.

Previous demonstrations of embedded 3D printing include (1) printing of collagen, alginate, and fibrin for creating femurs and branched coronary arteries [146], (2) human aortic endothelial cells for creating vascular networks [147], (3) silicone for the fabrication of soft robots [148], and (4) droplets, for the manipulation of selected individual droplets in droplet-based biological assays [149].

Besides bioprinting, embedded 3D printing was also explored in the field of polymer sciences and electronics. For instance, Karyappa *et al.* developed a method known as embedded ink writing, based on the concept of embedded 3D printing, for the printing of curable liquid polysiloxane within a liquid immiscible with the resins (e.g., ethanol, isopropanol, and methanol) [150]. The surrounding media allowed the structures of the patterned polysiloxane within the liquid to be maintained until curing, and the embedding liquid was subsequently removed by evaporation [150]. Embedded ink writing allowed for the direct writing of polysiloxane resins, which could be used to make microfluidic devices, flexible wearables, and soft actuators [150]. Similarly, embedded 3D printing has been demonstrated for thermoplastics [151, 152].

Emerging works have been explored to apply embedded 3D printing for food production. The use of yield stress fluids as a support bath for embedded 3D food printing was first demonstrated in a recent work by Yang *et al.*, who demonstrated how embedded 3D food printing could be used to overcome current constraints of 3D food printing technologies in attainable geometries of the printed foods [153]. The printing bath was made of biocompatible and edible hydrogels with shear-thinning behaviour, enabling the creation of freeform food shapes with fluid-like inks. The printed construct could then be subsequently freed by washing out the support bath or left inside the curable bath. Yang *et al.* also detailed the considerations for choosing the appropriate support bath and discussed their method of preparation (Figure 9.16) [153], and compared two kinds of support baths for different food ink characteristics, namely, gelatin and Carbopol. The former is obtained from animal tissues and is commonly the primary ingredient for producing jelly [153], while the latter is a known polyacrylic acid powder possessing the chemical name carbomer [147, 154], which is often used as a thickening agent in cosmetics,

(a)

(b)

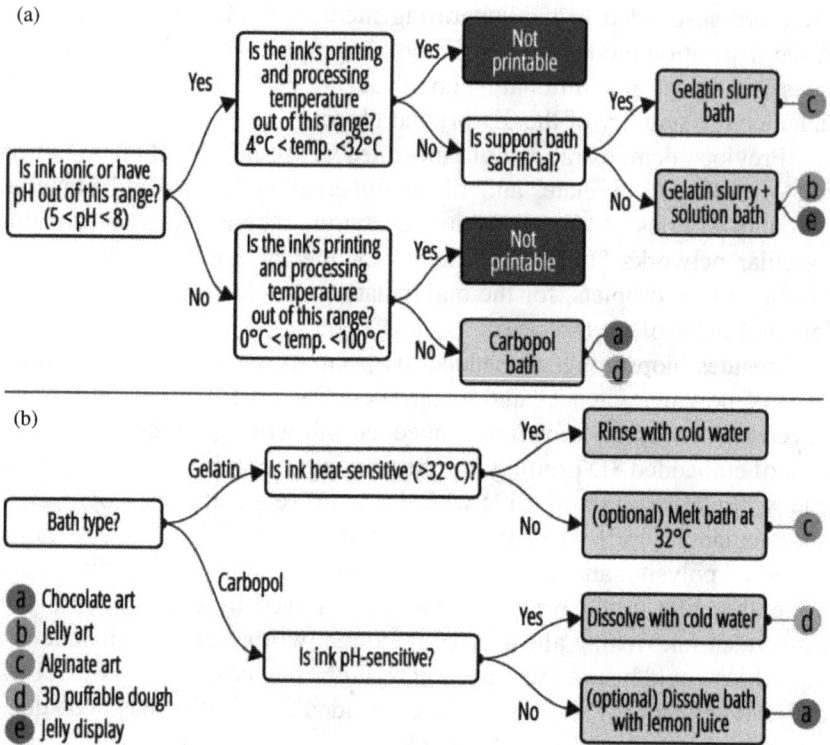

Figure 9.16.   Guidelines highlighting the steps for choosing (a) support bath types, and (b) release methods [153]. Reproduced from [153], which is licensed under CC BY 4.0.

pharmaceuticals, and food packaging [153]. Carbopol is also widely used for embedded 3D printing of non-edible materials.

To make the support bath for either medium, gelatin was prepared as a block of gel before being blended into fine particles to produce a slurry. Carbopol powder was mixed in alkaline water to generate a 1% solution, which has the consistency of hair gel [155]. Both the gelatin and Carbopol support baths offer omnidirectional floatation to the extruded ink.

The pH values of the food ink and printing temperature are two key aspects to consider when selecting the appropriate support bath, as both support baths respond differently to these variables. Carbopol baths are stable over a large range of temperatures. However, the pH value and presence of ionic compounds (e.g., salts) can affect their capacity to remain gelated. On the other hand, gelatin is suitable as a support bath for inks with a wide range of pH values and is relatively unaffected by the presence of ionic substances. However, its low melting point (~32 °C) prevents it from being used as a support bath for inks that need extrusion above 32 °C. Besides, the gelatin support bath employed in bioprinting is often sacrificial; it can be destroyed if it is heated above 32 °C. In 3D food printing, however, there is an option to retain the gelatin support bath as part of the foods [153]. Currently, the two support baths of gelatin and Carbopol can support the printing of a large array of materials. However, other materials (i.e., ionic inks that need more than 32 °C extrusion or thermal processing, materials that need to be treated at below 0 °C or above 100 °C) are incompatible with the existing approach [153].

The methods to separate the printed product from the support bath are also dependent on the material property of the ink. The stimuli to which the support bath is sensitive can help in the release process as long as they are not reactive with the printed foods [153]. Heat can be used to dissolve the gelatin bath, while the Carbopol bath can be liquefied with acidic juice to release the printed constructs. Cold distilled water is used to rinse away the support baths if the procedures are deemed to be incompatible with the printed material.

Luo *et al.* used the following applications to demonstrate the fidelity, freeform design, and interaction of food created using fluidic materials and printed with a consumer-grade printer modified with a syringe pump and a commercial modular printer [155]. These works included freeform chocolate printing and jelly art produced by embedded 3D printing.

- Freeform chocolate printing

Freeform chocolate structures can be printed within the Carbopol support bath, with the reservoir of the extruder being heated at 35 °C to prevent the chocolate ink from hardening (Figure 9.17). Embedded 3D printing of chocolate helps to overcome the issue of the melted chocolate's inherent fluidity, which makes it hard to support its printed structure by conventional direct ink writing.

- Jelly art

Jelly art, also referred to as *wagashi* (Japanese-style confection), is demonstrated with embedded 3D printing of food inks within gelatin. Since gelatin is hydrophilic, the possibilities of using colourants in the printing inks are limited to water-insoluble compounds; otherwise, the dye will spread into the bath. To overcome this issue, the authors used fibre-laden carrot and chocolate syrup-based alginate inks for printing within a gelatin bath containing calcium chloride (Figure 9.18) [153].

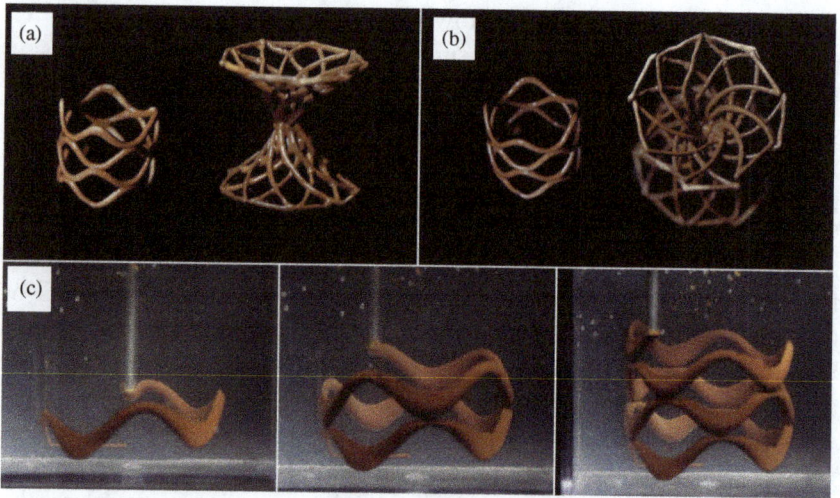

Figure 9.17.  Freeform embedded 3D printing of chocolate: (a) side view, (b) top view, and (c) printing process [153]. Reproduced from [153], which is licensed under CC BY 4.0.

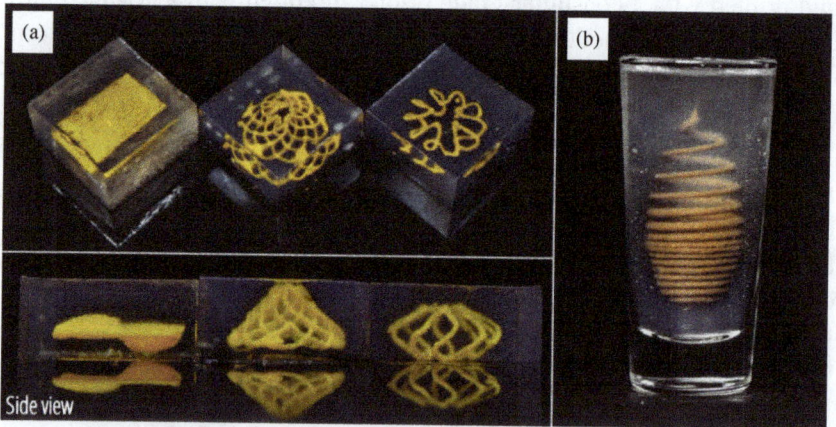

Figure 9.18.  Freeform embedded 3D printing of jelly art (*wagashi*) with (a) alginate-carrot ink, and (b) chocolate syrup ink [153]. Reproduced from [153], which is licensed under CC BY 4.0.

- Flower jellies

Flower jellies are a type of delicate dessert made with a flower-shaped jelly floating within another transparent jelly. In another related work, Miyatake *et al.* designed a flower jelly printer, which is a customized printer and design software for producing flower jellies [156]. They also developed the slit injection printing technique, where a petal-shaped knife creates a slit in the base jelly and injects the coloured jelly inside, taking the shape of the petal-shaped knife upon the retraction of a knife (Figure 9.19). Unlike most works in embedded 3D printing, the slit injection technique injects the coloured jelly directly into a firm bulk jelly [156]. This process is distinct from the use of granular gels in other demonstrations of embedded 3D printing [146]. By printing into a firm jelly, the 3D-printed jelly and surrounding jelly can be consumed without the firm jelly losing its aesthetic shape (Figure 9.20). This work introduces

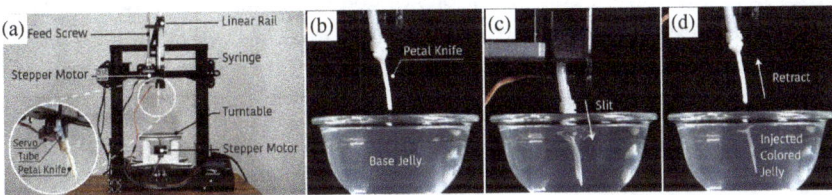

Figure 9.19. Flower jelly printer and the printing process: (a) the printer set-up, and (b) to (d) the slit injection sequence. Firstly, the petal-shaped knife creates a slit inside the base jelly. Subsequently, the knife is retracted while injecting the coloured jelly [156]. Reproduced from [156], which is licensed under CC BY-SA 4.0.

Figure 9.20. A showcase of the flower jellies and other printed patterns using the flower jelly printer [156]. Reproduced from [156], which is licensed under CC BY-SA 4.0.

new possibilities for greater personalization, fabrication, and consumption of 3D-printed gel-based foods in the future.

### 9.3.3  *Outlook*

Current works with 3D food printing using advanced fluidic technologies have yielded improvements to the additive manufacturing of food using conventional direct ink writing with a single nozzle. Among the microfluidics-enhanced extrusion 3D printing methods, printing with coaxial printheads is the most explored in 3D food printing. In the previous sections, it was demonstrated that coaxial printing can be used for the fabrication of filaments and fibres within 3D-printed foods to achieve desirable properties in terms of their texture, appearance, and shape. Future works for microfluidic-enhanced extrusion 3D food printing may explore the use of microfluidic chips to achieve multi-material, multi-ordered 3D food printing using varying configurations of channels (e.g., branched, coaxial) and fluid-control components (e.g., valve, mixer). Such microchannels should enable simultaneous, parallel, and alternate extrusion of multiple food inks with increased throughput to address the scalability for 3D food printing.

The adoption of embedded 3D printing for food production remains relatively new and pioneering, albeit it is a useful method for the fabrication of 3D-printed foods. In addition to the fabrication of complex and intricate foods for aesthetics and achieving desired mouthfeels, embedded 3D food printing enables the freeform fabrication of overhang features without a support structure. Embedded 3D printing allows the printing of inks with a larger range of viscosities, suggesting that rheological modifiers such as food thickeners may not be necessary to enhance their printability. Embedded 3D food printing possesses certain limitations [153]. Firstly, none of the support bath formulations (i.e., gelatin and Carbopol) can support ionic food inks that require the process to be at freezing and boiling temperatures. Secondly, the use of hydrogel support baths also necessitates the use of water-insoluble colour pigments within the printing inks, lest they spread throughout the support bath during printing [153]. Future research may investigate the possibilities of a non-hydrogel-based support bath to broaden the operating conditions of embedded 3D printing. Gelatin is animal-derived, and the investigation of alternative materials that meet certain dietary requirements may also help dissimilate the technology.

## 9.4 The Role of Data and Artificial Intelligence in Future 3D Food Printing

Data and artificial intelligence (AI) have been playing a bigger role in various areas like Industry 4.0, Robotics, and Fintech. Well-being is another area that has increasing demand to utilize the strength of AI to benefit society and, in particular, create personalized nutrition for a wide range of people, from "healthy" (e.g., sportsmen, soldiers) to "normal" individuals. Sportsmen are known to take specially designed nutritional food to boost their performance. Diet, on the other hand, is also critical for the well-being of normal people because it can help with chronic disease management and improve an individual's health. This is an area of research that involves several disciplines and combines findings from one's health history and lifestyle to provide personalized nutrition.

Food is critical because it provides us with energy and nutrients. Nutrient management, especially for sportsmen and the elderly, is a growing market with increasing importance. In addition, people are willing to pay to enjoy delicious food and are much less resistant to 3D-printed food as compared to regular medicine. Thus, there is huge potential for personalized nutrition, which 3D food printing can deliver, as one of its strengths is customization.

In order to achieve personalized nutrition via 3D printing, information extraction of the nutrients in food as well as the health and dietary requirements of any particular person is the first step and will be covered. After which, the 3D food printer can utilize AI and customize the nutrients in the dish for the person. There are, however, issues in establishing a large database for this purpose, which the challenges will be discussed in the next section.

### 9.4.1 *Nutrition Information Extraction*

Food and nutrition knowledge has been increasing and there is increasing interest to extract even more knowledge from raw food and nutrition-related data [157]. However, there is a wide range of food around the world and it is challenging to predict the exact nutritional value of food based solely on recipes. Ispirova *et al.* [157] have proposed performing machine learning on a large corpus of recipe data, where domain knowledge for data clustering can be used by machine learning to estimate the nutritional content of any designed dish based on its recipe.

With such capabilities now being developed by various researchers around the world, there is high potential that a 3D food printer can utilize these capabilities and automatically customize a dish with personalized nutrition. However, to achieve this, such a system needs to know a person's health history and dietary requirements.

Personalized and precision nutrition requires information about the person. Several sensors have been proposed for monitoring purposes. For example, a smart dining table that is capable of measuring the weight of food could also monitor the number of calories a person consumes [158]. In addition, camera-based solutions [159] could utilize images to estimate the amount of food intake, while sensors like accelerometers could monitor physiological processes relating to the food consumption process, like hand-to-mouth gestures, biting, chewing, and swallowing, to estimate the calorie intake [160]. A review of the existing methods, which include both sensors and mobile phones apps, have been nicely summarized by Sempionatto *et al.* [161].

Precision nutrition will require information about the person, and biomarkers have been proposed by several researchers to be part of this feedback information to allow AI to output a customized nutrition meal. Environmental genomics has also been investigated to offer more cost-effective monitoring as compared to conventional biomonitoring [162]. In addition, machine learning can be combined with microbial metabarcoding to enable improved classification and accuracy.

Accurate estimation of the energy requirement is important. The total daily energy expenditure in a day is the sum of various components like: (1) resting energy expenditure, (2) diet-induced thermogenesis, (3) energy expenditure due to physical activities, and (4) energy spent on growth for children [163]. A review of nutrition approaches that utilize machine learning is available in [164].

## 9.4.2 *Personalized Nutrition Intervention and Challenges*

General guidelines on diet and lifestyle have been provided by different health promotion agencies in various countries. However, individually customized interventions have increasingly been considered the direction to go next. Milani *et al.* [164] have shown promising results in applying AI to help pediatrics track obesity among children. AI is used to predict the glycemic response, which can then be fed into the system to recommend a personalized meal.

A prediction of glycemic responses has also been proposed to modify elevated postprandial blood glucose and its metabolic consequences [165]. High blood glucose levels are a concern in many countries, as the issues that diabetes can cause is significant. Different people present different post-meal blood glucose responses. Hence, personalized diets that can demonstrate lower postprandial responses have been proposed, which illustrates the potential impact of customized nutrition. With AI now being able to predict human responses as well as recommend personalized nutrition, 3D food printing can potentially be used to output a dish that meets a person's precise nutrition requirements.

To implement a larger scale and sustainable personalized nutrition diet, a few challenges need to be overcome. Verma *et al.* [166] succinctly summarized these challenges: (1) limitations in reductionist approaches, (2) a need for personalized nutrition infrastructure, (3) data standardization, and (4) data sparsity and the need for improved methods.

Personalized nutrition is still relatively new, and guiding principles are required for its implementation. Thus, Adams *et al.* [167] have proposed a set of 10 guiding principles that include: (1) clear definitions of potential users and beneficiaries, (2) usage of validated diagnostic methods and measures, (3) maintenance of data quality and relevance, (4) data-driven recommendations from validated models and algorithms, (5) the design of personalized nutrition studies around validated individual health or function needs and outcomes, (6) provision of rigorous scientific evidence for an effect on health or function, (7) deliverance of user-friendly tools, (8) alignment with population-based recommendations (for healthy individuals), (9) transparency about potential effects, and (10) protection of individual data privacy and responsiblities.

With the advancement of AI as well as food engineering technologies like the 3D food printer, there is a need to model them as a cyber-physical system. Cyber-physical systems have been proposed in several papers and are key to more nutritious and sustainable paths for novel food systems [168]. Athough these systems have been around for many years, they are still relatively new when it comes to the food industry. Efforts in this area are still being made, and there is huge growth potential for cyber-physical systems in the field of personalized nutrition.

# References

[1]  Khosravani, M. R. and Reinicke, T. (2020). On the environmental impacts of 3D printing technology, *Applied Materials Today*, 20, p. 100689.

[2]  Reis, G. G., Heidemann, M. S., Matos, K. H. O. D. and Molento, C. F. M. (2020). Cell-based meat and firms' environmental strategies: new rationales as per available literature, *Sustainability*, 12, p. 9418.

[3]  Gebler, M., Schoot Uiterkamp, A. J. M. and Visser, C. (2014). A global sustainability perspective on 3D printing technologies, *Energy Policy*, 74, pp. 158–167.

[4]  Kouzani, A. Z., Adams, S., Whyte, D. J., Oliver, R., Hemsley, B., Palmer, S. and Balandin, S. (2017). 3D printing of food for people with swallowing difficulties, *KnE Engineering*, 2, pp. 23–29.

[5]  Ramachandraiah, K. (2021). Potential development of sustainable 3D-printed meat analogues: a review, *Sustainability*, 13, p. 938.

[6]  Sun, J., Zhou, W., Huang, D., Fuh, J. Y. H. and Hong, G. S. (2015). An overview of 3D printing technologies for food fabrication, *Food and Bioprocess Technology*, 8, pp. 1605–1615.

[7]  Mattick, C. S., Landis, A. E., Allenby, B. R. and Genovese, N. J. (2015). Anticipatory life cycle analysis of *in vitro* biomass cultivation for cultured meat production in the United States, *Environmental Science & Technology*, 49, pp. 11941–11949.

[8]  Tuomisto, H. L. and Teixeira de Mattos, M. J. (2011). Environmental impacts of cultured meat production, *Environmental Science & Technology*, 45, pp. 6117–6123.

[9]  Oonincx, D. G. A. B. and de Boer, I. J. M. (2012). Environmental impact of the production of mealworms as a protein source for humans — a life cycle assessment, *PLOS One*, 7, p. e51145.

[10]  Ghosh, S., Gillis, A., Sheviryov, J., Levkov, K. and Golberg, A. (2019). Towards waste meat biorefinery: extraction of proteins from waste chicken meat with non-thermal pulsed electric fields and mechanical pressing, *Journal of Cleaner Production*, 208, pp. 220–231.

[11]  Jayathilakan, K., Sultana, K., Radhakrishna, K. and Bawa, A. S. (2012). Utilization of byproducts and waste materials from meat, poultry and fish processing industries: a review, *Journal of Food Science and Technology*, 49, pp. 278–293.

[12]  Govindharaj, M., Roopavath, U. K. and Rath, S. N. (2019). Valorization of discarded Marine Eel fish skin for collagen extraction as a 3D printable blue biomaterial for tissue engineering, *Journal of Cleaner Production*, 230, pp. 412–419.

[13]  Parodi, A., De Boer, I. J. M., Gerrits, W. J. J., Van Loon, J. J. A., Heetkamp, M. J. W., Van Schelt, J. and Van Zanten, H. H. E. (2020).

Bioconversion efficiencies, greenhouse gas and ammonia emissions during black soldier fly rearing — a mass balance approach, *Journal of Cleaner Production*, 271, p. 122488.

[14] Kaipia, R., Dukovska-Popovska, I. and Loikkanen, L. (2013). Creating sustainable fresh food supply chains through waste reduction, *International Journal of Physical Distribution & Logistics Management*, 43, pp. 262–276.

[15] Qiu, F., Hu, Q. and Xu, B. (2020). Fresh agricultural products supply chain coordination and volume loss reduction based on strategic consumer, *International Journal of Environmental Research and Public Health*, 17, p. 7915.

[16] Choudhury, D., Tseng, T. W. and Swartz, E. (2020). The business of cultured meat, *Trends in Biotechnology*, 38, pp. 573–577.

[17] Zhang, J., Liu, L., Jiang, Y., Faisal, S., Wei, L., Cao, C. and Wang, Q. (2019). Converting peanut protein biomass waste into "double green" meat substitutes using a high-moisture extrusion process: a multiscale method to explore a process for forming a meat-like fibrous structure, *Journal of Agricultural and Food Chemistry*, 67, pp. 10713–10725.

[18] Jiang, G., Ameer, K., Kim, H., Lee, E.-J., Ramachandraiah, K. and Hong, G.-P. (2020). Strategies for sustainable substitution of livestock meat, *Foods*, 9, p. 1227.

[19] Teo, C. C., Tan, S. N., Yong, J. W., Hew, C. S. and Ong, E. S. (2010). Pressurized Hot Water Extraction (PHWE), *Journal of Chromatography A*, 1217, pp. 2484–2494.

[20] Ong, E. S., Pek, C. J. N., Tan, J. C. W. and Leo, C. H. (2020). Antioxidant and cytoprotective effect of quinoa (Chenopodium quinoa Willd.) with Pressurized Hot Water Extraction (PHWE). *Antioxidants*, 9, p. 1110. doi. org/1110.3390/antiox9111110.

[21] Ong, E. S., Oh, C. L. Y., Tan, J. C. W., Foo, S. Y. and Leo, C. H. (2021). Pressurized hot water extraction of okra seeds reveals antioxidant, antidiabetic and vasoprotective activities, *Plants*, 10, p. 1645. https://doi. org/1610.3390/plants10081645

[22] Todd, R. and Baroutian, S. (2017). A techno-economic comparison of subcritical water, supercritical $CO_2$ and organic solvent extraction of bioactives from grape marc, *Journal of Cleaner Production*, 158, pp. 349–358.

[23] Lee, C. P., Takahashi, M., Arai, S., Lee, C. K. L. and Hashimoto, M. (2021). 3D printing of Okara ink: the effect of particle size on the printability, *ACS Food Science & Technology*, 1, pp. 2053–2061.

[24] Zimon, D., Madzik, P. and Domingues, P. (2020). Development of key processes along the supply chain by implementing the ISO 22000 standard, *Sustainability*, 12, p. 61.

[25] Murphy, S. V., De Coppi, P. and Atala, A. (2020). Opportunities and challenges of translational 3D bioprinting, *Nature Biomedical Engineering*, 4, pp. 370–380.

[26] Jia, A., Chua, C. K. and Mironov, V. (2016). A perspective on 4D bioprinting, *International Journal of Bioprinting*, 2, pp. 3–5.

[27] Lee, A. Y., An, J. and Chua, C. K. (2017). Two-way 4D printing: a review on the reversibility of 3D-printed shape memory materials, *Engineering*, 3, pp. 663–674.

[28] Lee, A. Y., An, J., Chua, C. K. and Zhang, Y. (2019). Preliminary investigation of the reversible 4D printing of a dual-layer component, *Engineering*, 5, pp. 1159–1170.

[29] Liu, Y., Zhang, W., Zhang, F., Lan, X., Leng, J., Liu, S., Jia, X., Cotton, C., Sun, B. and Gu, B. (2018). Shape memory behavior and recovery force of 4D printed laminated Miura-origami structures subjected to compressive loading, *Composites Part B: Engineering*, 153, pp. 233–242.

[30] Zhu, P., Yang, W., Wang, R., Gao, S., Li, B. and Li, Q. (2018). 4D printing of complex structures with a fast response time to magnetic stimulus, *ACS Applied Materials & Interfaces*, 10, pp. 36435–36442.

[31] Momeni, F., Liu, X. and Ni, J. (2017). A review of 4D printing, *Materials & Design*, 122, pp. 42–79.

[32] Pei, E. (2014). 4D Printing: dawn of an emerging technology cycle, *Assembly Automation*, 4, pp. 310–314.

[33] Khoo, Z. X., Teoh, J. E. M., Liu, Y., Chua, C. K., Yang, S., An, J., Leong, K. F. and Yeong, W. Y. (2015). 3D printing of smart materials: a review on recent progresses in 4D printing, *Virtual and Physical Prototyping*, 10, pp. 103–122.

[34] Alshebly, Y. S., Nafea, M., Ali, M. S. M. and Almurib, H. A. (2021). Review on recent advances in 4D printing of shape memory polymers, *European Polymer Journal*, 159, p. 110708.

[35] Tibbits, S., McKnelly, C., Olguin, C., Dikovsky, D. and Hirsch, S. (2014). 4D printing and universal transformation.

[36] Wang, W., Yao, L., Zhang, T., Cheng, C.-Y., Levine, D. and Ishii, H. (2017). Transformative appetite: shape-changing food transforms from 2D to 3D by water interaction through cooking. In *Proceedings of the 2017 CHI Conference on Human Factors in Computing Systems*, pp. 6123–6132.

[37] Tao, Y., Do, Y., Yang, H., Lee, Y.-C., Wang, G., Mondoa, C., Cui, J., Wang, W. and Yao, L. (2019). Morphlour: personalized flour-based morphing food induced by dehydration or hydration method. In *Proceedings of the 32nd Annual ACM Symposium on User Interface Software and Technology*, pp. 329–340.

[38] He, C., Zhang, M. and Devahastin, S. (2020). Investigation on spontaneous shape change of 4D printed starch-based purees from purple sweet

potatoes as induced by microwave dehydration, *ACS Applied Materials & Interfaces*, 12, pp. 37896–37905.

[39] Levy, M. (2019). Student Winner Speculative Design Award Core77 Design Awards 2019. Retrieved from https://designawards.core77.com/speculative-design/84449/Neo-Fruit

[40] He, C., Zhang, M. and Guo, C. (2020). 4D printing of mashed potato/purple sweet potato puree with spontaneous color change, *Innovative Food Science & Emerging Technologies*, 59, p. 102250.

[41] Ghazal, A. F., Zhang, M. and Liu, Z. (2019). Spontaneous color change of 3D printed healthy food product over time after printing as a novel application for 4D food printing, *Food and Bioprocess Technology*, 12, pp. 1627–1645.

[42] Shanthamma, S., Preethi, R., Moses, J. and Anandharamakrishnan, C. (2021). 4D printing of sago starch with turmeric blends: a study on pH-triggered spontaneous color transformation, *ACS Food Science & Technology*, 1, pp. 669–679.

[43] Yao, L., Ou, J., Cheng, C.-Y., Steiner, H., Wang, W., Wang, G. and Ishii, H. (2015). BioLogic: natto cells as nanoactuators for shape changing interfaces. In *Proceedings of the 33rd Annual ACM Conference on Human Factors in Computing Systems*, pp. 1–10.

[44] Institute for Quality and Efficiency in Health Care (IQWiG). (2011). How does our sense of taste work? Retrieved from https://www.ncbi.nlm.nih.gov/books/NBK279408/

[45] Kan, V., Vargo, E., Machover, N., Ishii, H., Pan, S., Chen, W. and Kakehi, Y. (2017). Organic primitives: synthesis and design of pH-reactive materials using molecular I/O for sensing, actuation, and interaction. In *Proceedings of the 2017 CHI Conference on Human Factors in Computing Systems*, pp. 989–1000.

[46] Spence, C. (2015). Eating with our ears: assessing the importance of the sounds of consumption on our perception and enjoyment of multisensory flavour experiences, *Flavour*, 4, pp. 1–14.

[47] Lu, R. (2013). *Instrumental Assessment of Food Sensory Quality*, "Principles of solid food texture analysis" (Elsevier), pp. 103–128.

[48] Pant, A., Lee, A. Y., Karyappa, R., Lee, C. P., An, J., Hashimoto, M., Tan, U.-X., Wong, G., Chua, C. K. and Zhang, Y. (2021). 3D food printing of fresh vegetables using food hydrocolloids for dysphagic patients, *Food Hydrocolloids*, 114, p. 106546.

[49] Lee, A. Y., Pant, A., Pojchanun, K., Lee, C. P., An, J., Hashimoto, M., Tan, U.-X., Leo, C. H., Wong, G. and Chua, C. K. (2021). Three-dimensional printing of food foams stabilized by hydrocolloids for hydration in dysphagia, *International Journal of Bioprinting*, 7, p. 393.

[50]  Zhang, Q., Zhang, M., Djeghlaf, L., Bataille, J., Gamby, J., Haghiri-Gosnet, A. M. and Pallandre, A. (2017). Logic digital fluidic in miniaturized functional devices: perspective to the next generation of microfluidic lab-on-chips, *Electrophoresis*, 38, pp. 953–976.

[51]  He, S., Joseph, N., Feng, S., Jellicoe, M. and Raston, C. L. (2020). Application of microfluidic technology in food processing, *Food & Function*, 11, pp. 5726–5737.

[52]  Michelon, M., Oliveira, D. R. B., de Figueiredo Furtado, G., de la Torre, L. G. and Cunha, R. L. (2017). High-throughput continuous production of liposomes using hydrodynamic flow-focusing microfluidic devices, *Colloids and Surfaces B: Biointerfaces*, 156, pp. 349–357.

[53]  Hou, X., Zhang, Y. S., Trujillo-de Santiago, G., Alvarez, M. M., Ribas, J., Jonas, S. J., Weiss, P. S., Andrews, A. M., Aizenberg, J. and Khademhosseini, A. (2017). Interplay between materials and microfluidics, *Nature Reviews Materials*, 2, pp. 1–15.

[54]  Calejo, J., Pinho, D., Galindo-Rosales, F. J., Lima, R. and Campo-Deaño, L. (2016). Particulate blood analogues reproducing the erythrocytes cell-free layer in a microfluidic device containing a hyperbolic contraction, *Micromachines*, 7, p. 4.

[55]  Nightingale, A. M., Beaton, A. D. and Mowlem, M. C. (2015). Trends in microfluidic systems for *in situ* chemical analysis of natural waters, *Sensors and Actuators B: Chemical*, 221, pp. 1398–1405.

[56]  Rackus, D. G., Shamsi, M. H. and Wheeler, A. R. (2015). Electrochemistry, biosensors and microfluidics: a convergence of fields, *Chemical Society Reviews*, 44, pp. 5320–5340.

[57]  Redman, E. A., Batz, N. G., Mellors, J. S. and Ramsey, J. M. (2015). Integrated microfluidic capillary electrophoresis-electrospray ionization devices with online MS detection for the separation and characterization of intact monoclonal antibody variants, *Analytical Chemistry*, 87, pp. 2264–2272.

[58]  Damiati, S., Kompella, U. B., Damiati, S. A. and Kodzius, R. (2018). Microfluidic devices for drug delivery systems and drug screening, *Genes*, 9, p. 103.

[59]  Chen, J., Xue, C., Zhao, Y., Chen, D., Wu, M.-H. and Wang, J. (2015). Microfluidic impedance flow cytometry enabling high-throughput single-cell electrical property characterization, *International Journal of Molecular Sciences*, 16, pp. 9804–9830.

[60]  Wen, N., Zhao, Z., Fan, B., Chen, D., Men, D., Wang, J. and Chen, J. (2016). Development of droplet microfluidics enabling high-throughput single-cell analysis, *Molecules*, 21, p. 881.

[61]  Peters, K. L., Corbin, I., Kaufman, L. M., Zreibe, K., Blanes, L. and McCord, B. R. (2015). Simultaneous colorimetric detection of improvised

explosive compounds using microfluidic paper-based analytical devices (μPADs), *Analytical Methods*, 7, pp. 63–70.

[62] Schroen, K., Bliznyuk, O., Muijlwijk, K., Sahin, S. and Berton-Carabin, C. C. (2015). Microfluidic emulsification devices: from micrometer insights to large-scale food emulsion production, *Current Opinion in Food Science*, 3, pp. 33–40.

[63] Kastner, E., Verma, V., Lowry, D. and Perrie, Y. (2015). Microfluidic-controlled manufacture of liposomes for the solubilisation of a poorly water soluble drug, *International Journal of Pharmaceutics*, 485, pp. 122–130.

[64] Teh, S.-Y., Lin, R., Hung, L.-H. and Lee, A. P. (2008). Droplet microfluidics, *Lab on a Chip*, 8, pp. 198–220.

[65] Hettiarachchi, S., Melroy, G., Mudugamuwa, A., Sampath, P., Premachandra, C., Amarasinghe, R. and Dau, V. (2021). Design and development of a microfluidic droplet generator with vision sensing for lab-on-a-chip devices, *Sensors and Actuators A: Physical*, 332, p. 113047.

[66] Costa, A. L. R., Gomes, A. and Cunha, R. L. (2017). Studies of droplets formation regime and actual flow rate of liquid-liquid flows in flow-focusing microfluidic devices, *Experimental Thermal and Fluid Science*, 85, pp. 167–175.

[67] Chen, X., Glawdel, T., Cui, N. and Ren, C. L. (2015). Model of droplet generation in flow focusing generators operating in the squeezing regime, *Microfluidics and Nanofluidics*, 18, pp. 1341–1353.

[68] Jing, T., Ramji, R., Warkiani, M. E., Han, J., Lim, C. T. and Chen, C.-H. (2015). Jetting microfluidics with size-sorting capability for single-cell protease detection, *Biosensors and Bioelectronics*, 66, pp. 19–23.

[69] Yu, Y., Fu, F., Shang, L., Cheng, Y., Gu, Z. and Zhao, Y. (2017). Bioinspired helical microfibers from microfluidics, *Advanced Materials*, 29, p. 1605765.

[70] Vijayan, S. and Hashimoto, M. (2019). 3D printed fittings and fluidic modules for customizable droplet generators, *RSC Advances*, 9, pp. 2822–2828.

[71] Martino, C., Berger, S., Wootton, R. C. and deMello, A. J. (2014). A 3D-printed microcapillary assembly for facile double emulsion generation, *Lab on a Chip*, 14, pp. 4178–4182.

[72] Meng, Z.-J., Wang, W., Liang, X., Zheng, W.-C., Deng, N.-N., Xie, R., Ju, X.-J., Liu, Z. and Chu, L.-Y. (2015). Plug-n-play microfluidic systems from flexible assembly of glass-based flow-control modules, *Lab on a Chip*, 15, pp. 1869–1878.

[73] Zhang, J. M., Aguirre-Pablo, A. A., Li, E. Q., Buttner, U. and Thoroddsen, S. T. (2016). Droplet generation in cross-flow for cost-effective 3D-printed "plug-and-play" microfluidic devices, *RSC Advances*, 6, pp. 81120–81129.

[74]  Boskovic, D. and Loebbecke, S. (2014). Synthesis of polymer particles and capsules employing microfluidic techniques, *Nanotechnology Reviews*, 3, pp. 27–38.

[75]  McDonald, J. C., Duffy, D. C., Anderson, J. R., Chiu, D. T., Wu, H., Schueller, O. J. and Whitesides, G. M. (2000). Fabrication of microfluidic systems in poly (dimethylsiloxane), *ELECTROPHORESIS*, 21, pp. 27–40.

[76]  Ushikubo, F., Birribilli, F., Oliveira, D. and Cunha, R. (2014). Y-and T-junction microfluidic devices: effect of fluids and interface properties and operating conditions, *Microfluidics and Nanofluidics*, 17, pp. 711–720.

[77]  Garstecki, M., Fuerstman M. J., Stone, H. A. and Whitesides G. M. (2006) Formation of droplets and bubbles in a microfluidic T-junction — scaling and mechanism of break-up, *Lab on a Chip*, 6, p. 437.

[78]  Tan, Y.-C., Cristini, V. and Lee, A. P. (2006). Monodispersed microfluidic droplet generation by shear focusing microfluidic device, *Sensors and Actuators B: Chemical*, 114, pp. 350–356.

[79]  Anna, S. L., Bontoux, N. and Stone, H. A. (2003). Formation of dispersions using "flow focusing" in microchannels, *Applied Physics Letters*, 82, pp. 364–366.

[80]  Galus, S. and Kadzińska, J. (2015). Food applications of emulsion-based edible films and coatings, *Trends in Food Science & Technology*, 45, pp. 273–283.

[81]  Đorđević, V., Balanč, B., Belščak-Cvitanović, A., Lević, S., Trifković, K., Kalušević, A., Kostić, I., Komes, D., Bugarski, B. and Nedović, V. (2015). Trends in encapsulation technologies for delivery of food bioactive compounds, *Food Engineering Reviews*, 7, pp. 452–490.

[82]  Skurtys, O. and Aguilera, J. (2008). Applications of microfluidic devices in food engineering, *Food Biophysics*, 3, pp. 1–15.

[83]  Comunian, T. A., Ravanfar, R., Alcaine, S. D. and Abbaspourrad, A. (2018). Water-in-oil-in-water emulsion obtained by glass microfluidic device for protection and heat-triggered release of natural pigments, *Food Research International*, 106, pp. 945–951.

[84]  Michelon, M., Huang, Y., de la Torre, L. G., Weitz, D. A. and Cunha, R. L. (2019). Single-step microfluidic production of W/O/W double emulsions as templates for β-carotene-loaded giant liposomes formation, *Chemical Engineering Journal*, 366, pp. 27–32.

[85]  Varshney, M., Li, Y., Srinivasan, B. and Tung, S. (2007). A label-free, microfluidics and interdigitated array microelectrode-based impedance biosensor in combination with nanoparticles immunoseparation for detection of Escherichia coli O157: H7 in food samples, *Sensors and Actuators B: Chemical*, 128, pp. 99–107.

[86] Kim, T.-H., Park, J., Kim, C.-J. and Cho, Y.-K. (2014). Fully integrated lab-on-a-disc for nucleic acid analysis of food-borne pathogens, *Analytical Chemistry*, 86, pp. 3841–3848.

[87] Oscar, S.-V., Fernando, O.-C. L. and Del Pilar, C.-M. M. (2017). Total polyphenols content in white wines on a microfluidic flow injection analyzer with embedded optical fibers, *Food Chemistry*, 221, pp. 1062–1068.

[88] Serex, L., Bertsch, A. and Renaud, P. (2018). Microfluidics: a new layer of control for extrusion-based 3D printing, *Micromachines*, 9, p. 86.

[89] Richard, C., Neild, A. and Cadarso, V. J. (2020). The emerging role of microfluidics in multi-material 3D bioprinting, *Lab on a Chip*, 20, pp. 2044–2056.

[90] du Chatinier, D. N., Figler, K. P., Agrawal, P., Liu, W. and Zhang, Y. S. (2021). The potential of microfluidics-enhanced extrusion bioprinting, *Biomicrofluidics*, 15, p. 041304.

[91] Ma, J., Wang, Y. and Liu, J. (2018). Bioprinting of 3D tissues/organs combined with microfluidics, *RSC Advances*, 8, pp. 21712–21727.

[92] Levato, R., Jungst, T., Scheuring, R. G., Blunk, T., Groll, J. and Malda, J. (2020). From shape to function: the next step in bioprinting, *Advanced Materials*, 32, p. 1906423.

[93] Ober, T. J., Foresti, D. and Lewis, J. A. (2015). Active mixing of complex fluids at the microscale, *Proceedings of the National Academy of Sciences*, 112, pp. 12293–12298.

[94] Stroock, A. D., Dertinger, S. K., Ajdari, A., Mezić, I., Stone, H. A. and Whitesides, G. M. (2002). Chaotic mixer for microchannels, *Science*, 295, pp. 647–651.

[95] Lee, C.-Y., Chang, C.-L., Wang, Y.-N. and Fu, L.-M. (2011). Microfluidic mixing: a review, *International journal of Molecular Sciences*, 12, pp. 3263–3287.

[96] Pati, F., Jang, J., Lee, J. W. and Cho, D.-W. (2015). *Essentials of 3D Biofabrication and Translation*, "Extrusion bioprinting" (Elsevier), pp. 123–152.

[97] Lee, C., Abelseth, E., De La Vega, L. and Willerth, S. (2019). Bioprinting a novel glioblastoma tumor model using a fibrin-based bioink for drug screening, *Materials Today Chemistry*, 12, pp. 78–84.

[98] Pi, Q., Maharjan, S., Yan, X., Liu, X., Singh, B., van Genderen, A. M., Robledo-Padilla, F., Parra-Saldivar, R., Hu, N. and Jia, W. (2018). Digitally tunable microfluidic bioprinting of multilayered cannular tissues, *Advanced Materials*, 30, p. 1706913.

[99] Hansen, C. J., Saksena, R., Kolesky, D. B., Vericella, J. J., Kranz, S. J., Muldowney, G. P., Christensen, K. T. and Lewis, J. A. (2013). High-throughput printing via microvascular multinozzle arrays, *Advanced Materials*, 25, pp. 96–102.

[100]  Skylar-Scott, M. A., Mueller, J., Visser, C. W. and Lewis, J. A. (2019). Voxelated soft matter via multimaterial multinozzle 3D printing, *Nature*, 575, pp. 330–335.
[101]  Costantini, M., Colosi, C., Święszkowski, W. and Barbetta, A. (2018). Co-axial wet-spinning in 3D bioprinting: state of the art and future perspective of microfluidic integration, *Biofabrication*, 11, p. 012001.
[102]  Onoe, H., Okitsu, T., Itou, A., Kato-Negishi, M., Gojo, R., Kiriya, D., Sato, K., Miura, S., Iwanaga, S. and Kuribayashi-Shigetomi, K. (2013). Metre-long cell-laden microfibres exhibit tissue morphologies and functions, *Nature Materials*, 12, pp. 584–590.
[103]  Ozbolat, I. T. and Hospodiuk, M. (2016). Current advances and future perspectives in extrusion-based bioprinting, *Biomaterials*, 76, pp. 321–343.
[104]  Zhu, K., Chen, N., Liu, X., Mu, X., Zhang, W., Wang, C. and Zhang, Y. S. (2018). A general strategy for extrusion bioprinting of bio-macromolecular bioinks through alginate-templated dual-stage crosslinking, *Macromolecular Bioscience*, 18, p. 1800127.
[105]  Zhang, Y. S., Arneri, A., Bersini, S., Shin, S.-R., Zhu, K., Goli-Malekabadi, Z., Aleman, J., Colosi, C., Busignani, F. and Dell'Erba, V. (2016). Bioprinting 3D microfibrous scaffolds for engineering endothelialized myocardium and heart-on-a-chip, *Biomaterials*, 110, pp. 45–59.
[106]  Gao, G., Lee, J. H., Jang, J., Lee, D. H., Kong, J. S., Kim, B. S., Choi, Y. J., Jang, W. B., Hong, Y. J. and Kwon, S. M. (2017). Tissue engineered bio-blood-vessels constructed using a tissue-specific bioink and 3D coaxial cell printing technique: a novel therapy for ischemic disease, *Advanced Functional Materials*, 27, p. 1700798.
[107]  Zhang, Y. S., Pi, Q. and van Genderen, A. M. (2017). Microfluidic bioprinting for engineering vascularized tissues and organoids, *Journal of Visualized Experiments*, 11, p. 55957.
[108]  Carmeliet, P. and Jain, R. K. (2000). Angiogenesis in cancer and other diseases, *Nature*, 407, pp. 249–257.
[109]  Liu, T., Liu, Q., Anaya, I., Huang, D., Kong, W., Mille, L. S. and Zhang, Y. S. (2021). Investigating lymphangiogenesis in a sacrificially bioprinted volumetric model of breast tumor tissue, *Methods*, 190, pp. 72–79.
[110]  Gong, J., Schuurmans, C. C., van Genderen, A. M., Cao, X., Li, W., Cheng, F., He, J. J., López, A., Huerta, V. and Manríquez, J. (2020). Complexation-induced resolution enhancement of 3D-printed hydrogel constructs, *Nature Communications*, 11, pp. 1–14.
[111]  Li, S., Liu, Y., Li, Y., Liu, C., Sun, Y. and Hu, Q. (2016). A novel method for fabricating engineered structures with branched micro-channel using hollow hydrogel fibers, *Biomicrofluidics*, 10, p. 064104.

[112] Zhang, Y., Yu, Y., Chen, H. and Ozbolat, I. T. (2013). Characterization of printable cellular micro-fluidic channels for tissue engineering, *Biofabrication*, 5, p. 025004.

[113] Gao, Q., He, Y., Fu, J.-z., Liu, A. and Ma, L. (2015). Coaxial nozzle-assisted 3D bioprinting with built-in microchannels for nutrients delivery, *Biomaterials*, 61, pp. 203–215.

[114] Gao, Q., Liu, Z., Lin, Z., Qiu, J., Liu, Y., Liu, A., Wang, Y., Xiang, M., Chen, B. and Fu, J. (2017). 3D bioprinting of vessel-like structures with multilevel fluidic channels, *ACS Biomaterials Science & Engineering*, 3, pp. 399–408.

[115] Attalla, R., Ling, C. and Selvaganapathy, P. (2016). Fabrication and characterization of gels with integrated channels using 3D printing with microfluidic nozzle for tissue engineering applications, *Biomed Microdevices*, 18, p. 17.

[116] Ghorbanian, S., Qasaimeh, M. A., Akbari, M., Tamayol, A. and Juncker, D. (2014). Microfluidic direct writer with integrated declogging mechanism for fabricating cell-laden hydrogel constructs, *Biomed Microdevices*, 16, pp. 387–395.

[117] Colosi, C., Shin, S. R., Manoharan, V., Massa, S., Costantini, M., Barbetta, A., Dokmeci, M. R., Dentini, M. and Khademhosseini, A. (2016). Microfluidic bioprinting of heterogeneous 3D tissue constructs using low-viscosity bioink, *Advanced Materials*, 28, pp. 677–684.

[118] Idaszek, J., Costantini, M., Karlsen, T. A., Jaroszewicz, J., Colosi, C., Testa, S., Fornetti, E., Bernardini, S., Seta, M. and Kasareło, K. (2019). 3D bioprinting of hydrogel constructs with cell and material gradients for the regeneration of full-thickness chondral defect using a microfluidic printing head, *Biofabrication*, 11, p. 044101.

[119] Costantini, M., Testa, S., Mozetic, P., Barbetta, A., Fuoco, C., Fornetti, E., Tamiro, F., Bernardini, S., Jaroszewicz, J. and Święszkowski, W. (2017). Microfluidic-enhanced 3D bioprinting of aligned myoblast-laden hydrogels leads to functionally organized myofibers *in vitro* and *in vivo*, *Biomaterials*, 131, pp. 98–110.

[120] Cooper, S., Maxwell, A., Kizana, E., Ghoddusi, M., Hardeman, E., Alexander, I., Allen, D. and North, K. (2004). C2C12 co-culture on a fibroblast substratum enables sustained survival of contractile, highly differentiated myotubes with peripheral nuclei and adult fast myosin expression, *Cell Motility and the Cytoskeleton*, 58, pp. 200–211.

[121] Kim, S. M., Kim, H. W. and Park, H. J. (2021). Preparation and characterization of *surimi*-based imitation crab meat using coaxial extrusion three-dimensional food printing, *Innovative Food Science & Emerging Technologies*, 71, p. 102711.

[122] Ko, H. J., Wen, Y., Choi, J. H., Park, B. R., Kim, H. W. and Park, H. J. (2021). Meat analog production through artificial muscle fiber insertion using coaxial nozzle-assisted three-dimensional food printing, *Food Hydrocolloids*, 120, p. 106898.

[123] Dong, X., Huang, Y., Pan, Y., Wang, K., Prakash, S. and Zhu, B. (2019). Investigation of sweet potato starch as a structural enhancer for three-dimensional printing of Scomberomorus niphonius *surimi*, *Journal of Texture Studies*, 50, pp. 316–324.

[124] Hunt, A., Getty, K. and Park, J. W. (2009). Roles of starch in *surimi* seafood: a review, *Food Reviews International*, 25, pp. 299–312.

[125] Yoo, B. and Lee, C. (1993). Rheological relationships between *surimi* sol and gel as affected by ingredients, *Journal of Food Science*, 58, pp. 880–883.

[126] Hamidi, A. and Tadesse, Y. (2019). Single step 3D printing of bioinspired structures via metal reinforced thermoplastic and highly stretchable elastomer, *Composite Structures*, 210, pp. 250–261.

[127] Yu, B., Ren, F., Zhao, H., Cui, B. and Liu, P. (2020). Effects of native starch and modified starches on the textural, rheological and microstructural characteristics of soybean protein gel, *International Journal of Biological Macromolecules*, 142, pp. 237–243.

[128] Hu, Y., Tian, J., Zou, J., Yuan, X., Li, J., Liang, H., Zhan, F. and Li, B. (2019). Partial removal of acetyl groups in konjac glucomannan significantly improved the rheological properties and texture of konjac glucomannan and κ-carrageenan blends, *International Journal of Biological Macromolecules*, 123, pp. 1165–1171.

[129] Vancauwenberghe, V., Verboven, P., Lammertyn, J. and Nicolaï, B. (2018). Development of a coaxial extrusion deposition for 3D printing of customizable pectin-based food simulant, *Journal of Food Engineering*, 225, pp. 42–52.

[130] Vancauwenberghe, V., Katalagarianakis, L., Wang, Z., Meerts, M., Hertog, M., Verboven, P., Moldenaers, P., Hendrickx, M. E., Lammertyn, J. and Nicolaï, B. (2017). Pectin based food-ink formulations for 3-D printing of customizable porous food simulants, *Innovative Food Science & Emerging Technologies*, 42, pp. 138–150.

[131] Fraeye, I., Colle, I., Vandevenne, E., Duvetter, T., Van Buggenhout, S., Moldenaers, P., Van Loey, A. and Hendrickx, M. (2010). Influence of pectin structure on texture of pectin–calcium gels, *Innovative Food Science & Emerging Technologies*, 11, pp. 401–409.

[132] Willats, W. G., Knox, J. P. and Mikkelsen, J. D. (2006). Pectin: new insights into an old polymer are starting to gel, *Trends in Food Science & Technology*, 17, pp. 97–104.

[133] Matsuura, T., Idota, N., Hara, Y. and Annaka, M. (2009). *Gels: Structures, Properties, and Functions,* "Dynamic light scattering study of pig vitreous body" (Springer), pp. 195–203.

[134] Fraeye, I., Duvetter, T., Doungla, E., Van Loey, A. and Hendrickx, M. (2010). Fine-tuning the properties of pectin–calcium gels by control of pectin fine structure, gel composition and environmental conditions, *Trends in Food Science & Technology,* 21, pp. 219–228.

[135] Jeon, W. Y., Yu, J. Y., Kim, H. W. and Park, H. J. (2021). Production of customized food through the insertion of a formulated nanoemulsion using coaxial 3D food printing, *Journal of Food Engineering,* p. 110689.

[136] Kim, H. W., Bae, H. and Park, H. J. (2017). Classification of the printability of selected food for 3D printing: Development of an assessment method using hydrocolloids as reference material, *Journal of Food Engineering,* 215, pp. 23–32.

[137] Joung, H. J., Choi, M. J., Kim, J. T., Park, S. H., Park, H. J. and Shin, G. H. (2016). Development of food-grade curcumin nanoemulsion and its potential application to food beverage system: antioxidant property and *in vitro* digestion, *Journal of Food Science,* 81, pp. N745–N753.

[138] Tønnesen, H. H., Másson, M. and Loftsson, T. (2002). Studies of curcumin and curcuminoids. XXVII. Cyclodextrin complexation: solubility, chemical and photochemical stability, *International Journal of Pharmaceutics,* 244, pp. 127–135.

[139] Cai, X., Du, X., Cui, D., Wang, X., Yang, Z. and Zhu, G. (2019). Improvement of stability of blueberry anthocyanins by carboxymethyl starch/xanthan gum combinations microencapsulation, *Food Hydrocolloids,* 91, pp. 238–245.

[140] Lu, Y., Mao, L., Hou, Z., Miao, S. and Gao, Y. (2019). Development of emulsion gels for the delivery of functional food ingredients: From structure to functionality, *Food Engineering Reviews,* 11, pp. 245–258.

[141] Kim, H. W., Lee, J. H., Park, S. M., Lee, M. H., Lee, I. W., Doh, H. S. and Park, H. J. (2018). Effect of hydrocolloids on rheological properties and printability of vegetable inks for 3D food printing, *Journal of Food Science,* 83, pp. 2923–2932.

[142] Alloncle, M. and Doublier, J.-L. (1991). Viscoelastic properties of maize starch/hydrocolloid pastes and gels, *Food Hydrocolloids,* 5, pp. 455–467.

[143] Zhao, J. and He, N. (2020). A mini-review of embedded 3D printing: supporting media and strategies, *Journal of Materials Chemistry B,* 8, pp. 10474–10486.

[144] Daly, A. C., Riley, L., Segura, T. and Burdick, J. A. (2020). Hydrogel microparticles for biomedical applications, *Nature Reviews Materials,* 5, pp. 20–43.

[145] Grosskopf, A. K., Truby, R. L., Kim, H., Perazzo, A., Lewis, J. A. and Stone, H. A. (2018). Viscoplastic matrix materials for embedded 3D printing, *ACS Applied Materials & Interfaces*, 10, pp. 23353–23361.

[146] Hinton, T. J., Jallerat, Q., Palchesko, R. N., Park, J. H., Grodzicki, M. S., Shue, H.-J., Ramadan, M. H., Hudson, A. R. and Feinberg, A. W. (2015). Three-dimensional printing of complex biological structures by freeform reversible embedding of suspended hydrogels, *Science Advances*, 1, p. e1500758.

[147] Bhattacharjee, T., Zehnder, S. M., Rowe, K. G., Jain, S., Nixon, R. M., Sawyer, W. G. and Angelini, T. E. (2015). Writing in the granular gel medium, *Science Advances*, 1, p. e1500655.

[148] Wehner, M., Truby, R. L., Fitzgerald, D. J., Mosadegh, B., Whitesides, G. M., Lewis, J. A. and Wood, R. J. (2016). An integrated design and fabrication strategy for entirely soft, autonomous robots, *Nature*, 536, pp. 451–455.

[149] Nelson, A. Z., Kundukad, B., Wong, W. K., Khan, S. A. and Doyle, P. S. (2020). Embedded droplet printing in yield-stress fluids, *Proceedings of the National Academy of Sciences*, 117, pp. 5671–5679.

[150] Karyappa, R., Ching, T. and Hashimoto, M. (2020). Embedded Ink Writing (EIW) of polysiloxane inks, *ACS Applied Materials & Interfaces*, 12, pp. 23565–23575.

[151] Karyappa, R., Ohno, A. and Hashimoto, M. (2019). Immersion precipitation 3D printing (ip 3DP), *Materials Horizons*, 6, pp. 1834–1844.

[152] Karyappa, R. and Hashimoto, M. (2021). Freeform polymer precipitation in microparticulate gels, *ACS Applied Polymer Materials*, 3, pp. 908–919.

[153] Yang, H., Luo, D., Qian, K. and Yao, L. (2021). Freeform fabrication of fluidic edible materials. In *Proceedings of the 2021 CHI Conference on Human Factors in Computing Systems*, pp. 1–10.

[154] Hajash, K., Sparrman, B., Guberan, C., Laucks, J. and Tibbits, S. (2017). Large-scale rapid liquid printing, *3D Printing and Additive Manufacturing*, 4, pp. 123–132.

[155] Luo, D., Yang, H., Khurana, M., Qian, K. and Yao, L. (2021). Demonstrating freeform fabrication of fluidic edible materials. In *Extended Abstracts of the 2021 CHI Conference on Human Factors in Computing Systems*, pp. 1–4.

[156] Miyatake, M., Narumi, K., Sekiya, Y. and Kawahara, Y. (2021). Demonstrating flower jelly printer for parametrically designed flower jelly. In *Extended Abstracts of the 2021 CHI Conference on Human Factors in Computing Systems*, pp. 1–4.

[157] Ispirova, G., Eftimov, T. and Seljak, B. K. (2020). Exploring knowledge domain bias on a prediction task for food and nutrition data.

In *2020 IEEE International Conference on Big Data (Big Data)*, IEEE, pp. 3563–3572.

[158] Chang, K.-h., Liu, S.-y., Chu, H.-h., Hsu, J. Y.-j., Chen, C., Lin, T.-y., Chen, C.-y. and Huang, P. (2006). The diet-aware dining table: observing dietary behaviors over a tabletop surface. In *International Conference on Pervasive Computing*, pp. 366–382.

[159] Martin, C. K., Han, H., Coulon, S. M., Allen, H. R., Champagne, C. M. and Anton, S. D. (2008). A novel method to remotely measure food intake of free-living individuals in real time: the remote food photography method, *British Journal of Nutrition*, 101, pp. 446–456.

[160] Fontana, J. M., Farooq, M. and Sazonov, E. (2021). *Wearable Sensors*, "Detection and characterization of food intake by wearable sensors" (Elsevier), pp. 541–574.

[161] Sempionatto, J. R., Montiel, V. R.-V., Vargas, E., Teymourian, H. and Wang, J. (2021). Wearable and mobile sensors for personalized nutrition, *ACS Sensors*, 6, pp. 1745–1760.

[162] Cordier, T., Lanzén, A., Apothéloz-Perret-Gentil, L., Stoeck, T. and Pawlowski, J. (2019). Embracing environmental genomics and machine learning for routine biomonitoring, *Trends in Microbiology*, 27, pp. 387–397.

[163] EFSA Panel on Dietetic Products, Nutrition and Allergies. (2013). Scientific opinion on dietary reference values for energy, *EFSA Journal*, 11, p. 3005.

[164] Milani, G. P., Silano, M., Mazzocchi, A., Bettocchi, S., De Cosmi, V. and Agostoni, C. (2021). Personalized nutrition approach in pediatrics: a narrative review, *Pediatric Research*, 89, pp. 384–388.

[165] Zeevi, D., Korem, T., Zmora, N., Israeli, D., Rothschild, D., Weinberger, A., Ben-Yacov, O., Lador, D., Avnit-Sagi, T. and Lotan-Pompan, M. (2015). Personalized nutrition by prediction of glycemic responses, *Cell*, 163, pp. 1079–1094.

[166] Verma, M., Hontecillas, R., Tubau-Juni, N., Abedi, V. and Bassaganya-Riera, J. (2018). Challenges in personalized nutrition and health, *Frontiers in Nutrition*, 5, p. 117.

[167] Adams, S. H., Anthony, J. C., Carvajal, R., Chae, L., Khoo, C. S. H., Latulippe, M. E., Matusheski, N. V., McClung, H. L., Rozga, M. and Schmid, C. H. (2020). Perspective: guiding principles for the implementation of personalized nutrition approaches that benefit health and function, *Advances in Nutrition*, 11, pp. 25–34.

[168] Smetana, S., Aganovic, K. and Heinz, V. (2021). Food supply chains as cyber-physical systems: a path for more sustainable personalized nutrition, *Food Engineering Reviews*, 13, pp. 92–103.

## Problems

1. Define sustainability and explain how the life cycle assessment works.
2. Describe how 3D printing contributes to sustainability and food security.
3. What is meant by 4D printing, and what dimensions are involved in this transformative process?
4. Give some examples of 4D printing and discuss a few future trends.
5. Describe different types of microfluidic devices.
6. What are the reasons for using microfluidic systems in extrusion-based bioprinting?
7. Describe some of the works involving the use of coaxial printheads for 3D food printing. What did they achieve?
8. What is a yield stress fluid? How are they important for embedded 3D printing?
9. What are some possible considerations for choosing gelatin or Carbopol as support baths in embedded 3D food printing?
10. 3D food printing can impact personalized nutrition. Who are the potential users that benefit and why?
11. What are examples of the information required by the AI system in order to recommend a personalized diet?

# Index

www.ingramcontent.com/pod-product-compliance
Lightning Source LLC
Chambersburg PA
CBHW061614220326
41598CB00026BA/3756